Robust H_∞ Team Formation Tracking Design Methods of Large-Scale UAV Networked Control Systems

This book introduces the centralized robust H_∞ team formation tracking control strategy of multi-unmanned aerial vehicle (multi-UAV) network system under intrinsic random fluctuation, time-varying delay and packet dropout in wireless communication, and external disturbance. A simple robust decentralized H_∞ proportional-integral-derivative (PID) reference tracking network control strategy is introduced for practical applications of team formation of large-scale UAV under control saturation constraint, external disturbance, and vortex coupling. It provides practical design procedures based on linear matrix inequalities (LMIs) solvable via LMI TOOLBOX in MATLAB®.

Features:

- Focuses on the stabilization of a QUAV under finite-time switching model control (SMC)
- Discusses robustness control design for formation tracking in UAV networks
- Introduces different robust centralized and decentralized H_∞ attack-tolerant observer-based reference team formation tracking control of large-scale UAVs
- Reviews practical case studies in each chapter to introduce the design procedures
- Includes design examples of team formation of 25 quadrotor UAVs and a team formation example of five hybrid quadrotor/biped robot sub-teams

This book is aimed at researchers and graduate students in [illegible] l engineering.

Robust H_∞ Team Formation Tracking Design Methods of Large-Scale UAV Networked Control Systems

Bor-Sen Chen

CRC Press
Taylor & Francis Group
Boca Raton London New York

CRC Press is an imprint of the
Taylor & Francis Group, an **informa** business

Designed cover image: Juanjo Tugores/shutterstock

First edition published 2025
by CRC Press
2385 NW Executive Center Drive, Suite 320, Boca Raton FL 33431

and by CRC Press
4 Park Square, Milton Park, Abingdon, Oxon, OX14 4RN

CRC Press is an imprint of Taylor & Francis Group, LLC

ISBN: 978-1-032-82937-1 (hbk)
ISBN: 978-1-032-90051-3 (pbk)
ISBN: 978-1-003-54596-5 (ebk)

DOI: 10.1201/9781003545965

Typeset in Times
by Apex CoVantage, LLC

Preface

In future 5G and 6G era, it is a landmark that we are going to the next industry revolution toward automation and large data exchange in manufacturing technology and smart cities. The involved processes include cyber-physical systems (CPS), cyber-social systems (CSS), the Internet of Things (IoT), Industrial Internet of Things (IIOT), cloud computing, and others. To meet the wide application of abovementioned needs in the next era, the unmanned aerial vehicle (UAV) technology has become a popular issue in industry and academia area due to its fast-growing market and vast application such as military, searching, rescuing delivery and wireless communication platform. Due to the wide prospect of new industry requirements and relative technologies, a large-scale UAV system connected by wireless communication network is expected for various fields and a large-scale UAV formation flight is a prospective important technology to construct a smart city in the future.

With the prospective widespread application of multi-UAV team formation system in the future, the formation fight design problem of multi-UAV network control system has recently become a hot topic. The common formation control method is to establish a control strategy for a team of UAVs to achieve their desired target tracking, such as switching interaction topologies method, lead-follower method with a consensus formation control algorithm, and interactive leader-follower configuration with leader to interact with the followers via virtual leader etc. The centralized control architecture has been usually embedded in conventional multi-UAV formation controls. In order to resist external disturbances such as wind gust and intrinsic fluctuation from impacting UAV components, a robust centralized H_∞ team formation control design of multi-UAV formation control design is needed. However, the number of UAVs in the multi-UAV system is almost less than five in most literature. For a large-scale UAV team formation system, robust centralized H_∞ control methods have the unsolved or unscaled difficulty due to computational complexity and even insolvability problem. Therefore, robust decentralized H_∞ team formation networked control scheme is not only comprehensive for actual flight and wireless communication environment but also scalable and feasible for controller design of the large-scale UAV team formation networked control system (NCS), while the number of UAVs in the team formation system will increase in future 5G and 6G era.

Further, robust decentralized H_∞ attack-tolerant observer-based team formation tracking control of large-scale quadrotor UAV system is also an important research topic while malicious attack signals invade via wireless channels in networked control system. In order to accomplish a more complex and difficult task such as search and rescue, an integration of local motion planning and robust decentralized H_∞ fault-tolerant team formation tracking control strategy of hybrid UAVs and biped robots team is necessary. For the practical industrial applications, the robust decentralized H_∞ proportional-integral-derivative (PID) team formation control for large-scale UAVs network control system is also a simple but useful design strategy to deserve more research works. These robust H_∞ control strategies and their applications to the networked control system of large-scale UAVs to accomplish some specific tasks are the mains topics of the book. Consequently, this book is useful for postgraduate students, engineers, and scientists in mechanical engineering, electrical engineering aerospace engineering, control engineering, and applied mathematics.

I have published more than 200 papers on robust H_∞ designs of control, signaling processing, and communication systems under intrinsic fluctuation and external disturbance. Further, I have published more than six papers about robust H_∞ team formation control of large-scale UAV networked control system. Recently, I have coordinated with six professors to involve in the main theme project on "Team Formation Flight Control of Large-Scale Quadrotor UAVs" supported by Ministry of Science and Technology of Taiwan and two companies. Therefore, I have some theoretical backgrounds and experiences in the robust H_∞ formation flight control of large-scale quadrotor UAVs network control system.

Last but not least, I would like to thank Senior Publisher Dr. Gagandeep Singh in Engineering and Environment for CRC Press, Taylor & Francis Group, for his enthusiastic help in the publication process. Finally, I would like to thank Ms. Ya Juan Peng for her careful typing.

About the Author

Bor-Sen Chen received B.S. degree in electrical engineering from Tatung Institute of Technology, Taipei, Taiwan, in 1970, and M.S. degree of geophysics from the National Central University, Chungli, Taiwan in 1973, and Ph.D degree from University of Southern California, Los Angeles, CA, USA, in 1982. From 1973 to 1987, he had been a lecturer, associate professor, and professor of Tatung Institute of Technology. From 1987, he has been a professor, chair professor, and Tsing Hua distinguished chair professor in the Department of Electrical Engineering of National Tsing Hua University, Hsinchu, Taiwan. His research interests include robust control theory and engineering design, robust signal processing and communication system design, systems biology and their applications. He has published more than 300 journal papers, including 130 papers in control, 70 papers in signal processing and communication, and 100 papers in systems biology. He has also published ten monographs. He was the recipient of numerous awards for his academic accomplishments in robust control, fuzzy control, H_∞ control, stochastic control, signal processing and systems biology, including four Outstanding Research Awards of National Science Council, Academic Award in Engineering from Ministry of Education, National Chair Professor of the Ministry of Education, Best Impact Award of IEEE Taiwan Section for his most SCI citations of IEEE members in Taiwan, and others. His current research interest focuses on the H_∞ team formation network tracking control of large-scale UAVs, large-scale biped robots and their team cooperation, deep neural network (DNN)-based control design of nonlinear dynamic systems, systems medicine design based on design specifications and deep learning schemes, and so on. He is a life fellow of IEEE.

1 An Introduction to Large-Scale Quadrotor UAV Networked Control System

1.1 AN OVERVIEW OF ROBUST TEAM FORMATION CONTROL OF LARGE-SCALE QUADROTOR UAV NETWORKED CONTROL SYSTEM

Along with the development in technology generation, automation system and the Internet of Things (IoT) have attracted more attention for creating a future smart city [1–2]. To meet the wide applications in the next era, the unmanned aerial vehicle (UAV) technology has become a popular issue in industrial and academic area due to its fast-growing market and vast applications such as agriculture, military, searching, rescuing, delivery, and wireless communication [3–9]. Due to the wide prospect of new industry requirements and relative technologies mentioned earlier, a large-scale quadrotor UAV system connected by the wireless communication network is expected for various fields, and a large-scale quadrotor UAV team formation flight is an important prospective technology to construct a smart city in future [10].

Among all types of UAVs, quadrotor UAVs have vertical take-off and landing (VTOL) capability [7]. In accordance with this maneuverability, quadrotor UAVs can follow a wider variety of trajectories than other types of UAVs. With four independent rotors, a quadrotor UAV has six degrees of freedom (DoF) and four independent control inputs [8]. Although quadrotor UAVs have many appealing advantages in practical applications, the power consumption of quadrotor UAVs during flight is a key issue that needs further consideration. Due to the increasing need to implement more complex tasks in recent times, the team formation tracking control problem by multiple quadrotor UAVs has become more appealing than the tracking control problem of a single quadrotor UAV in the future [9].

At present, generally common team formation methods of UAVs include lead-follower (L-F) method and virtual structure (V-S) method [9,10]. Among the above-mentioned methods, the V-S method is considered to be the most recognized method for team formation tracking control designs of multiple quadrotor UAVs. The reason is that while the leader is virtualized, it can avoid the problem of the L-F approach, where the desired team formation shape cannot be maintained once the leader collides during the flight process.

DOI: 10.1201/9781003545965-1

Recently, some linear and nonlinear control schemes have been proposed to deal with attitude control and position control of quadrotor UAV. For example, proportional-integral-derivative (PID) attitude control was studied on the basis of the linearized system of quadrotor UAV at some operation points [16–22]. An adaptive control was proposed by the immersion and invariance methodology [3] and the reconfiguration control scheme [4]. A quadrotor UAV control system was augmented with the multi-layer neural network, and then the weights of neural networks were adjusted online as an adaptive law [4]. Further, a backstepping approach was also proposed for quadrotor UAV as a nonlinear control strategy [5], with the corresponding back stepping controller being divided into three subcontrollers, that is, attitude controller, altitude controller, and position controller. A combining integral sliding mode with backstepping sliding model controller in a double-loop control structure was proposed for quadrotor UAVs [5]. Some other robust tracking control schemes were also proposed for quadrotor UAVs [6–8]: a robust signal compensator was introduced for tracking control under the effect of nonlinearities and uncertainties [7]; a centralized H_∞ multi-quadrotor UAV fuzzy tracking control is proposed in References [9,12,13]; a decentralized H_∞ fuzzy tracking control of large-scale quadrotor UAV network control system was given in References [14,15]. A DNN-based H_∞ reference tracking control via deep learning scheme is proposed for quadrotor UAV [27]. A robust H_∞ fuzzy control of multi-quadrotor UAV team formation network control system is introduced in References [23,28]. A decentralized H_∞ attack-tolerant quadrotor UAV reference tracking network control under external disturbance, measurement noise and malicious attack signals is also introduced in References [28,29]. A robust decentralized H_∞ attack-tolerant hybrid quadrotor UAVs and biped robots team formation networked control system is proposed by reinforcement-learning-based DNN scheme in [29].

In general, the control designs of large-scale team formation system are divided into two classes: (i) centralized control and (ii) decentralized control of the team formation control strategy are employed. For the centralized formation control strategy, all quadrotor UAVs are coordinated by a central controller (e.g., a single computing unit) by using all information from all quadrotor UAVs. However, with an increase in the magnitude of quadrotor UAVs to a large scale, the computational complexity of implementing the centralized formation control strategy will be large and even lead the formation control design to be infeasible. As shown in References [11,12], for the centralized robust H_∞ V-S team formation T-S fuzzy control strategy of multi-quadrotor UAV network system, it can be seen that five quadrotor UAVs are almost the limit in the simulation examples.

Recently, with the technique advancement of wireless communication network [12,15], the control signals of physical plants are not needed to calculate by themselves, but by remote computing units and then transmitted to physical plants via wireless network channels from the ground control system (GCS). In the team formation of large-scale quadrotor UAVs, network control system (NCS) can be applied to the formation tracking control via wireless network channels in future 5G and 6G era. Specifically, NCS is divided into two parts: the local side (quadrotor UAVs) and the remote side (computing unit, i.e., controller). When the state information of quadrotor UAVs are measured and then transmitted to the GCS, control commands are calculated based on control strategy and then sends control input back to the quadrotor UAVs. This can effectively reduce power consumption and provide flexibility and maintainability during tasks. Although the network-based control strategy can optimize the calculation process of control commands, it also creates some problems that need to be dealt with. First, due to the uncertainty of quality of service (QoS) in wireless network communication, packet dropout may occur [12,15]. When packets are lost from the transmission, signal information cannot travel over the wireless network communication channel and therefore the receiver will not be able to receive any signal from the sender [15]. Second, if the signal is affected by the network-induced delay, it cannot be delivered to the receiver on time. Accordingly, the entire NCS is delay-dependent and the corresponding control problem is tougher than a delay-free network.

Recently, due to the rapid development of wireless network-based communication technology and widespread application in recent years, network security has become a major issue, which has attracted the attention of many researchers and research institutions. Moreover, hackers may attack NCS to interfere with the transmission of information or even cause the NCS to crash. These malicious attack signals can interfere with the correctness of transmission signal in wireless network channels, thus disrupting the operation of the entire network system. Since these malicious attack signals are unavailable to designer, the impact of malicious attack on NCS of quadrotor UAV team formation is not easily eliminated. To deal with this problem, an attack-tolerant control strategy has been developed to compensate the impact of malicious attack signals on the NCS of quadrotor UAV [29].

Except malicious attacks, during the practical application, unexpected external disturbances will inevitably interfere with quadrotor UAVs, for example, air flow and loading. Additionally, quadrotor UAVs may suffer unknown coupling effects from other quadrotor UAVs in the team formation system such as trailing vortex coupling from other neighboring quadrotor UAVs. In a nonideal environment, the formation control design of the quadrotor UAVs should consider the external disturbance, measurement noise, and trailing vortex coupling of other quadrotor UAVs in the team. Otherwise, it may lead to instability or even collision, especially for a large-scale quadrotor UAVs team formation system. Based on the above discussion, the attenuation of these effects should be further considered in the team formation tracking control scheme to improve the team formation tracking control performance of the NCS quadrotor UAVs. Hence, we should deal with malicious attack signal estimation and attack-tolerant tracking control problems for NCS of large-scale team formation quadrotor UAVs under malicious attack, external disturbance, and trailing vortex coupling.

In the previous designs of quadrotor UAV systems, the external disturbances or internal fluctuations are always ignored for the convenience of control design. However, in the realistic quadrotor UAV system, it will inevitably suffer from unavailable external disturbance and unknown intrinsic fluctuation. For example, the effect of the air flow to quadrotor UAV system can be regarded as unknown external disturbance. Also, the continuous fluctuation from motors and discontinuous jump from the sudden voltage jump of electric circuit or sensors [24,25] can be described as a kind of stochastic internal fluctuation of quadrotor UAV dynamic system. Therefore, the flight control scheme should be further developed for the quadrotor UAV NCS with intrinsic continuous and discontinuous random fluctuation as well as external disturbance to improve the robust flight control performance to achieve the desired trajectory tracking in practical application. In addition, the coupling effect caused from other UAVs in a large-scale quadrotor UAV system should be also considered especially for the practical implementation of robust team formation tracking control of large-scale quadrotor UAV team formation system. Due to the network requirement to save transmission frequency band and power consumption in smart city, the event-triggered mechanism is adopted, and sometimes network-included delay and packet dropout are both considered in the large-scale UAV NCS too.

1.2 THE SCOPE OF THE BOOK

This book focuses on the robust H_∞ team formation tracking design of large-scale quadrotor UAV network control system (NCS) under intrinsic random fluctuation, external disturbance, sensor measurement noise, interconnected coupling and malicious attack. First, the robust centralized H_∞ team formation tracking control designs of multi-quadrotor UAV NCS are introduced, and then decentralized H_∞ team formation tracking control designs of large-scale quadrotor UAVs NCS are discussed. Finally, the decentralized H_∞ attack-tolerant observer-based team formation tracking control of hybrid quadrotor UAVs and biped robots are described. These important topics about team formation NCS of large-scale quadrotor UAVs will be sequentially discussed in each chapter of this book in detail. Their outlines are given in the following paragraphs.

In Chapter 2, based on a robust lead-follower tracking control scheme of multi-agent systems, a robust centralized H_∞ team formation tracking control strategy of multi-quadrotor UAV NCS under Wiener and Poisson random fluctuation is introduced [11]. First, the lead-follower tracking error networked control system is constructed by arranging a team NCS of quadrotor UAVs into a leader-follower formation. Then a robust H_∞ centralized tracking controller is proposed to efficiently attenuate the effect of intrinsic random fluctuation and external disturbance in the team formation NCS of UAVs. With the help of Itô-Lévy formula, the optimal robust H_∞ team formation control design problem of multi-quadrotor UAVs NCS is transformed to a Hamilton-Jacobi inequality (HJI)-constrained optimization problem. In order to avoid solving the complicated HJI constraint, the T-S fuzzy interpolation method is employed to interpolate several local linearized systems at different operation points to approximate each highly nonlinear quadrotor UAV system in the team. Therefore, the HJI-constrained optimization problem can be transformed to an equivalent linear matrix inequalities (LMIs)-constrained optimization problem, which can be efficiently solved by the help of LMI toolbox in MATLAB. A design example of a team formation NCS of five quadrotor UAVs is given to illustrate the design procedure of the proposed centralized H_∞ team formation tracking control method and to validate the effectiveness of the proposed method.

The robust H_∞ event-triggered team formation tracking control design of multi-quadrotor-UAVs in a networked control system (NCS) is introduced in Chapter 3 [12]. In order to describe the realistic quadrotor UAV NCS, the intrinsic continuous Wiener random fluctuation, intrinsic discontinuous Poisson random fluctuation, external disturbance and time-varying transmission delays of wireless communication network are formulated in the nonlinear stochastic jump diffusion quadrotor UAV NCS. By combining the event-triggered multi-quadrotor UAV NCS with reference dynamic model into an augmented nonlinear stochastic system, the robust H_∞ event-triggered team formation tracking problem can be transformed to an HJI-constrained optimization problem. In order to overcome the difficulties in solving the HJI-constrained optimization problem for practical applications, the T-S fuzzy techniques are employed to interpolate local linearized models at some operation points to efficiently approximate nonlinear multi-quadrotor UAV NCS by a set of local linearized quadrotor NCSs. Thus the HJI-constrained optimization problem for the robust H_∞ event-triggered formation control of multi-quadrotor UAV NCS can be transformed to a LMIs-constrained optimization problem and can be easily solved by the convex optimization technique with the help of LMI toolbox in MATLAB. Finally, a simulation example is provided to validate the effectiveness of the proposed robust H_∞ event-triggered team formation tracking control for the multi-quadrotor UAV NCS.

In order for more practical applications, a simple decentralized H_∞ proportional-integral-derivative (PID) team formation tracking control strategy is introduced in Chapter 4 for large-scale quadrotor UAVs NCS under external disturbance, intrinsic random fluctuation and trailing vortex coupling from neighboring quadrotor UAVs. Based on the structure of virtual lead and followers, the reference trajectory of each UAV is generated by the combination of virtual lead trajectory with a specific time-varying formation offset to form a desired flight formation shape. Then, each quadrotor UAV can employ an optimal H_∞ decentralized PID controller to efficiently attenuate the effect of external disturbance and trailing vortex coupling from the neighboring quadrotor UAVs on the team formation reference tracking performance simultaneously. In order to solve a complex nonlinear HJI for PID control design of each UAV, the T-S fuzzy method is used to interpolate several local linearized UAVs at some operation points to approximate the nonlinear stochastic quadrotor UAV system so that the HJI can be transformed to a set of LMIs. Then, using a novel variable transformation method, the optimal decentralized H_∞ PID team formation tracking control problem of large-scale quadrotor UAVs can be independently designed in terms of a set of independent LMIs-constrained optimization problems for each quadrotor UAV. Further, in order to simplify the design procedure in a single run, the large amount of parameters in PID controller of decentralized H_∞ team formation tracking of each quadrotor UAV can be efficiently obtained by a conventional convex optimization algorithm without any real-time parameter tunning procedure for PID control design. Finally, a simulation example of team formation reference tracking control of 25 quadrotor UAVs with external disturbance and vortex coupling is given to validate the team formation tracking performance of the decentralized H_∞ PID tracking control method in comparison with the decentralized H_∞ T-S fuzzy tracking control scheme in Chapter 4 [21].

In Chapter 5, a robust decentralized H_∞ formation tracking control scheme for a large-scale quadrotor UAV team formation is discussed. NCS is introduced to overcome a non-scalable or even infeasible design problem due to high computational complexity by the centralized team formation control. First, the effects of wind external disturbance, intrinsic random fluctuation, coupling from neighbor quadrotor UAVs, time-varying network-induced delay, and packet dropout are considered in the design procedure of large-scale team formation tracking quadrotor UAV NCS. Moreover, an event-triggered mechanism is embedded in NCS to reduce the number of transmitted control signals to save the network traffic especially for a team formation of large-scale quadrotor UAVs. In order to efficiently attenuate the effect of the above uncertainties, a robust H_∞ decentralized tracking control strategy is employed for team formation control design of large-scale UAV NCS. The robust decentralized H_∞ virtual leader-follower tracking control design problem of large-scale UAV team formation NCS can be transformed to an independent HJI-constrained optimization problem for each UAV in the team, which can be effectively solved by a corresponding LMIs-constrained optimization problem by the T-S fuzzy interpolation method. Finally, a large-scale team formation of 25 quadrotor UAVs is provided to validate the effectiveness of the proposed robust decentralized H_∞ team formation control of large-scale quadrotor UAV NCS under environmental wind gust, intrinsic random fluctuation, coupling from neighbor quadrotor UAVs and event-triggered mechanism.

In previous chapters, in order to avoid solving HJI in the H_∞ team formation tracking control design of a large-scale quadrotor UAV NCS, the T-S fuzzy interpolation method is employed to interpolate several local linearized quadrotor UAVs at some operation points to approximate highly nonlinear quadrotor UAV NCS. This not only increases the design complexity but also leads to a heavy computational loading of H_∞ team formation control law of each quadrotor UAV in the formation team. Further, malicious attacks will bully via wireless communication channel in quadrotor UAV NCS. Therefore in Chapter 6, a robust decentralized H_∞ attack-tolerant observer-based team formation network control of large-scale quadrotor UAVs NCS under external disturbance, measurement noise, couplings from neighboring quadrotor UAVs, and malicious attack on actuator and sensor of NCS via wireless communication channel is introduced [21]. At first, we constructed a smoothed model of attack signals to describe their behavior. Then by embedding the smoothed model in the system state of each quadrotor UAV, we can simultaneously estimate the attack signals and the system state of each quadrotor UAV through a traditional Luenberger for the efficient robust decentralized H_∞ attack-tolerant observer-based team formation control of large-scale quadrotor UAVs [26]. For the design of robust decentralized H_∞ attack-tolerant observer-based team formation control of large-scale quadrotor UAVs, a very difficult independent nonlinear partial differential observer/controller-coupled Hamilton-Jacobi-Issac equation (HJIE) must be solved for robust H_∞ observer and controller design of each quadrotor UAV [29]. At present, there are no analytical and numerical methods to resolve HJIE. In this chapter, an HJIE-reinforcement learning-based deep neural network (DNN) is trained to directly solve the observer/controller coupled HJIE for the robust decentralized H_∞ attack-tolerant observer-based team formation tracking control of each quadrotor UAV. Since the system model of quadrotor UAV and HJIE of H_∞ attack-tolerant observer-based team formation tracking control strategy have been employed for the HJIE-reinforcement Adam learning algorithm-based DNN training, compared to the traditional DNN big data-driven training, we save a lot of training data and time to achieve the robust decentralized H_∞ attack-tolerant observer-based team formation tracking control design. As the deep Adam algorithm converges, it could be shown that the proposed HJIE-reinforcement DNN-based decentralized H_∞ attack-tolerant observer-based control scheme can achieve the theoretical result. Finally, simulation results of a team formation of 25 quadrotor UAVs NCS are presented in comparison with Chapter 5 to verify the effectiveness of the proposed method in this chapter.

In addition to the number increment of quadrotor UAVs in the team formation NCS, a heterogeneous multi-agent system (HMAS) combining with various types of unmanned vehicles (UVs) is also valued in many tasks, such as search and rescue (S&R), battlefield, logistics and transportation, and surveillance; compared with homogeneous MAS, it can adapt to a wide variety of tasks because each agent has different aptitudes. To construct an unmanned HMAS, three required key capabilities are perception, decision-making, and control [30]. Perception is to obtain information through the sensor (e.g., localization or computer vision), decision-making is to make decision through the actuator. Therefore, in Chapter 7, to make robust decentralized H_∞ team formation tracking control of large-scale quadrotor UAV NCS for a wider variety of application scenarios, an integrating local motion planning and robust decentralized fault-tolerant tracking control is introduced for a hybrid quadrotor UAVs and biped robots team system (URTS) for the purpose of search and rescue (S&R). In this chapter, a system architecture of performing S&R task for each agent in URTS is introduced at first to explain how to integrate reference trajectory planning and tracking control in URTS for S&R usage. In order to optimally allocate tasks to each agent in URTS, a task allocation problem is also investigated in this chapter. In order to optimally plan a path for each agent in URTS to reach these allocated task locations, a path planning problem is also formulated. To deal with complex S&R terrain, the path planning problem is decomposed into three steps: (i) global path planning, (ii) behavior decision, and (iii) local motion planning. Through such decomposition, some roadmap-based path planning algorithms can be applied to the global planning of agents in URTS. By the behavior decision, we can decide what behavior to follow the path planning of agents in URTS. By the decision, we can decide what behavior to follow for the global path according to the terrain environment. Next, we focus on the local motion planning problem of flying behavior for each quadrotor UAV and walking behavior for each robot, and then the tracking control problem for quadrotor UAVs and biped robots in the hybrid team system. The relationship between local motion planning and tracking control, that is, the transformation of the reference trajectory, is also explored in detail in this chapter [30].

Then, by converting the dynamics models of quadrotor UAV and biped robot into an unified agent dynamic model, the robust H_∞ decentralized observer-based feedforward reference tracking FTC strategy is proposed for each agent in URTS. A novel smoothed signal model of fault signal is embedded into the linearized agent system to achieve the active fault-tolerant control through observer estimation. Then, the design of robust H_∞ decentralized observer-based feedforward reference tracking fault-tolerant control strategy of URTS is transformed into LMI—constrained optimization problem of each agent in the hybrid team formation system, which can be solved by a two-step design procedure with the help of MATLAB LMI toolbox. The robust H_∞ decentralized observer-based feedforward reference tracking fault-tolerant control design problem of each UAV and biped robot in URTS can be effectively solved. Finally, several simulation results are used to demonstrate the integration of local motion planning with the S&R tasks of hybrid URTS and to verify the effectiveness of the proposed robust H_∞ decentralized observer-based feedforward reference tracking fault-tolerant control strategy of hybrid URTS under the external disturbance and the actuator and sensor fault.

1.3 CONCLUSION

In this chapter, an overview of team formation tracking control of multi-quadrotor UAV NCS is given to provide readers some background of robust team formation control of multi-quadrotor UAV NCS at first. Then the centralized and decentralized H_∞ robust team formation NCSs of large-scale quadrotor UAVs are discussed. Finally, the scope of the book is also given chapterwise for readers to get an insight into the evolution from centralized H_∞ team formation control strategy of multi-quadrotor UAV NCS to decentralized H_∞ attack-tolerant observer-based team formation NCS of large-scale quadrotor UAVs under external disturbance, measurement noise, trailing vortex coupling, and malicious attacks via wireless communication channel as well as from homogeneous quadrotor UAVs team to hybrid quadrotor UAVs and biped robots team to adapt to a wide variety of application scenarios. Finally, the theoretical background of robust H_∞ team formation control strategies of multiple quadrotor UAVs via network control systems and their applications in future 5G and 6G era are given in detail in the following chapters.

REFERENCES

[1] N. H. Motlagh, M. Bagaa, and T. Taleb, "UAV-based IoT platform: A crowd surveillance use case," *IEEE Communications Magazine*, vol. 55, no. 2, pp. 128–134, Feb. 2017.

[2] M. Wollschlaeger, T. Sauter, and J. Jasperneite, "The future of industrial communication: Automation networks in the era of the internet of things and industry 4.0," *IEEE Industrial Electronics Magazine*, vol. 11, no. 1, pp. 17–27, Mar. 2017.

[3] B. Zhao, B. Xian, Y. Zhang, and X. Zhang, "Nonlinear robust adaptive tracking control of a quadrotor UAV via immersion and invariance methodology," *IEEE Transactions on Industrial Electronics*, vol. 62, no. 5, pp. 2891–2902, May 2015.

[4] M. Bisheban and T. Lee, "Geometric adaptive control with neural networks for a quadrotor in wind fields," *IEEE Transactions on Control Systems Technology*, vol. 29, no. 4, pp. 1533–1548, July 2021.

[5] D. J. Almakhles, "Robust backstepping sliding mode control for a quadrotor trajectory tracking application," *IEEE Access*, vol. 8, pp. 5515–5525, Dec. 2020.

[6] J. Sun, Y. Wang, Y. Yu, and C. Sun, "Nonlinear robust compensation method for trajectory tracking control of quadrotors," *IEEE Access*, vol. 7, pp. 26766–26776, Mar. 2019.

[7] A. C. Satici, H. Poonawala, and M. W. Spong, "Robust optimal control of quadrotor UAVs," *IEEE Access*, vol. 1, pp. 79–93, May 2013.

[8] D. Shi, Z. Wu, and W. Chou, "Generalized extended state observer based high precision attitude control of quadrotor vehicles subject to wind disturbance," *IEEE Access*, vol. 6, pp. 32349–32359, May 2018.

[9] H. Du, W. Zhu, G. Wen, Z. Duan, and J. Lu, "Distributed formation control of multiple quadrotor aircraft based on nonsmooth consensus algorithms," *IEEE Transactions on Cybernetics*, vol. 49, no. 1, pp. 342–353, Jan. 2019.

[10] Z. Hou and I. Fantoni, "Interactive leader–follower consensus of multiple quadrotors based on composite nonlinear feedback control," *IEEE Transactions on Control Systems Technology*, vol. 26, no. 5, pp. 1732–1743, Sep. 2018.

[11] B. S. Chen, C. P. Wang, and M. Y. Lee, "Stochastic robust team tracking control of multi-UAV networked system under Wiener and Poisson random fluctuations," *IEEE Transactions on Cybernetics*, vol. 51, no. 12, pp. 5486–5799, Dec. 2020.

[12] M. Y. Lee, B. S. Chen, Y. Chang, and C. L. Hwang, "Stochastic robust team formation tracking design of multi-VTOL-UAV networked control system in smart city under time-varying delay and random fluctuation," *IEEE Access*, vol. 8, pp. 131310–131326, July 2020.

[13] H. Luan, Y. Xu, D. Liu, Z. Du, H. Qian, X. Liu, and X. Tong, "Energy efficient task cooperation for multi-UAV networks: A coalition formation game approach," *IEEE Access*, vol. 8, pp. 149372–149384, Apr. 2020.

[14] J. Choi, Y. Song, S. Lim, C. Kwon, and H. Oh, "Decentralized multi-subgroup formation control with connectivity preservation and collision avoidance," *IEEE Access*, vol. 8, pp. 71525–71534, Apr. 2020.

[15] M. Y. Lee, B. S. Chen, C. Y. Tsai, and C. L. Hwang, "Stochastic H_∞ robust decentralized tracking control of large-scale team formation UAV network system with time-varying delay and packet dropout under interconnected couplings and wiener fluctuations," *IEEE Access*, vol. 9, pp. 41976–41997, Mar. 2021.

[16] M. H. Amoozgar, A. Chamseddine, and Y. Zhang, "Fault-tolerant fuzzy gain-scheduled PID for a quadrotor helicopter tested in the presence of actuator faults," *IFAC Proceedings*, vol. 45, no. 3, pp. 282–287, 2012.

[17] F. Goodarzi, D. Lee, and T. Lee, "Geometric nonlinear PID control of a quadrotor UAV on SE(3)," in *Proceedings of European Control Conferences*, Zurich, Switzerland, June 2013, pp. 3845–3850.

[18] A. Alaimo, V. Milazzo, C. L. R. Milazzo, and A. Ricciardello, "PID controller applied to hexacopter flight," *Journal of Intelligent and Robotic Systems*, vol. 73, no. 1–4, pp. 261–270, Oct. 2013.

[19] B. Kada and Y. Ghazzawi, "Robust PID controller design for an UAV flight control system," in *Proceedings WCECS2011*, San Francisco, vol. II 1–6. ISSN: 2078–0966 (Online), 2011.

[20] J. Moreno-Valenzuela, R. Perez-Alcocer, M. Guerrero-Medina, and A. Dzul, "Nonlinear PID-type controller for quadrotor trajectory tracking," *IEEE/ASME Transactions on Mechatronics*, vol. 23, no. 5, pp. 2436–2447, Oct. 2018.

[21] B. S. Chen, Y. C. Liu, M. Y. Lee, and C. L. Hwang, "Decentralized H_∞ PID team formation tracking control of large-scale quadrotor UAVs under external disturbance and vortex coupling," *IEEE Access*, vol. 10, pp. 108169–108184, Oct. 2022.

[22] C. S. Tseng and B. S. Chen, "Multiobjective PID control design in uncertain robotic systems using neural network elimination scheme," *IEEE Transactions on Systems, Man, and Cybernetics—Part A: Systems and Humans*, vol. 31, no. 6, pp. 632–644, Nov. 2001.

[23] M. A. S. Teixeira, F. Neves-Jr, A. Koubaa, L. V. R. De Arruda, and A. S. De Oliveira, "A quadral-fuzzy control approach to flight formation by a fleet of unmanned aerial vehicles," *IEEE Access*, vol. 8, pp. 64366–64381, Apr. 2020.

[24] W. Zhang and B. S. Chen, "State feedback H_∞ control for a class of nonlinear stochastic systems," *SIAM Journal on Control and Optimization*, vol. 44, no. 6, pp. 1973–1991, 2006.

[25] W. Zhang, L. Xie, and B. S. Chen, *Stochastic H2/H_∞ Control: A Nash Game Approach*, Boca Raton, FL: CRC Press, 2017.

[26] B. S. Chen, M. Y. Lee, and X. H. Chen, "Security-enhanced filter design for stochastic systems under malicious attack via smoothed signal model and multiobjective estimation method," *IEEE Transactions on Signal Processing*, vol. 68, pp. 4971–4986, Aug. 2020.

[27] B. S. Chen, M. Y. Lee, and T. H. Lin, "DNN-based H_∞ control scheme of nonlinear time-varying dynamic systems with external disturbance and its application to UAV tracking design," *IEEE Access*, vol. 9, pp. 69635–69653, May 2021.

[28] B. S. Chen, M. Y. Lee, T. H. Lin, and W. Zhang, "Robust state/fault estimation and fault-tolerant control in discrete-time T-S fuzzy systems: An embedded smoothing signal model approach," *IEEE Transactions on Cybernetics*, vol. 52, no. 7, pp. 6886–6900, July 2022.

[29] B. S. Chen and P. C. Chao, "Robust decentralized H_∞ attack-tolerant observer-based team formation network control of large-scale quadrotor UAVs: HJIE-reinforcement learning-based deep neural network method," *IEEE Access*, vol. 11, pp. 65810–65833.

[30] B. S. Chen and T. W. Hung, "Integrating local motion planning and robust decentralized fault-tolerant tracking control for search and rescue task of hybrid UAVs and biped robots team system," *IEEE Access*, vol. 11, pp. 45888–45909.

2 Robust H_∞ Team Formation Tracking Control of Stochastic Multi-Quadrotor UAV Networked Control System

2.1 INTRODUCTION

The quadrotor unmanned aerial vehicle (UAV) technology has been attracting more and more researchers and institutes due to its fast-growing market and vast application area which is not limited to military [1,2]. A single quadrotor UAV has many limitations on the applications such as load capacity and battery power. Compared to a single quadrotor UAV, a formation of multiple quadrotor UAVs can accomplish more complex and challenging tasks at higher efficiency with multiple capabilities. The use of multiple aircrafts has been investigated in cooperating tasks [3]. With the decreasing manufacturing cost, recently, multi-quadrotor UAV formations are becoming popular and practical in many applications such as environmental monitoring, battlefield surveillance, target search and rescue, intruder detection and interception, exploration of unsecured regions, etc [4,5].

Quadrotor UAV is one of the popular rotor UAVs, a typical under-actuated system with six degrees of freedom (DoF) and four independent control inputs. Although this underactuated characteristics can effectively reduce the manufacturing difficulty of the system, due to the nonlinearity of coupling system variables in the quadrotor UAV dynamic system, the control system design quadrotor UAV become a great challenge [6–8]. At present, some linear and nonlinear control schemes have been proposed to deal with the attitude control and position control of the quadrotor UAV. Proportion integration differentiation (PID) attitude control and linear quadratic regulator (LQR) attitude control were studied in Reference [9]. Sliding mode control [10] and back-stepping control [11] were developed to be easy for practical implementations. Some other control schemes have also been proposed such as adaptive control [12], predictor-based control [13], and dynamic inversion control [14].

Recently, due to the advances in the data sensing and communication technology, the networked control system (NCS) of quadrotor UAVs is proposed to achieve the desired goal of the team formation by exchanging the information among the

DOI: 10.1201/9781003545965-2

UAVs through the wireless communication network. In general, multiple UAVs operating in synchronization need to form a formation [15]. The common formation control method is to establish the control strategies for each quadrotor UAV to achieve their desired goal tracking formation. At present, there exist several formation control theories, such as the leader-follower formation control law [16], virtual structure formation control law [17], behavior-based formation control method [18], and graph theory formation approach [19]. Among the methods mentioned above, the leader-follower formation control approach has become one of the well-recognized and most popular formation control strategies for the synchronization problem of networked control system. The idea of the leader-follower formation scheme is one of the quadrotor UAV to be selected as the leader in response to each follower for tracking the reference path and guiding the team. The other quadrotor UAVs, which are called the followers, are required to track the trajectory of the leader with corresponding formation shape. Since the leader-follower scheme only needs to specify the leader quadrotor UAV information to represent the network team behavior, the team formation tracking design problem of multi-quadrotor UAV NCS can be simplified under the leader-follower scheme. To achieve the leader-follower formation, the error dynamic system technique that has been employed for synchronization design problem [20] or consensus design problem [21,22] is modified for the team formation tracking control design of stochastic multi-quadrotor UAV NCS in this chapter.

In the previous studies of UAV dynamic system, the external disturbances or internal fluctuations are usually ignored for the convenience of control design. However, in the realistic quadrotor UAV dynamic system, it will inevitably suffer from the unknown external disturbances and internal fluctuation. For example, the effect of the airflow on the quadrotor UAV system can be considered as an unknown external disturbance. Also the continuous fluctuation from the motors and the discontinuous jump from the sudden voltage jump of electric circuit or sensors [23,24] can be modeled as a kind of stochastic internal fluctuation in the dynamic system. Hence, the robust flight control scheme should be further developed for the quadrotor UAV networked control system with the ability to tolerate intrinsic continuous and discontinuous random fluctuations as well as external disturbances to improve the robust flight control performance for the desired trajectory tracking in the practical application. Based on the Itô-Lévy formula [25,26], the continuous Wiener process and discontinuous Poisson process are introduced to model the intrinsic continuous and discontinuous random fluctuations in the robust H_∞ team formation tracking design process of multi-quadrotor UAV networked control system, respectively.

In this chapter, the robust H_∞ multi-quadrotor UAV team formation reference tracking control design strategy is proposed to guarantee that the multi-UAV networked control system gradually converges to the desired attitude and reference path (target) with a specific team formation shape under the effect of intrinsic random fluctuation and external disturbance [27]. With the help of Itô-Lévy formula, the optimal H_∞ multi-UAV team tracking controller design is transformed to an equivalent Hamilton-Jacobi inequality (HJI)-constrained optimization problem. Since the HJI constraint is a nonlinear partial differential inequality constraint, there does not exist any numerical or analytic method to efficiently solve the HJI constraint. To simplify the HJI-constrained optimization problem for the optimal H_∞ robust leader-follower formation tracking control design of the multi-quadrotor UAV team formation networked control system, the Takagi-Sugeno (T-S) fuzzy interpolation method is proposed to approximate the nonlinear stochastic system with the interpolation of several local stochastic linearized systems [28,29]. Also, the HJI-constrained optimization problem for the optimal H_∞ robust leader-follower team formation tracking design of the multi-quadrotor UAV team formation NCS can be transformed to a linear matrix inequality (LMI)-constrained optimization problem which can be easily solved with the help of MATLAB LMI toolbox.

The main aims of this chapter are summarized as follows:

(i) A robust stochastic nonlinear multi-quadrotor UAV team formation reference tracking problem can be formulated as a stochastic leader-follower H_∞ reference tracking design problem with the consideration of the corruption of continuous Wiener fluctuations, discontinuous Poisson jump fluctuations, and external disturbances in the more realistic task of multi-quadrotor UAV team formation.

(ii) A stochastic nonlinear reference tracking error dynamic system is introduced for designing H_∞ robust controller for the stochastic nonlinear multi-quadrotor UAV team formation system to efficiently track the desired attitude and path in a leader-follower formation for more practical applications.

(iii) The optimal H_∞ team formation tracking control design problem of the stochastic optimal H_∞ nonlinear multi-quadrotor UAV networked control system is transformed to an equivalent HJI-constrained optimization problem. Moreover, with the help of T-S fuzzy interpolation method [31], the HJI-constrained optimization problem of leader-follower optimal H_∞ multi-quadrotor UAV networked team formation tracking control system is transformed to an equivalent linear matrix inequalities (LMIs)-constrained optimization problem so that the leader-follower robust H_∞ multi-quadrotor UAVs team formation tracking control design problem can be solved efficiently with the help of MATLAB LMI toolbox to simplify the design procedure.

Notation: A^T: the transpose of matrix A; $A \geq 0 (A > 0)$: symmetric positive semi-definite (symmetric positive definite) matrix A, respectively; I_n: n-dimensional identity matrix; $\|\mathbf{x}\|_2$: the Euclidean norm for the given vector $\mathbf{x} \in \mathbb{R}^n$; $C^2(\mathbb{R}^n)$: the class of functions $f(\mathbf{x})$ which is twice continuously differentiable with respect to $\mathbf{x}$; $f_\mathbf{x}$: the gradient column vector of continuously differentiable function $\mathbf{f}(\mathbf{x}) \in \mathbb{R}^1$ with respect to $\mathbf{x} \in \mathbb{R}^n$ (i.e., $\frac{\partial f(\mathbf{x})}{\partial \mathbf{x}}$); $f_{\mathbf{xx}}$: the Hessian matrix with elements of second partial derivatives of twice continuously differentiable function $\mathbf{f}(\mathbf{x}) \in \mathbb{R}^1$ with respect to $\mathbf{x} \in \mathbb{R}^n$, (i.e., $\frac{\partial^2 f(\mathbf{x})}{\partial \mathbf{x}^2}$); $\mathcal{L}^2_{\mathcal{F}}(\mathbb{R}^+, \mathbb{R}^n)$: the space of nonanticipative stochastic processes $y(t) \in \mathbb{R}^n$ with respect to an increasing $\sigma-$algebras $\mathcal{F}_t (t \geq 0)$ satisfying $\|\mathbf{y}(t)\|_{\mathcal{L}^2(\mathbb{R}^+;\mathbb{R}^n)} \triangleq \mathbf{E}\left\{\int_0^\infty \mathbf{y}^T(t)\mathbf{y}(t)dt\right\}^{\frac{1}{2}} < \infty$; E: the expectation operator.

2.2 DYNAMIC MODEL OF QUADROTOR UAV SYSTEM

In References [6,7], the position of the mass center of the quadrotor is denoted by the vector $\Xi = [x, y, z]^T$, as shown in Figure 2.1. This position vector is expressed relatively with respect to an inertial frame (i) associated with the unit vector basis (e_1, e_2, e_3). The attitude is denoted by $\Theta = [\phi, \theta, \psi]$. These three angles are the Euler angles, that is, roll $(-\frac{\pi}{2} < \phi < \frac{\pi}{2})$, pitch $(-\frac{\pi}{2} < \theta < \frac{\pi}{2})$, and yaw $(-\pi < \psi < \pi)$ that define the orientation vector of the quadrotor in space expressed in body frame (*B*) fixed to the body of the quadrotor UAV and associated with vector basis (e_{b1}, e_{b2}, e_{b3}). The linear velocity and acceleration of the translational system are given respectively as $\dot{\Xi} = [\dot{x}, \dot{y}, \dot{z}]$ and $\ddot{\Xi} = [\ddot{x}, \ddot{y}, \ddot{z}]$. The angular velocity and acceleration of the attitude system are given by $\dot{\Theta} = [\dot{\phi}, \dot{\theta}, \dot{\psi}], \ddot{\Theta} = [\ddot{\phi}, \ddot{\theta}, \ddot{\psi}]$, respectively. The quadrotor UAV dynamic equation will described by two subsystems corresponding to translational motion (referring to the position of the mass center of the UAV) and angular motion (referring to the attitude of the quadrotor UAV). With the Newton-Euler method, the *i*th quadrotor UAV dynamic of the multi-quadrotor UAV networked control system (NCS) can be described as follows:

$$
\begin{pmatrix}
\dot{x}_1^i(t) \\ \dot{x}_2^i(t) \\ \dot{y}_1^i(t) \\ \dot{y}_2^i(t) \\ \dot{z}_1^i(t) \\ \dot{z}_2^i(t) \\ \dot{\phi}_1^i(t) \\ \dot{\phi}_2^i(t) \\ \dot{\theta}_1^i(t) \\ \dot{\theta}_2^i(t) \\ \dot{\psi}_1^i(t) \\ \dot{\psi}_2^i(t)
\end{pmatrix}
=
\begin{pmatrix}
x_2^i(t) \\
-\frac{K_x^i}{m^i} x_2^i(t) + (\cos\phi_1^i(t)\sin\theta_1^i(t)\cos\psi_1^i(t) \\
+\sin\phi_1^i(t)\sin\psi_1^i(t))\frac{F^i(t)}{m^i} + v_x^i(t) \\
y_2^i(t) \\
-\frac{K_y^i}{m^i} y_2^i(t) + (\cos\phi_1^i(t)\sin\theta_1^i(t)\sin\psi_1^i(t) \\
-\sin\phi_1^i(t)\cos\psi_1^i(t))\frac{F^i(t)}{m^i} + v_y^i(t) \\
z_2^i(t) \\
-\frac{K_z^i}{m^i} z_2^i(t) - g + \cos\phi_1^i(t)\cos\theta_1^i(t)\frac{F^i(t)}{m^i} + v_z^i(t) \\
\phi_2^i(t) \\
\frac{J_y^i - J_z^i}{J_x^i}\theta_1^i(t)\psi_1^i(t) - \frac{K_\phi^i}{J_x^i}\phi_2^i(t) + \frac{1}{J_x^i}\tau_\phi^i(t) + v_\phi^i(t) \\
\theta_2^i(t) \\
\frac{J_z^i - J_x^i}{J_y^i}\phi_1^i(t)\psi_1^i(t) - \frac{K_\theta^i}{J_y^i}\theta_2^i(t) + \frac{1}{J_y^i}\tau_\theta^i(t) + v_\theta^i(t) \\
\psi_2^i(t) \\
\frac{J_x^i - J_y^i}{J_z^i}\phi_1^i(t)\theta_1^i(t) - \frac{K_\psi^i}{J_z^i}\psi_2^i(t) + \frac{1}{J_z^i}\tau_\psi^i(t) + v_\psi^i
\end{pmatrix}
\tag{2.1}
$$

where $x_1^i(t), y_1^i(t), z_1^i(t) \in \mathbb{R}^1$ are the location of the ith quadrotor UAV in Cartesian coordinate space with respect to the inertial frame, $\phi_1^i(t), \theta_1^i(t), \psi_1{}^i(t) \in \mathbb{R}^1$ are the attitude of the ith quadrotor UAV consisted of the Euler angles of the quadrotor UAV with respect to the inertial frame. The total thrust $F^i(t) \in \mathbb{R}^1$ and the rotational force $\tau_\phi^i(t), \tau_\theta^i(t), \tau_\psi^i(t) \in \mathbb{R}^1$ of the ith quadrotor UAV are produced by four rotors. g is the gravity constant. $v_x^i(t), v_y^i(t), v_z^i(t)$ are the external disturbances of the ith UAV in the three translation dynamics. $v_\phi{}^i(t)$, $v_\theta^i(t)$, $v_\psi^i(t)$ are the external disturbances of the ith UAV caused by the unexpected rotation force in raw, pitch, yaw dynamics. $m^i \in \mathbb{R}^+$ denotes the total mass of the ith UAV; $J_x^i, J_y^i, J_z^i \in \mathbb{R}^+$ are the moments of inertia of the ith UAV; $K_x^i, K_y^i, K_z^i, K_\phi^i, K_\theta^i, K_\psi^i \in \mathbb{R}^+$ represent the aerodynamic damping coefficient of the ith quadrotor UAV; and $l \in \mathbb{R}^+$ denotes the distance between the epicenter of UAV and the rotor axis. Thus, the ith nonlinear quadrotor UAV system of the N quadrotor UAVs team can be written as follows:

$$\dot{X}_i(t) = f_i(X_i(t)) + g_i(X_i(t))U_i(t) + \mathbf{v}_i(t) \tag{2.2}$$

$$\begin{aligned}\mathbf{X}_i(t) = [&x_1^i(t), x_2^i(t), y_1^i(t), y_2{}^i(t), z_1^i(t), z_2^i(t),\\ &\phi_1^i(t), \phi_2^i(t), \theta_1^i(t), \theta_2{}^i(t), \psi_1^i(t), \psi_2^i(t)]^T\end{aligned}$$

$$U_i(t) = \left[F^i(t), \tau_\phi^i(t), \tau_\theta^i(t), \tau_\psi^i(t)\right]^T$$

$$\mathbf{v}_i(t) = \left[0, v_x^i(t), 0, v_y^i(t), 0, v_z{}^i(t), 0, v_\phi^i(t), 0, v_\theta^i(t), 0, v_\psi^i(t)\right]^T$$

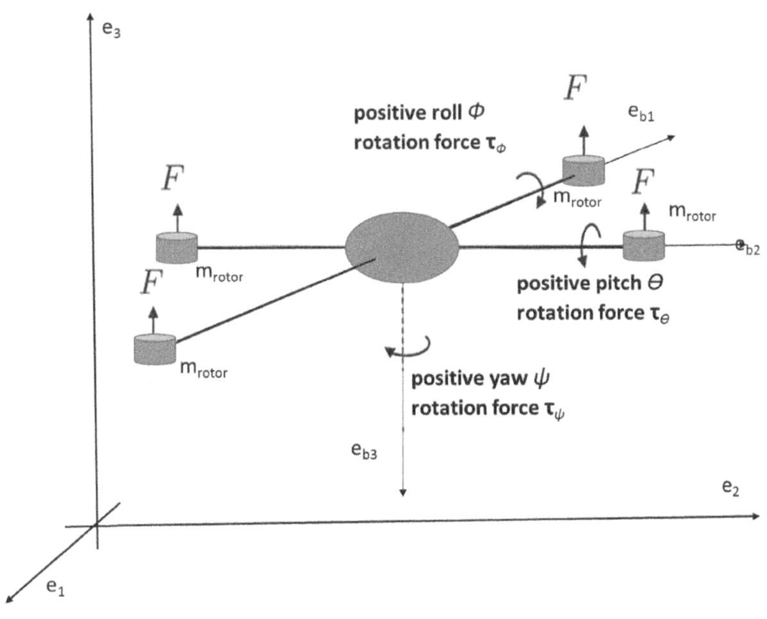

FIGURE 2.1 The quadrotor UAV dynamic model with the attitude and position in the inertial frame and the body frame (e_{b1}, e_{b2}, e_{b3}).

$$f_i\left(\mathbf{X}_i(t)\right)=[x_2^i(t),-\frac{K_x^i}{m^i}x_2^i(t),y_2^i(t),-\frac{K_y^i}{m^i}y_2^i(t),z_2^i(t),-g-\frac{K_z^i}{m^i}z_2^i(t),$$
$$\phi_2^i(t),\frac{J_y^i-J_z^i}{J_x^i}\theta_1^i(t)\psi_1^i(t)-\frac{K_\phi^i}{J_x^i}\phi_2^i(t),\theta_2^i(t),\frac{J_z^i-J_x^i}{J_y^i}\phi_1^i(t)$$
$$\times\psi_1^i(t)-\frac{K_\theta^i}{J_x^i}\theta_2^i(t),\psi_2^i(t),\frac{J_x^i-J_y^i}{J_z^i}\phi_1^i(t)\theta_1^i(t)-\frac{K_\psi^i}{J_z^i}\psi_2^i(t)]^T$$

$$g_i(\mathbf{X}_i(t))=\begin{bmatrix} 0 & 0 & 0 & 0 \\ (\cos\phi_1^i(t)\sin\theta_1^i(t)\cos\psi_1^i(t) & & & \\ +\sin\phi_1^i(t)\sin\psi_1^i(t))\frac{1}{m^i} & 0 & 0 & 0 \\ 0 & 0 & 0 & 0 \\ (\cos\phi_1^i(t)\sin\theta_1^i(t)\sin\psi_1^i(t) & & & \\ -\sin\phi_1^i(t)\cos\psi_1^i(t))\frac{1}{m^i} & 0 & 0 & 0 \\ 0 & 0 & 0 & 0 \\ \cos\phi_1^i(t)\cos\theta_1^i(t)\frac{1}{m^i} & 0 & 0 & 0 \\ 0 & 0 & 0 & 0 \\ 0 & \frac{1}{J_x^i} & 0 & 0 \\ 0 & 0 & 0 & 0 \\ 0 & 0 & \frac{1}{J_y^i} & 0 \\ 0 & 0 & 0 & 0 \\ 0 & 0 & 0 & \frac{1}{J_z^i} \end{bmatrix}$$

2.3 PROBLEM DESCRIPTION OF ROBUST H$_\infty$ TEAM FORMATION TRACKING CONTROL OF MULTI-QUADROTOR UAV NETWORKED CONTROL SYSTEM

In this chapter, the leader-follower architecture in Reference [34] is employed to deal with the multi-quadrotor UAV team tracking control problem. Under the concept of leader-follower architecture, the multi-UAV networked control system can be separated into one leader system and several follower systems. Since the leader has its own desired target and each follower is asked to track the leader's trajectories with a specific distance, for the convenience of design, a specific error-tracking augmented system is constructed via combining the leader's networked control system and the follower-networked control systems with their corresponding desired tracking path and attitude. To efficiently reduce the effect of external disturbance in the team tracking path on the desired target tracking, a robust controller is proposed to ensure the team formation tracking performance of the nonlinear multi-quadrotor UAV system.

2.3.1 Leader-Follower Quadrotor UAV Team Formation Networked Control System

Consider a connected team network consisting of N quadrotor UAVs and let $H = \{h_1, h_2, \cdots, h_N\}$ denote the index set of quadrotor UAVs. Without loss of generality, the h_1 UAV is assigned as leader (it can receive and transmit information) and the other UAVs are assigned as followers (they can only receive information). Hence, the set of multiple UAVs is partitioned into a set of leader H_L and a set of followers H_F, that is, $H = H_L \cup H_F = \{h_1\} \cup \{h_2, \cdots, h_N\}$

The dynamic of each follower quadrotor UAV $h_i \in H_F$ based on the interactions with the leader quadrotor UAV is given as:

$$\dot{X}_i(t) = f_i(X_i(t)) + g_i(X_i(t))U_i(t) + \mathbf{v}_i(t) \tag{2.3}$$

for i = 2, 3, . . ., N, and the dynamic of the leader quadrotor UAV $h_1 \in H_L$ is represented as:

$$\dot{X}_1(t) = f_1(X_1(t)) + g_1(X_1(t))U_1(t) + \mathbf{v}_1(t) \tag{2.4}$$

which will be controlled to track the desired trajectory X_d generated by the following reference model:

$$\dot{X}_d(t) = A_r X_d(t) + r(t) \tag{2.5}$$

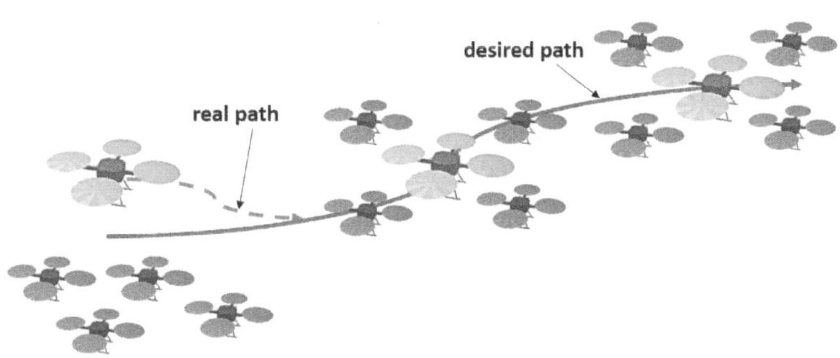

FIGURE 2.2 The leader-follower team formation of the multi-quadrotor UAV networked control system to track a desired reference trajectory. The yellow quadrotor UAV is the leader and the blue quadrotor UAVs are the followers.

Remark 2.1 (i) The team formation tracking control of the multi-quadrotor UAV networked control system is to specify the control input $U_1(t)$ of leader quadrotor UAV h_1 to track the desired trajectory $X_d(t)$ in (2.5) and to specify the control input $U_i(t)$ of each follower to track the trajectory of the leader quadrotor UAV in a specified leader-follower formation. (ii) The reference model introduced in (2.5) is to generate the desired tracking trajectory of the leader quadrotor UAV. The matrix A_r denotes a specific asymptotically stable matrix. The vector $X_d(t)$ denotes reference state trajectories which include the desired temporary position, velocity, angle and angle velocity of leader. The vector $r(t)$ denotes the bounded reference input. At the steady state, $X_d(t) = -A_r^{-1}r(t)$. If we choose $A_r = -I$, the desired trajectory $X_d(t)$ is equivalent to the reference input $r(t)$ at the steady state. In this case, $X_d(t)$ will converge to $r(t)$ at the steady state.

Aggregating the states of all followers into a vector $X_F(t) \in \mathbb{R}^{12(N-1)}$ and the states of the leader quadrotor UAV into a vector $X_L(t) \in \mathbb{R}^{12}$, the follower quadrotor UAV networked control system in (2.3) and the leader quadrotor UAV networked control system in (2.4) can be augmented as the following dynamic system:

$$\dot{\mathbf{X}}(t) = f_0(\mathbf{X}(t)) + g_0(\mathbf{X}(t))\mathbf{u}(t) + \mathbf{v}_0(t) \tag{2.6}$$

where

$$X(t) = [X_1^T(t), X_2^T(t), \cdots, X_N^T(t)]^T, f_0(X(t)) = [f_1^T(X_1(t)), f_2^T(X_2(t)), \cdots,$$
$$f_N{}^T(X_N(t))]^T, g_0(X(t)) = diag\{g_1(X_1(t)), g_2(X_2(t)), \cdots, g_N(X_N(t))\},$$
$$u(t) = [U_1^T(t), U_2^T(t), \cdots, U_N{}^T(t)]^T.$$

2.3.2 The Robust Leader-Follower Team Formation Tracking Controller Design of Multi-Quadrotor UAV Networked Control System

In order to arrange the quadrotor UAVs to a leader-follower formation e_d, we introduce a nonlinear tracking error dynamic of leader-follower team formation tracking system. By shifting the origin of nonlinear tracking error dynamic system to the leader-follower formation e_d, the robust leader-follower team formation tracking problem can be transformed to a H_∞ regulation control problem which can be easily solved.

Let $e_1(t) = X_1(t) - X_d(t)$, $e_i(t) = X_i(t) - X_1(t)$, for $i = 2, \cdots, N$, the tracking error dynamic of the multi-quadrotor UAV leader-follower team formation tracking networked control system is introduced as follows:

$$\dot{\mathbf{e}}(t) = \bar{f}(\mathbf{e}(t)) + \bar{g}(\mathbf{e}(t))\mathbf{U}(t) + \bar{\mathbf{v}}(t) \tag{2.7}$$

$$where\ \ \mathbf{e}(t) = \begin{pmatrix} X_d(t) \\ e_1(t) \\ e_2(t) \\ \vdots \\ e_N(t) \end{pmatrix} = \begin{pmatrix} X_d(t) \\ X_1(t) - X_d(t) \\ X_2(t) - X_1(t) \\ \vdots \\ X_N(t) - X_1(t) \end{pmatrix},$$

$$\bar{f}(\mathbf{e}(t)) = \begin{pmatrix} A_r X_d(t) \\ \bar{f}_1(e_1(t)) \\ \bar{f}_2(e_2(t)) \\ \vdots \\ \bar{f}_N(e_N(t)) \end{pmatrix} = \begin{pmatrix} A_r X_d(t) \\ f_1(X_1(t)) - A_r X_d(t) \\ f_2(X_2(t)) - f_1(X_1(t)) \\ \vdots \\ f_N(X_N(t)) - f_1(X_1(t)) \end{pmatrix},$$

$$\bar{g}(\mathbf{e}(t)) = \begin{pmatrix} 0 & 0 & 0 & \cdots & 0 \\ 0 & \bar{g}_1(e_1(t)) & 0 & \cdots & 0 \\ 0 & -\bar{g}_1(e_1(t)) & \bar{g}_2(e_2(t)) & & \vdots \\ \vdots & \vdots & & \ddots & \vdots \\ 0 & -\bar{g}_1(e_1(t)) & \cdots & \cdots & \bar{g}_N(e_N(t)) \end{pmatrix}$$

$$\mathbf{U}(t) = \begin{pmatrix} 0 \\ U_1(t) \\ U_2(t) \\ \vdots \\ U_N(t) \end{pmatrix}, \bar{\mathbf{v}}(t) = \begin{pmatrix} r(t) \\ \mathbf{v}_1(t) - r(t) \\ \mathbf{v}_2(t) - \mathbf{v}_1(t) \\ \vdots \\ \mathbf{v}_N(t) - \mathbf{v}_1(t) \end{pmatrix}$$

We define the target of the multi-quadrotor UAVs leader-follower formation vector as $e_d = [0, 0, e_{d2}^T, \ldots, e_{dN}^T]^T$. To arrange the UAVs in the leader-follower formation, the origin of nonlinear multi-quadrotor UAV tracking error dynamic system in (2.7) should be shifted to the e_d. In this situation, if the shifted nonlinear multi-quadrotor UAV tracking error dynamic system in (2.7) is robustly stabilized at the origin, then the robust asymptotic stability of $\tilde{\mathbf{e}}(t) = \mathbf{e}(t) - e_d$ is equivalent to asymptotically achieving the leader-follower formation (target) e_d.

Then, we get the following shifted nonlinear multi-quadrotor UAV tracking error dynamic system as follows:

$$\dot{\tilde{\mathbf{e}}}(t) = \tilde{f}(\tilde{\mathbf{e}}(t)) + \tilde{g}(\tilde{\mathbf{e}}(t))\mathbf{U}(t) + \bar{\mathbf{v}}(t) \tag{2.8}$$

where $\tilde{f}(\tilde{\mathbf{e}}(t)) = \bar{f}(\tilde{\mathbf{e}}(t) + e_d), \tilde{g}(\tilde{\mathbf{e}}(t)) = \bar{g}(\tilde{\mathbf{e}}(t) + e_d)$ for the simplicity of notation.

Thus, the origin $\tilde{\mathbf{e}}(t) = 0$ of the nonlinear multi-quadrotor UAV team tracking error dynamic system in (2.8) is at the desired steady state (target of leader-follower formation) e_d of the original nonlinear multi-quadrotor UAV tracking error dynamic system in (2.7), that is, the multi-quadrotor UAV team formation problem in (2.6) with the desired steady formation e_d is transformed to the stabilization problem of the shifted nonlinear multi-UAV error dynamic system in (2.8).

In practice, the quadrotor UAV networked control system always suffers from the parametric fluctuations due to the rotors, rigid bodies, electrical circuits or sensors. To mimic the internal fluctuation of real quadrotor UAV networked control system in the team tracking process, the dynamical model in (2.8) should be modified with continuous and discontinuous intrinsic random processes. In the following, the Wiener process $W_i(t) \in \mathbb{R}^1$ can be deemed as a continuous random fluctuation on the information transmitting via the uncertain network channel. On the other hand, the Poisson counting processes $N_i(t) \in \mathbb{R}^1$ are discontinuous abruptly random fluctuations and it is used to mimic the sudden voltage jump of electrical circuits and sensors during the flight, which will cause discontinuous changes of the system in (2.8).

Let $(\Omega, \mathcal{F}, \mathcal{F}_t, \mathcal{P})$ be a filtration probability space, Ω denotes the sample space, $\mathcal{P}$ is the probability measure, $\mathcal{F}_t$ is σ-algebra generated by the following two mutually independent stochastic processes: (i) the Wiener process $W_i(s)\}_{i=1}^N$; (ii) the Poisson counting processes $N_i(s)_{i=1}^N$ for $s \le t$, $\mathcal{F} = \{\mathcal{F}_t : t \ge 0\}$.

The stochastic nonlinear leader-follower multi-quadrotor UAV networked tracking error system is described as follows:

$$d\tilde{\mathbf{e}}(t) = (\tilde{f}(\tilde{\mathbf{e}}(t)) + \tilde{g}(\tilde{\mathbf{e}}(t))\mathbf{U}(t) + \overline{\mathbf{v}}(t))dt \\ + \sum_{i=1}^{N} \tilde{\sigma}_i(\tilde{\mathbf{e}}(t))dW_i(t) + \sum_{i=1}^{N} \tilde{\Gamma}_i(\tilde{\mathbf{e}}(t))dN_i(t) \tag{2.9}$$

where $\tilde{f}: \mathbb{R}^{12(N+1)} \to \mathbb{R}^{12(N+1)}$, $\tilde{g}: \mathbb{R}^{12(N+1)} \to \mathbb{R}^{12(N+1)\times 4(N+1)}$, $\tilde{\sigma}_i: \mathbb{R}^{12(N+1)} \to \mathbb{R}^{12(N+1)}$, and $\tilde{\Gamma}_i: \mathbb{R}^{12(N+1)} \to \mathbb{R}^{12(N+1)}$ are nonlinear Borel measurable continuous functions, which are satisfied with Lipschitz continuity. The input vector $\mathbf{U}(t) \in \mathcal{L}^2_{\mathcal{F}}(\mathbb{R}^+; \mathbb{R}^{4(N+1)})$ is the control force produced by the four rotors of the UAVs. Wiener process $W_i(t)$ is the continuous but non-differentiable stochastic process, and $\tilde{\sigma}_i(\tilde{\mathbf{e}}(t))dW_i(t)$ denotes the effect of continuous stochastic intrinsic noise caused by the ith quadrotor UAV in the team tracking process. Poisson counting process $N_i(t)$ is the discontinuous changes of the system at time instant t caused by an incident, and $\tilde{\Gamma}_i(\tilde{\mathbf{e}}(t))dN_i(t)$ denotes the effect of discontinuous stochastic intrinsic noise caused by the ith quadrotor UAV in the team tracking process. Assume $W_i(t)$ and $N_j(t)$ are independent for $i = 1,2,...,N$ and $j = 1,2,...,N$.

Remark 2.2 Some important properties of Wiener process and Poisson jump process in (2.9) are given as follows [25,26]:

1) $\mathbf{E}\{W_i(t)\} = \mathbf{E}\{dW_i(t)\} = 0, \mathit{for}\ i = 1,\cdots,N.$

2) $\mathbf{E}\{dW_i(t)dW_j(t)\} = \begin{cases} dt, & \text{for i=j.} \\ 0, & \text{for i} \neq \text{j}' \end{cases},$

3) $\mathbf{E}\{dN_i(t)\} = \lambda_i dt$, where the finite scalar number $\lambda_i > 0$ is the Poisson jump intensity, $\mathit{for}\ i = 1,\cdots,N.$

The analysis of stochastic dynamical system in (2.9) is different from the deterministic case due to the indifferentiable property of these two stochastic processes. For Wiener process, it is continuous but not differentiable at every time instant. For Poisson jump process, there exists discontinuity when jump occurs. The indifferentiable property of these two stochastic processes in stochastic dynamical system is unable to be analyzed by deterministic method. Fortunately, with the help of Itô-Lévy formula, the stochastic differential property of the function of $\tilde{\mathbf{e}}(t)$ in (2.9) is able to be analyzed. Before Theorem 2.1, Itô-Lévy formula is proposed in the following lemma:

Lemma 2.1 Let the Lyapunov function $V(\cdot) \in C^2(\mathbb{R}^{12(N+1)})$ with $V(0)=0$ and $V(\cdot) \geq 0$. For the stochastic nonlinear system in (2.9), the Itô-Lévy formula of $V(\tilde{e}(t))$ is given as follows:

$$dV(\tilde{\mathbf{e}}(t)) = [V_{\tilde{\mathbf{e}}}^T(\tilde{f}(\tilde{\mathbf{e}}(t)) + \tilde{g}(\tilde{\mathbf{e}}(t))\mathbf{U}(t) + \overline{\mathbf{v}}(t)) + \frac{1}{2}\sum_{i=1}^{N} \tilde{\sigma}_i^T(\tilde{\mathbf{e}}(t))V_{\tilde{\mathbf{e}}\tilde{\mathbf{e}}}\tilde{\sigma}_i(\tilde{\mathbf{e}}(t))]dt + \sum_{i=1}^{N} V_{\tilde{\mathbf{e}}}^T\tilde{\sigma}_i(\tilde{\mathbf{e}}(t))dW_i(t) + \sum_{i=1}^{N}\{V(\tilde{\mathbf{e}}(t) + \tilde{\Gamma}_i(\tilde{\mathbf{e}}(t))) - V(\tilde{\mathbf{e}}(t))\}dN_i(t) \tag{2.10}$$

where $V_{\tilde{\mathbf{e}}} = \dfrac{\partial V(\tilde{\mathbf{e}}(t))}{\partial \tilde{\mathbf{e}}(t)}$ and $V_{\tilde{\mathbf{e}}\tilde{\mathbf{e}}} = \dfrac{\partial^2 V(\tilde{\mathbf{e}}(t))}{\partial \tilde{\mathbf{e}}^2(t)}$

Also, another useful lemma could provide as an indirect method for the proof of the following theorems.

Lemma 2.2 [30] For any matrix M_1 and M_2 with appropriate dimensions, we have the following inequality:

$$M_1^T M_2 + M_2^T M_1 \leq M_1^T Z M_1 + M_2^T Z^{-1} M_2 \tag{2.11}$$

where Z is any positive-definite symmetric matrix.

Since the asymptotic tracking is one of the most important issues in the leader–follower team formation tracking control of multi-quadrotor UAV networked control system, therefore, in the following theorem, a controller is constructed to guarantee the asymptotic tracking performance of multi-quadrotor UAV networked tracking system in (2.9). Also, the design of the controller for the multi-quadrotor UAV networked tracking system in (2.9) is transformed to an equivalent HJI problem.

Theorem 2.1 Consider the following stochastic nonlinear multi-UAV tracking error dynamic system in (2.9) without the external disturbances, that is, $\overline{\mathbf{v}}(t) = 0$.

$$d\tilde{\mathbf{e}}(t) = (\tilde{f}(\tilde{\mathbf{e}}(t)) + \tilde{g}(\tilde{\mathbf{e}}(t))\mathbf{U}(t))dt + \sum_{i=1}^{N}\tilde{\sigma}_i(\tilde{\mathbf{e}}(t))dW_i(t) + \sum_{i=1}^{N}\tilde{\Gamma}_i(\tilde{\mathbf{e}}(t))dN_i(t) \tag{2.12}$$

If there exist a feedback controller $\mathbf{U}(t) = K(\tilde{\mathbf{e}}(t))$ and a Lyapunov function $V(\cdot) \in C^2(\mathbb{R}^{12(N+1)})$ with $V(0)=0$ and $V(x) \geq 0$, $\forall x \in \mathbb{R}^{12(N+1)}$, satisfying the following inequalities

$$m_1 \|\tilde{\mathbf{e}}(t)\|_2^2 \leq V(\tilde{\mathbf{e}}(t)) \leq m_2 \|\tilde{\mathbf{e}}(t)\|_2^2, \tag{2.13}$$

where $m_1, m_2 > 0$, and the following HJI

$$V_{\tilde{\mathbf{e}}}^T(\tilde{f}(\tilde{\mathbf{e}}(t)) + \tilde{g}(\tilde{\mathbf{e}}(t))\mathbf{K}(\tilde{\mathbf{e}}(t)) + \frac{1}{2}{}_i^T(\tilde{\mathbf{e}}(t))V_{\tilde{\mathbf{e}}\tilde{\mathbf{e}}}\tilde{\sigma}_i(\tilde{\mathbf{e}}(t)) + \sum_{i=1}^{N}\lambda_i\{V(\tilde{\mathbf{e}}(t) + \tilde{\Gamma}_i(\tilde{\mathbf{e}}(t))) - V(\tilde{\mathbf{e}}(t))\} < 0 \tag{2.14}$$

then the stochastic nonlinear multi-quadrotor UAV team networked control system in (2.9) could achieve the asymptotic tracking in probability, that is, $\tilde{\mathbf{e}}(t) \to 0$ or $\bar{\mathbf{e}}(t) \to e_d$ in probability.

Proof. Please refer to Appendix 2A.

In general, the effect of external disturbance is unavoidable during the team tracking process for the multi-quadrotor UAV networked control system. Since the external disturbances are unpredictable, a robust H_∞ tracking controller is constructed to efficiently attenuate the effect of external disturbances $\bar{\mathbf{v}}(t)$ on the team tracking performance of the stochastic nonlinear multi-quadrotor UAV networked control system. The definition of the robust H_∞ team formation tracking performance is defined as follows:

$$J_\infty(\mathbf{U}(t)) = \sup_{\substack{\bar{\mathbf{v}}(t)\in \\ \mathcal{L}^2_{\mathcal{F}}(\mathbb{R}^+;\mathbb{R}^{\bar{N}})}} \frac{E\{\int_0^{t_f} [\tilde{\mathbf{e}}^T(t)Q\tilde{\mathbf{e}}(t)]dt - V(\tilde{\mathbf{e}}_0)\}}{E\{\int_0^{t_f} \bar{\mathbf{v}}^T(t)\bar{\mathbf{v}}(t)dt\}} \tag{2.15}$$

where the term $-V(\tilde{\mathbf{e}}_0)$ is used to deduct the effect of initial condition $\tilde{\mathbf{e}}_0$ for some Lyapunov functions $V(\tilde{\mathbf{e}}(t))$, $V(\cdot) \in C^2(\mathbb{R}^{12(N+1)})$, $V(\cdot) > 0$. t_f is the terminal time of tracking control, Q is a positive definite weighting matrix, and $\bar{N} = 12(N+1)$. The design objective is to specify $\mathbf{U}(t)$ to attenuate the effect of any finite energy external disturbance on the leader-follower team formation tracking of the multi-quadrotor UAV networked control system under a prescribed level ρ^2. If one can specify a control $U^*(t)$ such that $J_\infty(U^*(t)) \le \rho^2$ for some prescribed $\rho > 0$, then the effect of any external disturbance $\bar{\mathbf{v}}(t)$ on the tracking error $\tilde{\mathbf{e}}(t)$ must be attenuated below a desired level ρ^2 from the viewpoint of energy, that is, the $\mathcal{L}_2$ gain from $\bar{\mathbf{v}}(t)$ to $\tilde{\mathbf{e}}(t)$ must be equal to or less than a prescribed value ρ^2.

Theorem 2.2 In the stochastic nonlinear multi-quadrotor UAV networked tracking system (2.9), if one could specify a nonlinear tracking controller $\mathbf{U}(t) = \mathbf{K}(\tilde{\mathbf{e}}(t))$, such that the following HJI has a positive solution $V(\tilde{\mathbf{e}}(t)) \in C^2(\mathbb{R}^{12(N+1)})$ with $V(0) = 0$ and $V(\tilde{\mathbf{e}}(t)) \ge 0$:

$$\begin{aligned} &\tilde{\mathbf{e}}^T(t)Q\tilde{\mathbf{e}}(t) + V_{\tilde{\mathbf{e}}}^T\tilde{f}(\tilde{\mathbf{e}}(t)) + V_{\tilde{\mathbf{e}}}^T\tilde{g}(\tilde{\mathbf{e}}(t))\mathbf{K}(\tilde{\mathbf{e}}(t)) + \frac{1}{4\rho^2}V_{\tilde{\mathbf{e}}}^T V_{\tilde{\mathbf{e}}} \\ &+ \frac{1}{2}\sum_{i=1}^{N}\tilde{\sigma}_i^T(\tilde{\mathbf{e}}(t))V_{\tilde{\mathbf{e}}\tilde{\mathbf{e}}}\tilde{\sigma}_i(\tilde{\mathbf{e}}(t)) + \sum_{i=1}^{N}\{V(\tilde{\mathbf{e}}(t) + \tilde{\Gamma}_i(\tilde{\mathbf{e}}(t))) - V(\tilde{\mathbf{e}}(t))\} \le 0 \end{aligned} \tag{2.16}$$

then the robust leader-follower H_∞ team tracking control performance in (2.15) of the stochastic nonlinear multi-quadrotor UAV networked control system in (2.9) is guaranteed with a prescribed level ρ^2 for all possible $\bar{\mathbf{v}}(t) \in \mathcal{L}^2_{\mathcal{F}}(\mathbb{R}^+, \mathbb{R}^n)$, the asymptotic tracking of the multi-quadrotor UAV networked control system in (2.9) can be achieved in probability with the proposed controller too, that is, $E[\tilde{\mathbf{e}}^T(t)\tilde{\mathbf{e}}(t)] \to 0$, as $t \to \infty$.

Proof. Please refer to Appendix 2B.

2.4 T-S FUZZY MODEL FOR STOCHASTIC MULTI-QUADROTOR UAV NETWORKED CONTROL SYSTEM

In general, the robust H$_\infty$ team tracking control problem for stochastic nonlinear multi-quadrotor UAV networked control system needs to solve an HJI in (2.16), which is difficult to be solved directly. Therefore, a T-S fuzzy model is introduced to approximate the nonlinear stochastic system by interpolating several local linear systems at some operation points. Subsequently, the controller design can be significantly simplified.

For the stochastic nonlinear multi-quadrotor UAV shifted team tracking error dynamic system in (2.9), the nonlinear terms $\tilde{f}(\tilde{\mathbf{e}}(t))$, $\tilde{g}(\tilde{\mathbf{e}}(t))$, $\tilde{\sigma}\left(\tilde{\mathbf{e}}(t)\right)$, and $\tilde{\Gamma}\left(\tilde{\mathbf{e}}(t)\right)$ consist of the functions of the leader's state $X_1(t)$ and the followers' state $X_i(t)$. Since all quadrotor UAVs are identical, we can use a T-S fuzzy model to approximate $\tilde{f}_i(\tilde{e}_i(t))$, $\tilde{g}_i(\tilde{e}_i(t))$, $\tilde{\sigma}_i\left(\tilde{e}_i(t)\right)$, and $\tilde{\Gamma}_i\left(\tilde{e}_i(t)\right)$ by several local linear functions of $\tilde{e}_i(t)$. The T-S fuzzy model of $\tilde{f}_i(\tilde{e}_i(t))$, $\tilde{g}_i(\tilde{e}_i(t))$, $\tilde{\sigma}_i\left(\tilde{e}_i(t)\right)$, and $\tilde{\Gamma}_i\left(\tilde{e}_i(t)\right)$ can be described by several fuzzy if–then rules.

The *kth* rule of T-S fuzzy model for stochastic nonlinear multi-quadrotor UAV shifted team tracking error dynamic system in (2.9) can be described as the following [31]:

Plant Rule k :

$$\begin{aligned}
&\textit{if } X_{1,1}(t) \textit{ is } G_{k,1},\cdots, X_{1,g}(t) \textit{ is } G_{k,g}, X_{i,1}(t) \textit{ is } G_{k,g+1},\cdots, X_{i,g}(t) \textit{ is } G_{k,2g},\\
&\text{then } \tilde{\mathrm{f}}_{\mathrm{i}}(\tilde{\mathrm{e}}_{\mathrm{i}}(\mathrm{t})) = \mathrm{A}_{\mathrm{ik}}\tilde{\mathrm{e}}_{\mathrm{i}}(\mathrm{t}), \tilde{\mathrm{g}}_{\mathrm{i}}(\tilde{\mathrm{e}}_{\mathrm{i}}(\mathrm{t})) = \mathrm{B}_{\mathrm{ik}},\ \tilde{\sigma}_{\mathrm{ij}}\left(\tilde{\mathrm{e}}_{\mathrm{i}}(\mathrm{t})\right) = \mathrm{C}_{\mathrm{ijk}}\tilde{\mathrm{e}}_{\mathrm{i}}(\mathrm{t}),\\
&\text{for j } = 1,2,\ldots,\mathrm{N},\ \tilde{\Gamma}_{\mathrm{ij}}\left(\tilde{\mathrm{e}}_{\mathrm{i}}(\mathrm{t})\right) = \mathrm{D}_{\mathrm{ijk}}\tilde{\mathrm{e}}_{\mathrm{i}}(\mathrm{t}),\ \text{for j } = 1,2,\ldots,\mathrm{N}, \textit{for } \mathrm{k} = 1,2,\ldots,L.
\end{aligned}$$

g is the number of the premise variable of each UAV. $G_{k,l}$ are the fuzzy sets of plant rule k for $l = 1,\cdots,2g$. $A_{ik}, B_{ik}, C_{ijk}, D_{ijk}$ are the local linearization system matrices of $\tilde{f}_i(\tilde{e}_i(t))$, $\tilde{g}_i(\tilde{e}_i(t))$, $\tilde{\sigma}_{ij}\left(\tilde{e}_i(t)\right)$, and $\tilde{\Gamma}_{ij}\left(\tilde{e}_i(t)\right)$ for $i, j = 1,2,\ldots,N$ at the corresponding fuzzy set. The premise variables $X_{1,1}(t)$, $X_{1,2}(t)$,. . ., $X_{1,N}(t)$ and $X_{i,1}(t)$, $X_{i,2}(t)$,. . ., $X_{i,N}(t)$ are the variables in the state vectors of the leader X_1 and the follower X_i, respectively, and the membership grades of the premise variables are defined as $\{G_{k,l}(X_{1,l}(t))\}_{l=1}^{2g}$

We define the the function $\mu_k(X_1, X_i)$ as [31]:

$$\mu_k(X_1, X_i) = \prod_{l=1}^{g} G_{k,l}(X_{1,l}(t)) \prod_{s=1}^{g} G_{k,g+s}(X_{i,s}(t))$$

where $G_{k,l}(X_{1,l}(t))$ denotes the grade of membership of leader $X_{1,l}(t)$ in $G_{k,l}$, and $G_{k,g+s}(X_{i,s}(t))$ denotes the grade of membership of follower $X_{i,s}(t)$ in $G_{k,g+s}$.

Assuming that $\mu_k(X_1, X_i) \geq 0$, $\sum_{k=1}^{L} \mu_k(X_1, X_i) > 0$,

for $k = 1,2,\cdots,L$, the interpolation functions $h_k(X_1, X_i)$ are defined as:

$$h_k(X_1,X_i)=\frac{\mu_k(X_1,X_i)}{\sum_{k=1}^{L}\mu_k(X_1,X_i)},\ with\ \sum_{k=1}^{L}h_k(X_1,X_i)=1,for\ k=1,2,\cdots,L$$

With the if-then rules above, the overall stochastic nonlinear multi-quadrotor UAV shifted team tracking error dynamic system can be denoted as:

$$\begin{aligned}d\tilde{\mathbf{e}}(t)=&\sum_{k=1}^{L}H_k(\bar{X})[\big(A_k\tilde{\mathbf{e}}(t)+B_k\mathbf{U}(t)+\bar{\mathbf{v}}(t)\big)dt+\sum_{i=1}^{N}C_{ik}\tilde{\mathbf{e}}(t)dW_i(t)\\&+\sum_{i=1}^{N}D_{ik}\tilde{\mathbf{e}}(t)dN_i(t)]\end{aligned}\tag{2.17}$$

where $H_k(\bar{X})=diag\{I_{12},h_k(X_1),h_k(X_1,X_2),\cdots,h_k(X_1,X_N),\ A_k=diag\{A_r,A_{1k},A_{2k},\cdots,A_{Nk}\},\ C_{jk}=diag\{0,\ C_{1jk},C_{2jk},\cdots,C_{Njk}\},$

$$D_{jk}=diag\{0,D_{1jk},D_{2jk},\cdots,D_{Njk}\},\ B_k=\begin{bmatrix}0&0&0&\cdots&0\\0&B_{1k}&0&\cdots&0\\0&-B_{1k}&B_{2k}&&\vdots\\\vdots&\vdots&&\ddots&\vdots\\0&-B_{1k}&\cdots&\cdots&B_{Nk}\end{bmatrix}\text{ for } j=1,2,...N.$$

Also, by the T-S fuzzy interpolation method, the *lth* control rule is given as follows:

Control Rule l:
$if\ X_{1,1}(t)\ is\ G_{l,1},\cdots,\ X_{1,g}(t)\ is\ G_{l,g},\ X_{i,1}(t)\ is\ G_{l,g+1},\cdots,\ X_{i,g}(t)\ is\ G_{l,2g},$
then $\mathrm{U}_1(\mathrm{t})=\mathrm{K}_{1l}\tilde{\mathrm{e}}_1(\mathrm{t})$for the leader $\mathrm{U}_i(\mathrm{t})=\mathrm{K}_{il}\tilde{\mathrm{e}}_i(\mathrm{t})$ for the follower $\mathrm{i}=2,3,\cdots,\mathrm{N}$

for $l=1,2,\cdots,L$, and the fuzzy tracking controllers for the leader quadrotor UAV and follower quadrotor UAVs are given as follows:

$$U_1(t)=\sum_{l=1}^{L}h_l(X_1,X_1)K_{1l}\tilde{e}_1(t)\tag{2.18}$$

$$U_i(t)=\sum_{l=1}^{L}h_l(X_1,X_i)K_{il}\tilde{e}_i(t),\ \text{for } i=2,3,\cdots,N\tag{2.19}$$

By the fuzzy controllers above, the overall stochastic nonlinear multi-quadrotor UAV shifted team tracking error dynamic system in (2.17) can be denoted as:

$$\begin{aligned}d\tilde{\mathbf{e}}(t)=&\sum_{k=1}^{L}H_k(\bar{X})\sum_{l=1}^{L}H_l(\bar{X})[\big(A_k\tilde{\mathbf{e}}(t)+B_kK_l\tilde{\mathbf{e}}(t)+\bar{\mathbf{v}}(t)\big)dt\\&+\sum_{i=1}^{N}C_{ik}\tilde{\mathbf{e}}(t)dW_i(t)+\sum_{i=1}^{N}D_{ik}\tilde{\mathbf{e}}(t)dN_i(t)]\end{aligned}\tag{2.20}$$

where $K_l = diag\{0, K_{1l}, K_{2l}, \cdots, K_{Nl}\}$

Remark 2.3 In Reference [31], Takagi and Sugeno have proposed the systematic method to build a T-S fuzzy model for nonlinear function by the system identification scheme, that is, the local system matrices can be identified by many kinds of identification methods such as least square estimation method. Moreover, many studies have proven that the T-S fuzzy model can approximate a continuous function with any degree of accuracy [32]. However, there are still some fuzzy approximation errors in (2.17) and (2.20). In this design, the fuzzy approximation error is state dependent and very complex and can be merged into the intrinsic random fluctuation and external disturbance and its effect could be efficiently attenuated by the proposed robust H$_\infty$ team formation tracking control strategy.

2.5 ROBUST H$_\infty$ FUZZY TEAM TRACKING CONTROL DESIGN OF MULTI-QUADROTOR UAV NETWORKED CONTROL SYSTEM

In the previous section, we have transformed the controller design for the multi-quadrotor UAV networked control system into an equivalent HJI-constrained optimization problem in (2.16). However, due to the difficulties in solving the HJI constraint, there is no method to efficiently solve the HJI problem efficiently. In this section, with the help of T-S fuzzy interpolation method, the controller design for multi-quadrotor UAV networked control system is transformed to an equivalent LMIs-constraint optimization problem which can be easily solved by using the MATLAB LMI toolbox. To prove the asymptotic tracking ability of the stochastic nonlinear multi-quadrotor UAV networked control system in (2.20) by the proposed fuzzy controller, a Lyapunov function for the team tracking system of (2.20) is selected as

$$V(t) = \tilde{\mathbf{e}}^T(t) P \tilde{\mathbf{e}}(t) \tag{2.21}$$

where $P = P^T > 0$ is the positive definite matrix to be designed.

Theorem 2.3 In the nonlinear multi-quadrotor UAV networked tracking system in (2.20), if we can find the matrices $W = W^T > 0$ and $\{Y_l\}_{l=1}^L$ as the solution of the following LMIs:

$$\begin{bmatrix} \Phi_{lk} & WC_{1k}^T & \cdots & WC_{Nk}^T & WD_{1k}^T & \cdots & WD_{Nk}^T \\ * & -W & 0 & 0 & 0 & 0 & 0 \\ * & * & \ddots & 0 & 0 & 0 & 0 \\ * & * & * & -W & 0 & 0 & 0 \\ * & * & * & * & -\lambda_1^{-1}W & 0 & 0 \\ * & * & * & * & * & \ddots & 0 \\ * & * & * & * & * & * & -\lambda_N^{-1}W \end{bmatrix} < 0 \tag{2.22}$$

for $k, l = 1, 2, \ldots, L$,

where $\Phi_{lk} = WA_k^T + A_k W + B_k Y_l + (B_k Y_l)^T + \sum_{i=1}^{N} \lambda_i (WD_{ik}^T + D_{ik} W)$, then the stochastic nonlinear multi-quadrotor UAV networked control system in (2.9) without external disturbance could achieve the asymptotic tracking in probability of the reference model in (2.5) by the proposed fuzzy controller $K_l = Y_l W^{-1}$ in (2.18) and (2.19) in a leader-follower team formation e_d.

Proof. Refer to Appendix 2C.

On the other hand, while the effect of external disturbance is taken into consideration, the design purpose of the fuzzy controller in this chapter is to specify the fuzzy controller in (2.18) and (2.19) for the multi-quadrotor UAV networked control system to achieve the robust H_∞ team formation tracking performance in (2.15) under the effect of the external disturbance. That is, the design objective is to specify $\mathbf{U}(t)$ to attenuate the effect of any external disturbance $\bar{\mathbf{v}}(t)$ on the tracking error $\tilde{\mathbf{e}}(t)$ below a desired level ρ^2 from the viewpoint of energy. Similar to Theorem 2.3, the robust fuzzy controller design problem is immediately transformed to an equivalent LMI-constrained optimization problem with the help of T-S fuzzy approximation approach.

Theorem 2.4 In the stochastic nonlinear multi-UAV networked team formation tracking error system (2.20), if we can find the matrices $P = P^T > 0$ and $\{Y_l\}_{l=1}^{L}$ as the solution of the following linear matrix inequalities (LMIs):

$$\begin{bmatrix} \Psi_{kl} & W & WC_{1k}^T & \cdots & WC_{Nk}^T & WD_{1k}^T & \cdots & WD_{Nk}^T \\ * & -Q^{-1} & 0 & 0 & 0 & 0 & & 0 \\ * & * & -W & & & & & 0 \\ * & * & * & \ddots & & & & 0 \\ * & * & * & * & -W & & & 0 \\ * & * & * & * & * & -\lambda_1^{-1} W & & 0 \\ * & * & * & * & * & * & \ddots & 0 \\ * & * & * & * & * & * & * & -\lambda_N^{-1} W \end{bmatrix} \leq 0 \tag{2.23}$$

for $k, l = 1, 2, \cdots, L$, where $W = P^{-1}$, $\Psi_{kl} = \frac{1}{\rho^2} I + A_k W + WA_k^T + B_k Y_l + Y_l^T B_k^T + \sum_{i=1}^{N} \lambda_i (WD_{ik}^T + D_{ik} W)$, then the H_∞ team formation tracking control performance in (2.15) is guaranteed with a prescribed level ρ^2 by the proposed fuzzy controller $K_l = Y_l W^{-1}$ in (2.18) and (2.19) for all possible $\bar{\mathbf{v}}(t) \in \mathcal{L}_{\mathcal{F}}^2(\mathbb{R}^+, \mathbb{R}^{12(N+1)})$. In the case $\bar{\mathbf{v}}(t) \in \mathcal{L}_{\mathcal{F}}^2(\mathbb{R}^+, \mathbb{R}^{12(N+1)})$, the asymptotic tracking is also achieved in probability by the proposed fuzzy controller (2.18) and (2.19), that is, $E[\tilde{\mathbf{e}}^T(t)\tilde{\mathbf{e}}(t)] \to 0$, as $t \to \infty$.

TABLE 2.1
Parameters of Quadrator UAV

Symbol	Value	Unit
m	2	kg
$J_x = J_y$	1.25	$Ns^2\ /\ rad$
J_z	2.2	$Ns^2\ /\ rad$
$K_x = K_y = K_z$	0.01	$Ns\ /\ m$
$K_\varphi = K_\theta = K_\psi$	0.012	$Ns\ /\ m$
l	0.2	m
b	2	Ns^2
d	5	Nms^2
g	9.81	$m\ /\ s^2$

Proof. Please refer to Appendix 2D.

In order to obtain a better robust H_∞ team formation tracking performance, the optimally robust H_∞ team formation tracking control problem of the multi-quadrotor UAV networked control system is formulated as the following minimization problem, i.e., we can achieve the optimal H_∞ robust team formation tracking performance in (2.15) to efficiently attenuate the effect of random fluctuations and external disturbances as possible.

$$\begin{gathered}\min_{W,\{Y_l\}_{l=1}^{L}} \rho^2 \\ \text{subject to } W = W^T > 0 \text{ and (2.23)}\end{gathered} \tag{2.24}$$

The minimization problem (2.24) subject to the LMIs in (2.23), is also called an eigenvalue problem (EVP) [35], which could be solved by decreasing ρ until no solution $W > 0$ for the LMIs in (2.23). Based on the analyses above, the design procedure for the H_∞ fuzzy team tracking control of stochastic multi-quadrotor UAV networked control system is summarized as follows:

The design procedure of optimal H_∞ robust leader-follower team formation tracking control for stochastic multi-UAV networked control system:

Step 1: Augment the leader quadrotor UAV dynamic networked control system in (2.4) and the follower quadrotor UAV dynamic systems in (2.3) to an augmented system in (2.6).

Step 2: Transform the augmented system to a team tracking error dynamic system in (2.8) between the leader and the desired trajectory with the relative position e_d among the follower quadrotor UAVs and the leader quadrotor UAV.

Step 3: Construct the T-S fuzzy system in (2.20) with T-S fuzzy identification method to approximate the nonlinear shifted team formation tracking error dynamic system in (2.9) by interpolating several local linearized systems. Then we can get the LMIs in (2.23) for the LMI-constrained optimization problem in (2.24) for the optimal H_∞ robust leader-follower team formation tracking design of multi-quadrotor UAVs.

Step 4: Solve the EVP in (2.24) to obtain W and Y_l, for all l (thus, $W = P^{-1}, K_l = Y_l W^{-1}$ can also be obtained) to obtain fuzzy controller gain K_l in each local linearized system of quadrotor UAV.

Step 5: Construct the fuzzy control law $\mathbf{U}(t) = \sum_{l=1}^{L} H_l(\bar{X})\bar{K}_l\tilde{\mathbf{e}}(t)$ by interpolating several local linearized fuzzy control laws. The fuzzy control law consists of the leader's quadrotor UAV control law (2.18) and the follower quadrotor UAVs' control law (2.19).

2.6 DESIGN EXAMPLE

To illustrate the design procedure and confirm the effectiveness of proposed robust controller, a robust H_∞ leader-follower team formation tracking control scheme of multi-quadrotor UAVs is proposed to deal with the formation of the practical quadrotor unmanned aerial vehicle (UAV) networked control system consisting of one leader quadrotor UAV and four follower quadrotor UAVs with continuous and discontinuous random fluctuations in this design example. The stochastic shifted team tracking error dynamic model of the quadrotor UAV team is shown in (2.9), and the parameters of each quadrotor UAV of the networked control system are listed in Table 2.1 [33].

To ensure the flight of quadrotor UAVs is smooth and fast in practical applications, the design of attitude reference is a very important issue. In the designing of the attitude reference, we should consider not only the distance to the desired position but also the velocity of the quadrotor UAV. For each quadrotor UAV, the desired roll reference and the desired pitch reference are respectively constrained by the distance $e_x = x - x_d$, $e_y = y - y_d$, $e_z = z - z_d$ and the velocity $\dot{e}_x = \dot{x} - \dot{x}_d$, $\dot{e}_y = \dot{y} - \dot{y}_d$, $\dot{e}_z = \dot{z} - \dot{z}_d$ while the desired yaw trajectory is set by the designer. The desired roll reference and desired pitch reference can be described as

$$\begin{aligned} \phi_d &= \sin^{-1}\left(\frac{m}{F(t)}(u_x \sin\psi_d - u_y \cos\psi_d)\right) \\ \theta_d &= \tan^{-1}\left(\frac{1}{u_z + g}(u_x \cos\psi_d + u_y \sin\psi_d)\right) \end{aligned} \tag{2.25}$$

where $F(t)$ denotes total thrust, $u_x = a_1 e_x + a_2 \dot{e}_x, u_y = b_1 e_y + b_2 \dot{e}_y$, $u_z = c_1 e_z + c_2 \dot{e}_z$ are the force vectors of x, y, z axes and $[a_1\ a_2\ b_1\ b_2\ c_1\ c_2] = [10\ 5\ 10\ 5\ 10\ 5]$. To make the control scheme much clear, the blocks diagram of control flowchart is shown in Figure 2.3. In this simulation we set the number of the followers to be four, and the team formation e_d, that is, the relative positions between the leader quadrotor UAV and four follower quadrotor UAVs are set to be $(0,0,0), (+0.1,+0.1,0), (+0.1,-0.1,0), (-0.1,+0.1,0)$, $(-0.1,-0.1,0)$, respectively.

In this simulation example, the Wiener process is used to model the system parameter variations of quadrotor UAV networked control system and compensate the approximation error of fuzzy model. On the other hand, there exist some discontinuous fluctuation in the quadrotor UAVs network system such as drop out of data transmission, deformation, and fault of mechanical elements in quadrotor UAV. Since the simulation has five independent quadrotor UAVs, including a leader quadrotor and four follower quadrotor UAVs, we set five independent Poisson jump processes $\{P_i(t)\}_{i=1}^5$ and five independent Wiener processes $\{W_i(t)\}_{i=1}^5$ in this stochastic multi-quadrotor UAV networked control system. The following stochastic functions are used to mimic the aforementioned fluctuations on the three position velocities and accelerations as well as three attitude angular velocities and angular accelerations in the quadrotor UAV networked control system:

$$\begin{aligned}
\tilde{\sigma}_1(\tilde{\mathbf{e}}(t)) &= 0.01*[0_{12\times1}, c_1(\tilde{\mathbf{e}}(t)), 0_{48\times1}]^T \\
\tilde{\sigma}_2(\tilde{\mathbf{e}}(t)) &= 0.01*[0_{24\times1}, c_2(\tilde{\mathbf{e}}(t)), 0_{36\times1}]^T \\
\tilde{\sigma}_3(\tilde{\mathbf{e}}(t)) &= 0.01*[0_{36\times1}, c_3(\tilde{\mathbf{e}}(t)), 0_{24\times1}]^T \\
\tilde{\sigma}_4(\tilde{\mathbf{e}}(t)) &= 0.01*[0_{48\times1}, c_4(\tilde{\mathbf{e}}(t)), 0_{12\times1}]^T \\
\tilde{\sigma}_5(\tilde{\mathbf{e}}(t)) &= 0.01*[0_{60\times1}, c_5(\tilde{\mathbf{e}}(t))]^T \\
\tilde{\Gamma}_1(\tilde{\mathbf{e}}(t)) &= 0.1*[0_{12\times1}, d_1(\tilde{\mathbf{e}}(t)), 0_{48\times1}]^T \\
\tilde{\Gamma}_2(\tilde{\mathbf{e}}(t)) &= 0.1*[0_{24\times1}, d_2(\tilde{\mathbf{e}}(t)), 0_{36\times1}]^T \\
\tilde{\Gamma}_3(\tilde{\mathbf{e}}(t)) &= 0.1*[0_{36\times1}, d_3(\tilde{\mathbf{e}}(t)), 0_{24\times1}]^T \\
\tilde{\Gamma}_4(\tilde{\mathbf{e}}(t)) &= 0.1*[0_{48\times1}, d_4(\tilde{\mathbf{e}}(t)), 0_{12\times1}]^T \\
\tilde{\Gamma}_5(\tilde{\mathbf{e}}(t)) &= 0.1*[0_{60\times1}, d_5(\tilde{\mathbf{e}}(t))]^T
\end{aligned}$$

where

$$\begin{aligned}
c_1(\tilde{\mathbf{e}}(t)) = [&0, -\frac{K_x}{m}x_2^1, 0, -\frac{K_y}{m}y_2^1, 0, -\frac{K_z}{m}z_2^1, 0, \frac{J_y - J_z}{J_x}\theta_1^1\psi_1^1 - \frac{K_\phi}{J_x}\phi_2^1, 0, \\
&0, \frac{J_z - J_x}{J_y}\phi_1^1\psi_1^1 - \frac{K_\theta}{J_y}\theta_2^1, 0, \frac{J_x - J_y}{J_z}\phi_1^1\theta_1^1 - \frac{K_\psi}{J_z}\psi_2^1]^T
\end{aligned}$$

$$\begin{aligned}
c_i(\tilde{\mathbf{e}}(t)) = [&0, -\frac{K_x}{m}(x_2^i - x_2^1), 0, -\frac{K_y}{m}(y_2^i - y_2^1), 0, -\frac{K_z}{m}(z_2^i - z_2^1), \\
&0, \frac{J_y - J_z}{J_x}(\theta_1^i - \theta_1^1)(\psi_1^i - \psi_1^1) - \frac{K_\phi}{J_x}(\phi_2^i - \phi_2^1), \\
&0, \frac{J_z - J_x}{J_y}(\phi_1^i - \phi_1^1)(\psi_1^i - \psi_1^1) - \frac{K_\theta}{J_y}(\theta_2^i - \theta_2^1), \\
&0, \frac{J_x - J_y}{J_z}(\phi_1^i - \phi_1^1)(\theta_1^i - \theta_1^1) - \frac{K_\psi}{J_z}(\psi_2^i - \psi_2^1)]^T
\end{aligned}$$

$$\begin{aligned}
d_1(\tilde{\mathbf{e}}(t)) = [&0, -\frac{K_x}{m}x_2^1, 0, -\frac{K_y}{m}y_2^1, 0, -\frac{K_z}{m}z_2^1, 0, \frac{J_y - J_z}{J_x}\theta_1^1\psi_1^1 - \frac{K_\phi}{J_x}\phi_2^1, 0, \\
&0, \frac{J_z - J_x}{J_y}\phi_1^1\psi_1^1 - \frac{K_\theta}{J_y}\theta_2^1, 0, \frac{J_x - J_y}{J_z}\phi_1^1\theta_1^1 - \frac{K_\psi}{J_z}\psi_2^1]^T
\end{aligned}$$

$$d_i\left(\tilde{\mathbf{e}}(t)\right)=[0,-\frac{K_x}{m}(x_2^i-x_2^1),0,-\frac{K_y}{m}(y_2^i-y_2^1),0,-\frac{K_z}{m}(z_2^i-z_2^1),$$
$$0,\frac{J_y-J_z}{J_x}(\theta_1^i-\theta_1^1)(\psi_1^i-\psi_1{}^1)-\frac{K_\phi}{J_x}(\phi_2^i-\phi_2^1),$$
$$0,\frac{J_z-J_x}{J_y}(\phi_1^i-\phi_1^1)(\psi_1^i-\psi_1{}^1)-\frac{K_\theta}{J_y}(\theta_2^i-\theta_2^1),$$
$$0,\frac{J_x-J_y}{J_z}(\phi_1^i-\phi_1^1)(\theta_1^i-\theta_1{}^1)-\frac{K_\psi}{J_z}(\psi_2^i-\psi_2^1)]^T$$
$$for\ i=2,\cdots,5$$

In this design example, the state variables of $X_{i,1}=\phi_1^i, X_{i,2}=\theta_1^i$ are chosen as premise variables and it is assumed these state variables are available, for $i=1,\cdots,5$. Since there are three fuzzy sets associated with each premise variable, in total, there are 6,561 fuzzy rules in the multi-quadrotor UAV team networked control system. The operation points of the T-S fuzzy model and their linearized local linear stochastic models are shown as follows:

The three operation points of $X_{i,1}$ are given as $op_1^1=-\pi/6$, $op_1^2=0, op_1^3=\pi/6$, for $i=1,2,...5$, and the other three operation points of $X_{i,2}$ are also given as $op_2^1=-\pi/6$, $op_2^2=0, op_2^3=\pi/6$, for $i=1,2,...5$.

The qth rule of this T-S fuzzy model for the shifted stochastic nonlinear multi-quadrotor UAV team tracking error dynamic system in (2.9) is described as follows:

$$if\ X_{i,1}\ is\ op_1^{q_{i,1}}\ and\ X_{i,2}\ is\ op_2^{q_{i,2}},\ for\ i=1,2,...5$$
$$then\ d\tilde{e}(t)=\left(A_q\tilde{e}(t)+B_qU_q(t)+\bar{\mathbf{v}}(t)\right)dt+\sum_{i=1}^{5}C_{iq}\tilde{e}(t)dW_i\left(t\right)$$
$$+\sum_{i=1}^{5}D_{iq}\tilde{e}(t)dN_i\left(t\right)$$

where A_q, B_q, C_{iq}, D_{iq}, for $i=1,2,...5$, are obtained by fuzzy system identification with the help of MATLAB system identification toolbox.

Suppose the H_∞ team formation tracking performance in (2.15) is used to design the multi-UAV networked control system with the following weighting matrices. In this simulation example, to achieve the desired team formation tracking performance, the designer considers the position tracking performance more important than the velocity tracking control performance. Therefore, for the proposed weighting matrices, the weighting on the position tracking of each quadrotor UAV is greater than the weighting on the velocity tracking of each quadrotor UAV. The weighting matrices in this simulation are given as follows:

$$Q = diag\{Q_1, Q_2, Q_3, Q_4, Q_5, Q_6\}$$
$$Q_i = 0.01diag\{1, 0.001, 1, 0.001, 1, 0.001, 1, 0.001, 1, 0.001, 1, 0.001\}$$
$$for\ i = 1, 2, \ldots 6.$$

In the reference model in (2.5), let $A_r = -I_{12\times12}$ and select the desired steady state position trajectory and desired yaw angle trajectory as:

$$x_d(t) = sin(0.5t), y_d(t) = \cos(0.5t),$$
$$z_d(t) = 0.8t\ , \psi_d(t) = 0.$$

The initial positions of the leader quadrotor UAV and the follower quadrotor UAVs are selected as:

$$(0,0,0),(+0.2,+0.2,0),(+0.2,-0.2,0),(-0.2,+0.2,0)(-0.2,-0.2,0).$$

The external disturbances are set to be zero mean Gaussian noises with unit variance.

The simulation results are shown in Figures 2.4 to 2.7. In Figure 2.4, the three-dimensional trajectories show that the actual flight path of the leader quadrotor UAV can track the trajectory of desired position quickly, and four follower quadrotor UAVs

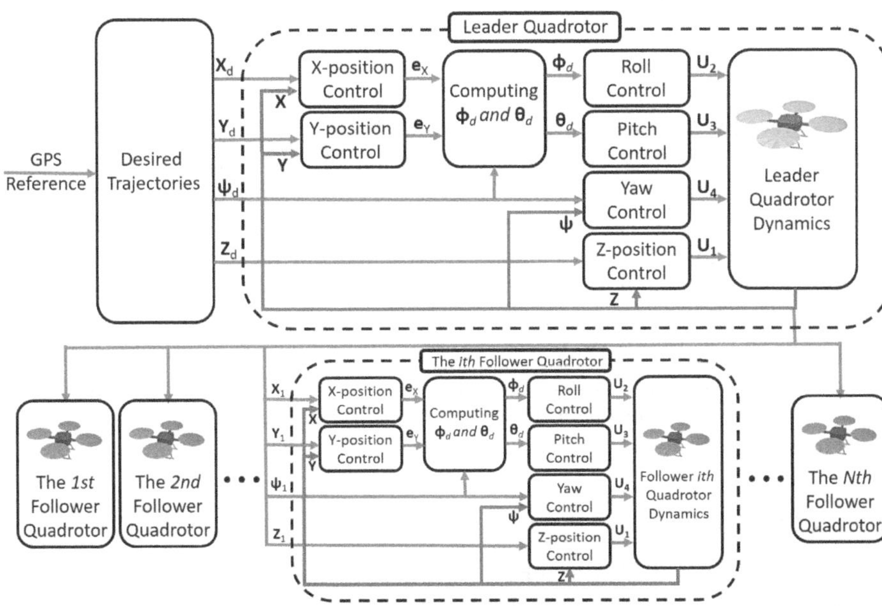

FIGURE 2.3 The structure of leader-follower team formation tracking control scheme for the multi-quadrotor UAV networked control system where the red line denotes the information flow inside the quadrotor UAV, and the blue line denotes the information flow between the quadrotor UAVs.

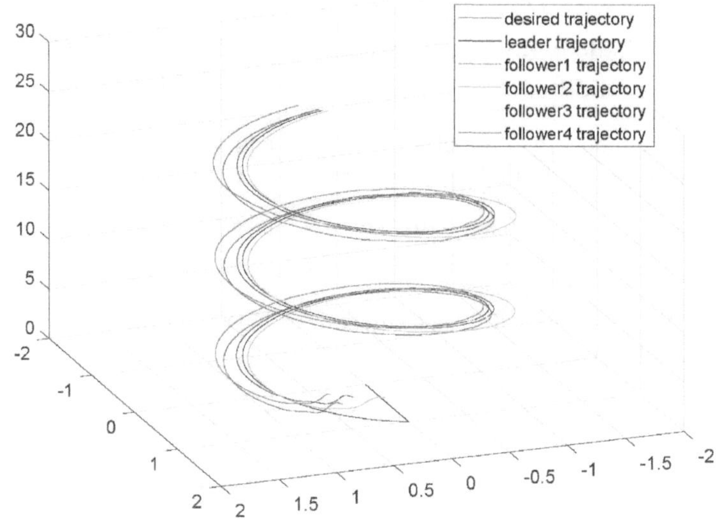

FIGURE 2.4 The 3D trajectories show that the actual flight path of the leader quadrotor UAV and the four follower quadrotor UAVs. In this figure, the leader quadrotor path trajectory converges to the desired trajectory quickly. Also, the leader-follower team formation of five quadrotor UAVs is achieved, that is, each follower quadrotor UAV successfully tracks the leader quadrotor UAV path trajectory with a specific formation shape.

can also track the trajectory of leader quadrotor UAV with the leader-follower formation quickly despite the external disturbances, intrinsic continuous Wiener fluctuation and discontinuous Poisson jump fluctuation. The team of five quadrotor UAVs, including a leader and four followers, can complete the task of spiral upward maneuvering flight under the reference model tracking scheme. It can be seen from Figures 2.5 and 2.6 that the quadrotor UAVs' state trajectories of velocity and angular velocity could achieve the robust team formation tracking performance under the influence of the continuous random Wiener fluctuations, discontinuous Poisson jump fluctuations, and external disturbances. In Figure 2.7, we show the trajectories of discontinuous random fluctuations due to Poisson jump counting process, and the continuous random fluctuations due to Weiner process for each quadrotor UAV system.

2.7 CONCLUSION

In this chapter, the stochastic robust H_∞ leader-follower team formation tracking control design is proposed to deal with the stochastic nonlinear multi-quadrotor UAV networked control system. To make the quadrotor UAVs be arranged in a fixed leader-follower team formation simultaneously, a shifted team formation tracking error dynamic system is introduced to simplify the design procedure. By the proposed nonlinear reference team formation tracking error dynamic system, the stochastic optimal H_∞ leader-follower team formation tracking control problem could

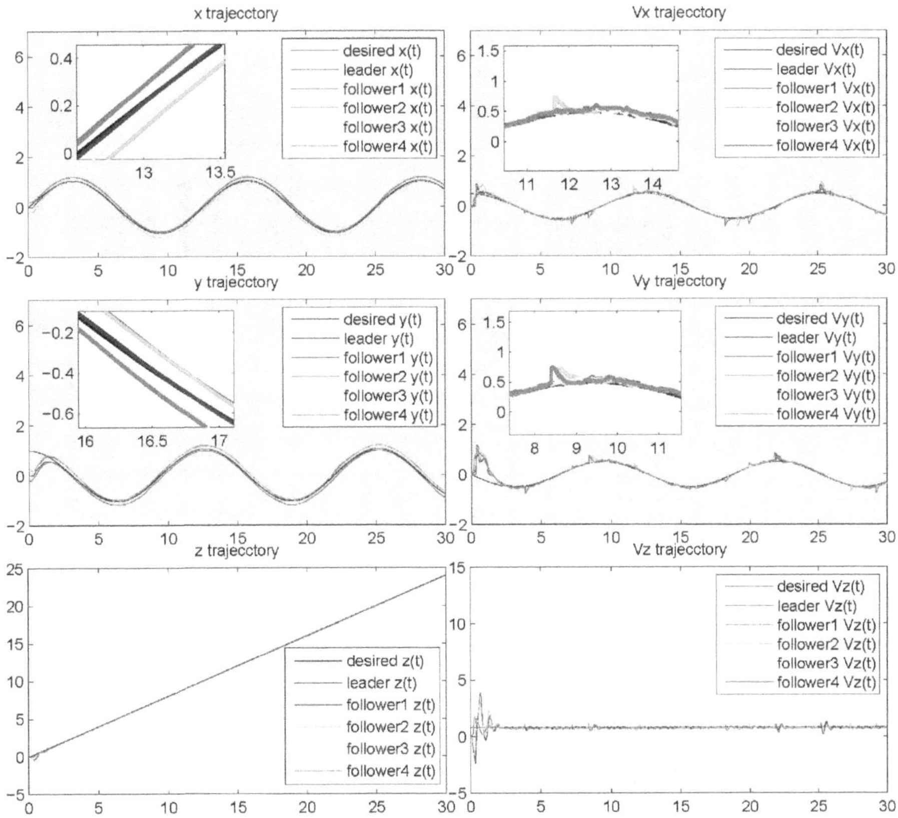

FIGURE 2.5 The position trajectories of the leader quadrotor UAV and four follower quadrotor UAVs. In this figure, the proposed controller can effectively ensure the robust team formation tracking performance of each quadrotor UAV to achieve their desired team formation tracking goal.

be transformed to an equivalent HJI-constrained optimization problem. Due to the difficulties in solving HJI constraint for the controller design, the HJI-constrained optimization problem is replaced by a LMIs-constrained optimization problem by the stochastic T-S fuzzy interpolation method of local linearized systems of quadrotor UAV. Hence, the stochastic robust H_∞ leader-follower team formation tracking control problem of the stochastic nonlinear multi-quadrotor UAV networked control system can be solved efficiently with the help of MATLAB LMI toolbox to simplify the design procedure. Finally, a design example of five quadrotor UAVs team is given to validate the effectiveness of the proposed robust H_∞ leader-follower team formation tracking control design of the stochastic nonlinear multi-quadrotor UAV networked control system. From the simulation results, the proposed stochastic robust H_∞ leader-follower team formation tracking control can make both the leader and the followers robustly track the desired trajectories of attitude and path in a fixed

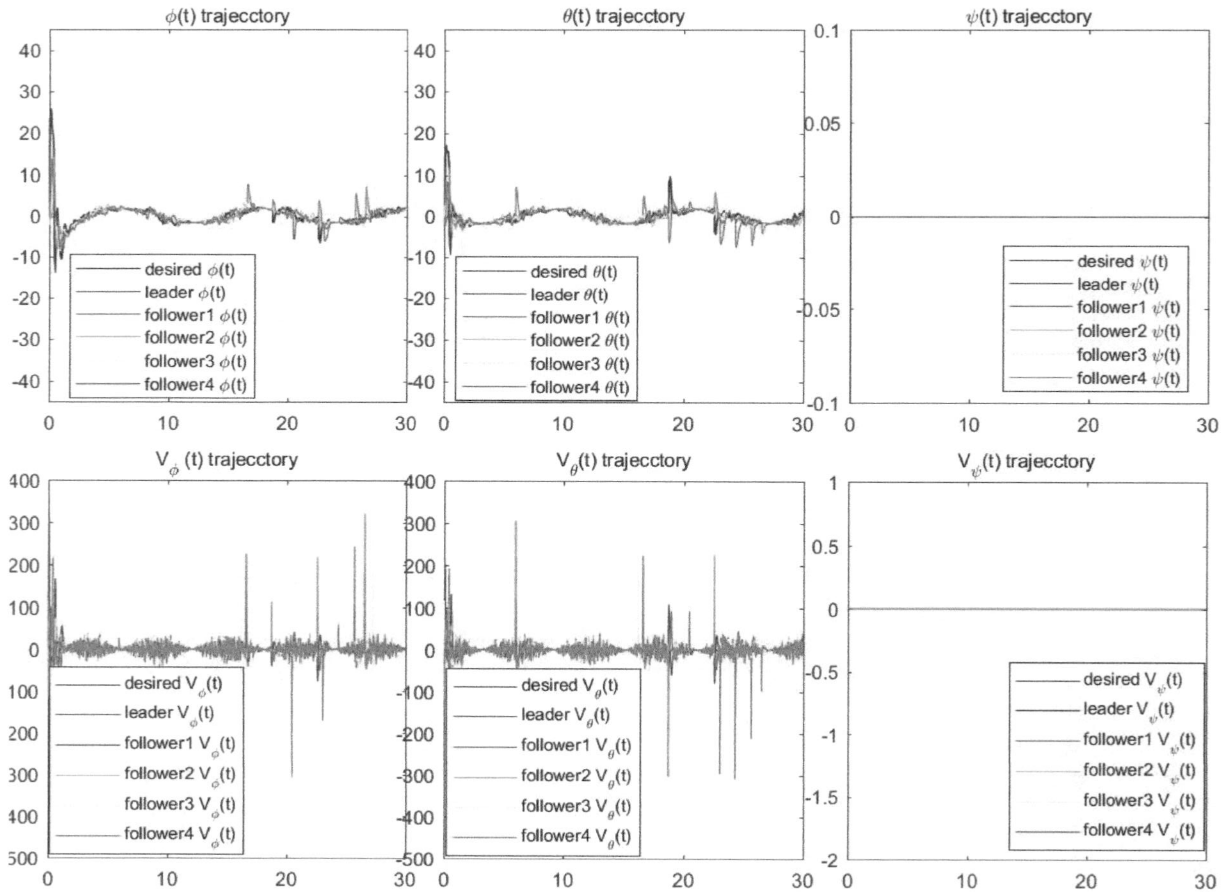

FIGURE 2.6 The attitude trajectories of the leader quadrotor UAV and four follower quadrotor UAVs. The intense changes of the attitude are caused by the change of position and velocity due to intrinsic fluctuations of Wiener and Poisson processes.

Poisson counting process of the quadrotor UAVs

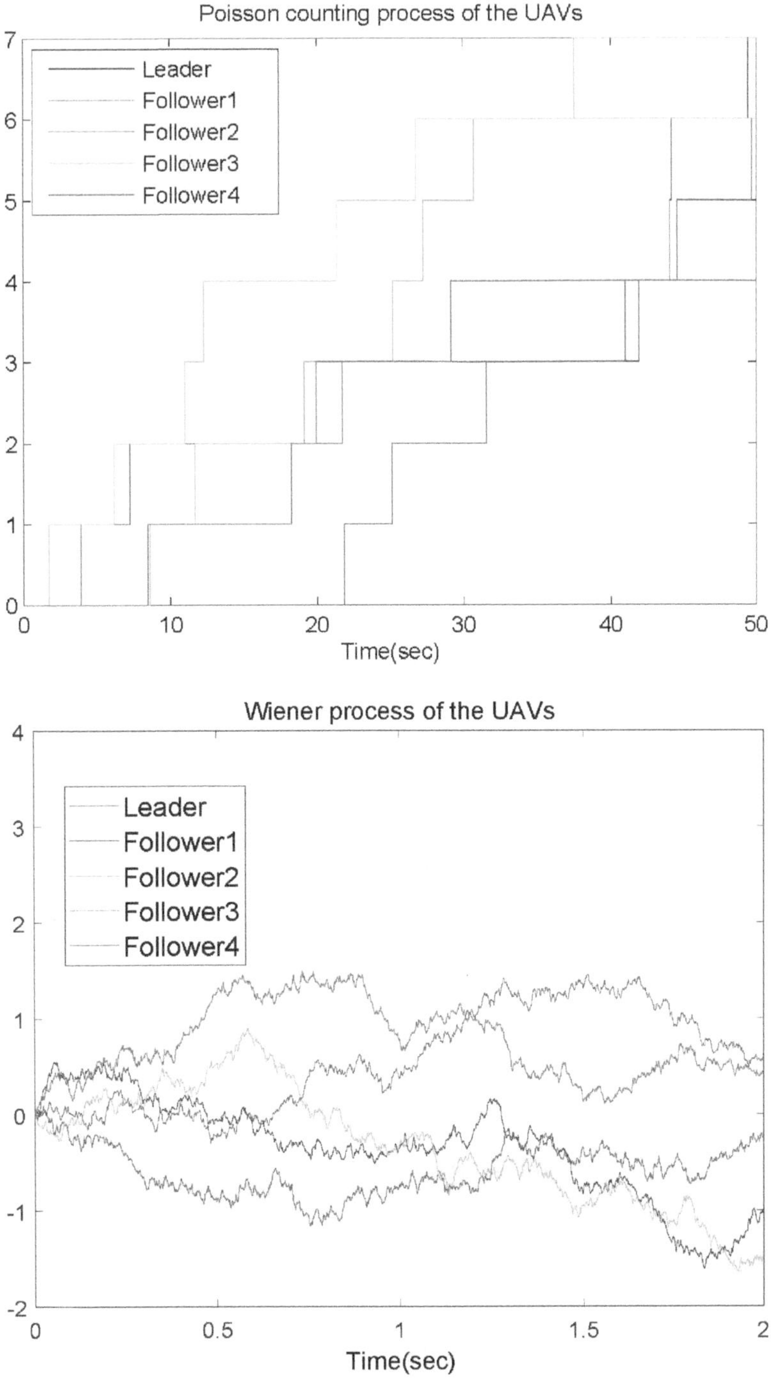

FIGURE 2.7 Stochastic random processes of the quadrotor UAVs. (a) Poisson counting processes of the quadrotor UAVs with intensity $\lambda i = 0.1$. (b) Wiener processes of the quadrotor UAVs.

leader-follower team formation shape under the unknown external disturbances and intrinsic system fluctuations. Under the concept of Internet of Things (IoT) in future smart city, the multi-quadrotor UAV networked control system can be connected with a ground computing unit to reduce the computational complexity of multi-UAV networked control system. In this framework, the ground computing unit can receive the information of multiple quadrotor UAVs and the suitable control strategy needs to be designed for the multiple quadrotor UAVs through the wireless network to accomplish the team formation task. Therefore, the effect of time-varying delay and event-triggered scheme of wireless communication network of multi-quadrotor UAV NCS will be considered in Chapter 3, and the effect of time-varying delay and packet dropout of wireless transmission channel will be discussed in Chapter 5 too. On the other hand, under the wireless network-based structure, the ground computing unit may receive channel interferences or hostile attack signals from the wireless network. Since the interferences and attack signals will deteriorate the performance of designed control strategy, the estimation and elimination of the aforementioned signals become a major issue of fault-tolerant design in the wireless network-based control problem. Hence, the Chapters 6 and 7 will focus on the network-based multi-quadrotor UAV team formation tracking problem and network-based multi-quadrotor UAV fault-tolerant control problem.

APPENDIX 2A: PROOF OF THEOREM 2.1

By applying Itô-Lévy formula in Lemma 2.1, we have:

$$\begin{aligned} E\{dV(\tilde{\mathbf{e}}(t))\} &= E\{V_{\tilde{\mathbf{e}}}^{T}(\tilde{f}(\tilde{\mathbf{e}}(t))+\tilde{g}(\tilde{\mathbf{e}}(t))\mathbf{K}(\tilde{\mathbf{e}}(t))) \\ &+\frac{1}{2}\sum_{i=1}^{N}\tilde{\sigma}_i(\tilde{\mathbf{e}}(t))V_{\tilde{\mathbf{e}}\tilde{\mathbf{e}}}\tilde{\sigma}_i(\tilde{\mathbf{e}}(t))+\sum_{i=1}^{N}\lambda_i\{V(\tilde{\mathbf{e}}(t)+\tilde{\Gamma}_i(\tilde{\mathbf{e}}(t))) \\ &-V(\tilde{\mathbf{e}}(t))\}\}dt \end{aligned} \tag{2A.1}$$

If the HJIE in (2.14) holds, then we can conclude:

$$E\{dV(\tilde{\mathbf{e}}(t))\}\le E\{-m_3\left\|\tilde{\mathbf{e}}(t)\right\|_2^2\}dt \tag{2A.2}$$

for some $m_3>0$. By using inequality (2.13), we get:

$$\frac{dE\{V(\tilde{\mathbf{e}}(t))\}}{dt}\le\frac{-m_3}{m_2}E\{V(\tilde{\mathbf{e}}(t))\} \tag{2A.3}$$

Based on the inequality in (2A.3), we have:

$$E\{V(\tilde{\mathbf{e}}(t))\}\le E\{V(\tilde{\mathbf{e}}(0))\}\exp\left(\frac{-m_3 t}{m_2}\right) \tag{2A.4}$$

and from (2A.4) it implies:

$$E\{\|\tilde{\mathbf{e}}(t)\|_2^2\} \le E\{\frac{V(\tilde{\mathbf{e}}(0))}{m_1}\}\exp\left(\frac{-m_3 t}{m_2}\right) \tag{2A.5}$$

so that,

$$\lim_{t\to\infty} E\{\|\tilde{\mathbf{e}}(t)\|_2^2\} = 0 \tag{2A.6}$$

Also, the asymptotically stable of stochastic nonlinear multi-quadrotor UAV shifted team formation tracking error dynamic system implies that $\overline{\mathbf{e}}(t) \to e_d$ in probability, i.e., the leader-follower team formation can be achieved. The proof is done.

APPENDIX 2B: PROOF OF THEOREM 2.2

It is clear that:

$$\begin{aligned} &E\{\int_0^{t_f} \tilde{\mathbf{e}}^T(t)Q\tilde{\mathbf{e}}(t)dt\} = E\{V(\tilde{\mathbf{e}}_0)\} - E\{V(\tilde{\mathbf{e}}(t_f))\} \\ &+E\{\int_0^{t_f} \tilde{\mathbf{e}}^T(t)Q\tilde{\mathbf{e}}(t)dt + dV(\tilde{\mathbf{e}}(t))\} \le E\{V(\tilde{\mathbf{e}}_0)\} \\ &+E\{\int_0^{t_f} \tilde{\mathbf{e}}^T(t)Q\tilde{\mathbf{e}}(t)dt + dV(\tilde{\mathbf{e}}(t))\} \end{aligned} \tag{2A.7}$$

By applying Lemma 2.1, we have:

$$\begin{aligned} &E\{\int_0^{t_f} \tilde{\mathbf{e}}^T(t)Q\tilde{\mathbf{e}}(t)dt\} \le E\{V(\tilde{\mathbf{e}}_0)\} + E\{\int_0^{t_f} \tilde{\mathbf{e}}^T(t)Q\tilde{\mathbf{e}}(t) \\ &+V_{\tilde{\mathbf{e}}}^T \tilde{f}(\tilde{\mathbf{e}}(t)) + V_{\tilde{\mathbf{e}}}^T \tilde{g}(\tilde{\mathbf{e}}(t))\mathbf{K}(\tilde{\mathbf{e}}(t)) + V_{\tilde{\mathbf{e}}}^T \overline{\mathbf{v}}(t) + \frac{1}{2}\sum_{i=1}^N \tilde{\sigma}_i^T(\tilde{\mathbf{e}}(t)) \\ &\times V_{\tilde{\mathbf{e}}\tilde{\mathbf{e}}}\tilde{\sigma}_i(\tilde{\mathbf{e}}(t)) + \sum_{i=1}^N \lambda_i\{V(\tilde{\mathbf{e}}(t) + \tilde{\Gamma}_i(\tilde{\mathbf{e}}(t))) - V(\tilde{\mathbf{e}}(t))\}dt\} \end{aligned} \tag{2A.8}$$

By using Lemma 2.2 with $M_1^T = \frac{1}{2}V_{\tilde{\mathbf{e}}}^T, M_2^T = \overline{\mathbf{v}}(t)$, and $Z = \rho^2 I$, the term $V_{\tilde{\mathbf{e}}}^T\overline{\mathbf{v}}(t)$ from (2A.8) can be bounded as follows:

$$V_{\tilde{\mathbf{e}}}^T\overline{\mathbf{v}}(t) = \frac{1}{2}V_{\tilde{\mathbf{e}}}^T\overline{\mathbf{v}}(t) + \frac{1}{2}\overline{\mathbf{v}}^T(t)V_{\tilde{\mathbf{e}}} \le \frac{1}{4\rho^2}V_{\tilde{\mathbf{e}}}^T V_{\tilde{\mathbf{e}}} + \rho^2\overline{\mathbf{v}}^T(t)\overline{\mathbf{v}}(t) \tag{2A.9}$$

Replacing the term $V_{\tilde{\mathbf{e}}}^T \bar{\mathbf{v}}(t)$ in (2A.8) by the inequality (2A.9), we get the following inequality:

$$\begin{aligned} E\{\int_0^{t_f} \tilde{\mathbf{e}}^T(t)Q\tilde{\mathbf{e}}(t)\} &\le E\{V(\tilde{\mathbf{e}}_0)\} + E\{\int_0^{t_f} \tilde{\mathbf{e}}^T(t)Q\tilde{\mathbf{e}}(t) + V_{\tilde{\mathbf{e}}}^T \tilde{f}(\tilde{\mathbf{e}}(t)) \\ &+ V_{\tilde{\mathbf{e}}}^T \tilde{g}(\tilde{\mathbf{e}}(t))\mathbf{K}(\tilde{\mathbf{e}}(t)) + \frac{1}{4\rho^2} V_{\tilde{\mathbf{e}}}^T V_{\tilde{\mathbf{e}}} + \rho^2 \bar{\mathbf{v}}^T(t)\bar{\mathbf{v}}(t) \\ &+ \frac{1}{2}\sum_{i=1}^{N} \tilde{\sigma}_i^T(\tilde{\mathbf{e}}(t)) V_{\tilde{\mathbf{e}}\tilde{\mathbf{e}}} \tilde{\sigma}_i(\tilde{\mathbf{e}}(t)) + \sum_{i=1}^{N} \lambda_i\{V(\tilde{\mathbf{e}}(t) + \tilde{\Gamma}_i(\tilde{\mathbf{e}}(t))) \\ &- V(\tilde{\mathbf{e}}(t))\}dt\} \end{aligned} \tag{2A.10}$$

If the following Hamilton-Jacobi inequality is satisfied:

$$\begin{aligned} &\tilde{\mathbf{e}}^T(t)Q\tilde{\mathbf{e}}(t) + V_{\tilde{\mathbf{e}}}^{\mathrm{T}} \tilde{f}(\tilde{\mathbf{e}}(t)) + V_{\tilde{\mathbf{e}}}^T \tilde{g}(\tilde{\mathbf{e}}(t))\mathbf{K}(\tilde{\mathbf{e}}(t)) + \frac{1}{4\rho^2} V_{\tilde{\mathbf{e}}}^T V_{\tilde{\mathbf{e}}} \\ &+ \frac{1}{2}\sum_{i=1}^{N} \tilde{\sigma}_i^T(\tilde{\mathbf{e}}(t)) V_{\tilde{\mathbf{e}}\tilde{\mathbf{e}}} \tilde{\sigma}_i(\tilde{\mathbf{e}}(t)) + \sum_{i=1}^{N} \lambda_i\{V(\tilde{\mathbf{e}}(t) + \tilde{\Gamma}_i(\tilde{\mathbf{e}}(t))) - V(\tilde{\mathbf{e}}(t))\} \le 0 \end{aligned} \tag{2A.11}$$

then we immediately have

$$E\{\int_0^{t_f} \tilde{\mathbf{e}}^T(t)Q\tilde{\mathbf{e}}(t)\} \le E\{V(\tilde{\mathbf{e}}_0)\} + E\{\int_0^{t_f} \rho^2 \bar{\mathbf{v}}^T(t)\bar{\mathbf{v}}(t)\} \tag{2A.12}$$

that is, $J_\infty(\mathbf{U}(t)) \le \rho^2$ is satisfied in (2.15).

If the external disturbance is of finite energy in the system, i.e., $\bar{\mathbf{v}}(t) \in \mathcal{L}_{\mathcal{F}}^2(\mathbb{R}^+, \mathbb{R}^{12(N+1)})$, then the r.h.s. of (2A.12) is finite, which shows the total energy of $\tilde{\mathbf{e}}(t)$ in the interval $[0,\infty]$ is bounded, and we have $E[\tilde{\mathbf{e}}^T(t)\tilde{\mathbf{e}}(t)] \to 0$, as $t \to \infty$, that is, the asymptotic tracking for the multi-quadrotor UAV networked control system is achieved; hence proved.

APPENDIX 2C: PROOF OF THEOREM 2.3

From the result of Theorem 2.2, if the HJI in (2.14) is satisfied, the stochastic nonlinear multi-quadrotor UAV networked control system achieves the asymptotic tracking in probability. However, the HJI is hard to be solved. We can transform the HJI to a set of LMIs by using the T-S fuzzy model proposed in Section 2.4. With the T-S fuzzy model in (2.20) and Lemma 2.2, we can get the following equalities:

$$V_{\tilde{\mathbf{e}}}^T \tilde{f}(\tilde{\mathbf{e}}(t)) = \sum_{k=1}^{L} H_k(\bar{X})\tilde{\mathbf{e}}^T(t)[A_k^T P + PA_k]\tilde{\mathbf{e}}(t) \tag{2A.13}$$

$$V_{\tilde{\mathbf{e}}}{}^T \tilde{g}(\tilde{\mathbf{e}}(t))\mathbf{U}(t) = \sum_{k=1}^{L} H_k(\bar{X}) \sum_{l=1}^{L} H_l(\bar{X}) \tilde{\mathbf{e}}^T(t) 2PB_k K_l \tilde{\mathbf{e}}(t)$$
$$= \sum_{k=1}^{L} H_k(\bar{X}) \sum_{l=1}^{L} H_l(\bar{X}) \tilde{\mathbf{e}}^T(t)[PB_k K_l + (PB_k K_l)^T]\tilde{\mathbf{e}}(t) \tag{2A.14}$$

$$\frac{1}{2}\sum_{i=1}^{N} \tilde{\sigma}_i(\tilde{\mathbf{e}}(t)) V_{\tilde{\mathbf{e}}\tilde{\mathbf{e}}} \tilde{\sigma}_i(\tilde{\mathbf{e}}(t))$$
$$= \sum_{i=1}^{N} (\sum_{k=1}^{L} H_k(\bar{X}) C_{ik} \tilde{\mathbf{e}}(t))^T P \times (\sum_{j=1}^{L} H_j(\bar{X}) C_{ij} \tilde{\mathbf{e}}(t))^T \tag{2A.15}$$
$$\leq \sum_{k=1}^{L} H_k(\bar{X}) \tilde{\mathbf{e}}^T(t) \sum_{i=1}^{N} C_{ik}{}^T P C_{ik}]\tilde{\mathbf{e}}(t)$$

$$\sum_{i=1}^{N} \lambda_i \{V(\tilde{\mathbf{e}}(t) + \tilde{\Gamma}_i(\tilde{\mathbf{e}}(t))) - V(\tilde{\mathbf{e}}(t)\}$$
$$= \sum_{i=1}^{N} \lambda_i \{(\tilde{\mathbf{e}}(t) + \tilde{\Gamma}_i(\tilde{\mathbf{e}}(t)))^T P \times (\tilde{\mathbf{e}}(t) + \tilde{\Gamma}_i(\tilde{\mathbf{e}}(t))) - \tilde{\mathbf{e}}^T(t) P \tilde{\mathbf{e}}(t)\}$$
$$= \sum_{i=1}^{N} \lambda_i \{(\tilde{\mathbf{e}}(t) + \sum_{k=1}^{L} H_k(\bar{X}) D_{ik} \times \tilde{\mathbf{e}}(t))^T P(\tilde{\mathbf{e}}(t) \tag{2A.16}$$
$$+ \sum_{k=1}^{L} H_k(\bar{X}) D_{ik} \tilde{\mathbf{e}}(t)) - \tilde{\mathbf{e}}^T(t) P \tilde{\mathbf{e}}(t)\}$$
$$\leq \sum_{k=1}^{L} H_k(\bar{X}) \tilde{\mathbf{e}}^T(t)[\sum_{i=1}^{N} \lambda_i \{D_{ik}^T P D_{ik} + D_{ik}^T P + P D_{ik}\}]\tilde{\mathbf{e}}(t)$$

By (2A.13)–(2A.16), the HJI in (2.14) can be replaced by:

$$\sum_{k=1}^{L}\sum_{l=1}^{L} H_k(\bar{X}) H_l(\bar{X}) \tilde{\mathbf{e}}^T(t)[A_k^T P + PA_k + PB_k K_l + (PB_k K_l)^T$$
$$+ \sum_{i=1}^{N} C_{ik}{}^T P C_{ik} + \sum_{i=1}^{N} \lambda_i (D_{ik}^T P D_{ik} + D_{ik}^T P + P D_{ik})]\tilde{\mathbf{e}}(t) < 0 \tag{2A.17}$$

If the following algebraic Riccati like inequalities are satisfied:

$$A_k^T P + PA_k + PB_k K_l + (PB_k K_l)^T + \sum_{i=1}^{N} C_{ik}{}^T P C_{ik}$$
$$+ \sum_{i=1}^{N} \lambda_i (D_{ik}^T P D_{ik} + D_{ik}^T P + P D_{ik}) < 0 \tag{2A.18}$$

for $k,l = 1,2,...,L$, then the HJI constraint in (2.14) the stochastic nonlinear multi-quadrotor UAV networked system without external disturbance could achieve the asymptotic tracking in probability.

By pre-multiplying and post-multiplying $W = P^{-1}$ to (2A.18), we have:

$$WA_k^T + A_kW + B_kY_l + (B_kY_l)^T + \sum_{i=1}^{N} WC_{ik}^{\ T} W^{-1} C_{ik} W$$
$$+ \sum_{i=1}^{N} \lambda_i (WD_{ik}^T W^{-1} D_{ik} W + WD_{ik}^T + D_{ik} W) < 0$$

for $k,l = 1,2,...,L$ with $Y_l = K_l W$. Then, by Schur complement [35], the above Riccati-like inequalities are equivalent to the LMIs in (2.22). This completes the proof.

APPENDIX 2D: PROOF OF THEOREM 2.4

From the result of Theorem 2.2, the H_∞ tracking performance (2.15) of stochastic nonlinear multi-quadrotor UAV networked tracking system (2.9) is achieved when the inequality (2.16) is satisfied. By substituting the Lyapunov function in (2.21) and the T-S fuzzy model in (2.20), we can get the following equalities:

$$V_{\tilde{\mathbf{e}}}^{\ T} \tilde{f}(\tilde{\mathbf{e}}(t)) = \sum_{k=1}^{L} H_k(\bar{X}) \tilde{\mathbf{e}}^T(t) [A_k^T P + PA_k] \tilde{\mathbf{e}}(t) \tag{2A.19}$$

$$\begin{aligned} V_{\tilde{\mathbf{e}}}^{\ T} \tilde{g}(\tilde{\mathbf{e}}(t)) \mathbf{U}(t) &= \sum_{k=1}^{L} H_k(\bar{X}) \sum_{l=1}^{L} H_l(\bar{X}) \tilde{\mathbf{e}}^T(t) 2PB_kK_l \tilde{\mathbf{e}}(t) \\ &= \sum_{k=1}^{L} H_k(\bar{X}) \sum_{l=1}^{L} H_l(\bar{X}) \tilde{\mathbf{e}}^T(t) [PB_kK_l + (PB_kK_l)^T] \tilde{\mathbf{e}}(t) \end{aligned} \tag{2A.20}$$

$$\begin{aligned} &\frac{1}{2} \sum_{i=1}^{N} \tilde{\sigma}_i \left(\tilde{\mathbf{e}}(t) \right) V_{\tilde{\mathbf{e}}\tilde{\mathbf{e}}} \tilde{\sigma}_i \left(\tilde{\mathbf{e}}(t) \right) \\ &= \sum_{i=1}^{N} \left(\sum_{k=1}^{L} H_k(\bar{X}) C_{ik} \tilde{\mathbf{e}}(t) \right)^T P \times \left(\sum_{j=1}^{L} H_j(\bar{X}) C_{ij} \tilde{\mathbf{e}}(t) \right)^T \\ &\le \sum_{k=1}^{L} H_k(\bar{X}) \tilde{\mathbf{e}}^T(t) [\sum_{i=1}^{N} C_{ik}^{\ T} P C_{ik}] \tilde{\mathbf{e}}(t) \end{aligned} \tag{2A.21}$$

$$\begin{aligned}
&\sum_{i=1}^{N}\lambda_i\{V(\tilde{\mathbf{e}}(t)+\tilde{\Gamma}_i(\tilde{\mathbf{e}}(t)))-V(\tilde{\mathbf{e}}(t))\} \\
&=\sum_{i=1}^{N}\lambda_i\{(\tilde{\mathbf{e}}(t)+\tilde{\Gamma}_i(\tilde{\mathbf{e}}(t)))^T P\times(\tilde{\mathbf{e}}(t)+\tilde{\Gamma}_i(\tilde{\mathbf{e}}(t)))-\tilde{\mathbf{e}}^T(t)P\tilde{\mathbf{e}}(t)\} \\
&=\sum_{i=1}^{N}\lambda_i\{(\tilde{\mathbf{e}}(t)+\sum_{k=1}^{L}H_k(\bar{X})D_{ik}\times\tilde{\mathbf{e}}(t))^T P(\tilde{\mathbf{e}}(t) \\
&+\sum_{k=1}^{L}H_k(\bar{X})D_{ik}\tilde{\mathbf{e}}(t))-\tilde{\mathbf{e}}^T(t)P\tilde{\mathbf{e}}(t)\} \\
&\le\sum_{k=1}^{L}H_k(\bar{X})\tilde{\mathbf{e}}^T(t)[\sum_{i=1}^{N}\lambda_i\{D_{ik}^T PD_{ik}+D_{ik}^T P+PD_{ik}\}]\tilde{\mathbf{e}}(t)
\end{aligned} \tag{2A.22}$$

By using (2A.19) to (2A.22), (2A.10) can be replaced by:

$$\begin{aligned}
&E\{\int_0^{t_f}\tilde{\mathbf{e}}^T(t)Q\tilde{\mathbf{e}}(t)\}\le E\{V(\tilde{\mathbf{e}}_0)\}+E\{\int_0^{t_f}[\sum_{k=1}^{L}\sum_{l=1}^{L}H_k(\bar{X}) \\
&\times H_l(\bar{X})\tilde{\mathbf{e}}^T(t)[Q+A_k^T P+PA_k+PB_kK_l+(PB_kK_l)^T+\frac{1}{\rho^2}PP \\
&+\sum_{i=1}^{N}C_{ik}{}^T PC_{ik}+\sum_{i=1}^{N}\lambda_i(D_{ik}^T PD_{ik}+D_{ik}^T P+PD_{ik})]\tilde{\mathbf{e}}(t)]dt\} \\
&+E\{\int_0^{t_f}\rho^2\bar{\mathbf{v}}^T(t)\bar{\mathbf{v}}(t)dt\}
\end{aligned} \tag{2A.23}$$

If the following algebraic Riccati-like inequalities are satisfied:

$$\begin{aligned}
&Q+A_k^T P+PA_k+PB_kK_l+(PB_kK_l)^T+\frac{1}{\rho^2}PP+\sum_{i=1}^{N}C_{ik}{}^T PC_{ik} \\
&+\sum_{i=1}^{N}\lambda_i(D_{ik}^T PD_{ik}+D_{ik}^T P+PD_{ik})\le 0,\ for\ k,l=1,2,\cdots,n
\end{aligned} \tag{2A.24}$$

then we immediately have

$$E\{\int_0^{t_f}\tilde{\mathbf{e}}^T(t)Q\tilde{\mathbf{e}}(t)\}\le E\{V(\tilde{\mathbf{e}}_0)\}+E\{\int_0^{t_f}\rho^2\bar{\mathbf{v}}^T(t)\bar{\mathbf{v}}(t)dt\}$$

, that is, $J_\infty(\mathbf{U}(t))\le\rho^2$ is satisfied.

Let $W = P^{-1}$ and $Y_l = K_l W$; then multiplying W to two sides of (2A.24), we get

$$\begin{aligned} & WQW + WA_k^T + A_k W + B_k Y_l + Y_l^T B_k^T + \frac{1}{\rho^2} I \\ & + \sum_{i=1}^{N} WC_{ik}{}^T W^{-1} C_{ik} W + \sum_{i=1}^{N} \lambda_i (WD_{ik}^T W^{-1} D_{ik} W + WD_{ik}^T + D_{ik} W) \leq 0 \end{aligned} \tag{2A.25}$$

By Schur complement [35], the Riccati-like inequalities in (2A.25) are equivalent to the LMIs in (2.23). The H_∞ team formation tracking control performance is achieved with a prescribed ρ^2.

If the external disturbance is of finite energy in the system, that is, $\overline{\mathbf{v}}(t) = \mathcal{L}_{\mathcal{F}}^2(\mathbb{R}_3^+, \mathbb{R}^{n+1})$, from (2A.24), the (2A.23) can be rewritten as follows:

$$E\{\int_0^{t_f} \tilde{\mathbf{e}}^T(t) Q \tilde{\mathbf{e}}(t) dt\} \leq E\{V(\tilde{\mathbf{e}}_0)\} + E\int_0^{t_f} \overline{V}(t) \overline{V}(t) dt \forall t_f \in [0, \infty] \tag{2A.26}$$

From (2A.26), since the right-hand side of (2A.26) is finite, we have $E[\tilde{\mathbf{e}}^T(t)\tilde{\mathbf{e}}(t)] \rightarrow 0$ as $t \rightarrow \infty$, that is, the asymptotic tracking for the multi-quadrotor UAV networked system is achieved too. This completes the proof.

REFERENCES

[1] M. A. Ma'sum, M. K. Arrofi, G. Jati, F. Arifin, M. N. Kurniawan, P. Mursanto, and W. Jatmiko, "Simulation of intelligent Unmanned Aerial Vehicle (UAV) For military surveillance," Proceedings of The IEEE 2013 International Conference on Advanced Computer Science and Information Systems (iCACSIS), Sanur Bali, Indonesia, 2013, pp. 161–168.

[2] D. H. Lyon, "A military perspective on small unmanned aerial vehicles," *IEEE Instrumentation and Measurement Magazine*, vol. 7, no. 3, pp. 27–31, Sep. 2004.

[3] N. Nigam, "The multiple unmanned air vehicle persistent surveillance problem: A review," *Machines*, vol. 2, no. 1, pp. 13–72, Jan. 2014.

[4] J. J. Acevedo, B. C. Arrue, I. Maza, and A. Ollero, "Distributed approach for coverage and patrolling missions with a team of heterogeneous aerial robots under communication constraints," *International Journal of Advanced Robotic Systems*, vol. 10, no. 28, pp. 13–72, Jan. 2014.

[5] J. J. Acevedo, B. C. Arrue, J. M. Diaz-Banez, I. Ventura, I. Maza, and A. Ollero, "One-to-one coordination algorithm for decentralized area partition in surveillance missions with a team of aerial robots," *Journal of Intelligent and Robotic Systems*, vol. 74, no. 1–2, pp. 269–285, Apr. 2014.

[6] P. Pounds, R. Mahony, and P. Corke, "Modelling and control of a large quadrotor robot," *Photoelectric System*, vol. 18, no. 7, pp. 691–699, Feb. 2010.

[7] T. Luukkonen, *Modelling and Control of Quadcopter*, Aalto University, School of Science, Espoo, Finland, Aug. 2011.

[8] T. Bresciani, *Modelling, Identification and Control of a Quadrotor Helicopter*, Lund University, Department of Automatic Control, Lund, Sweden, Oct. 2008.

[9] L. M. Argentim, W. C. Rezende, P. E. Santos, and R. A. Aguiar, "PID LQR and LQR-PID on a quadcopter platform," *IEEE International Conference on Industrial Informatics, Electronics and Vision*, Dhaka, Bangladesh, May 2013, pp. 1–6.
[10] M. Herrera, W. Chamorro, A. Gomez, and O. Camacho, "Sliding mode control: An approach to control a quadrotor," in *IEEE Asia-Pacific Conference on Computer Aided System Engineering*, Quito, Ecuador, July 2015, pp. 3–4.
[11] T. Madani and A. Benallegue, "Backstepping control for a quadrotor helicopter," in *IEEE International Conference on Intelligent Robots and Systems*, Beijing, China, Jan. 2006, pp. 3255–3260.
[12] S. S. Ge, E. Hong, and T. H. Lee, "Robust adaptive control of nonlinear systems with unknown time delays," *Automatica*, vol. 41, no. 7, pp. 1181–1190, July 2005.
[13] R. Lozano, P. Castillo, P. Garcia, and A. Dzul, Robust prediction-based control for unstable delay systems: Application to the yaw control of a mini-helicopter," *Automatica*, vol. 40, no. 4, pp. 603–612, Apr. 2004.
[14] A. Das, K. Subbarao, and F. Lewis, "Dynamic inversion with zero-dynamics stabilisation for quadrotor control," *IET Control Theory and Applications*, vol. 3, no. 3, pp. 303–314, Mar. 2009.
[15] D. J. Bennet, C. R. McInnes, M. Suzuki, and K. Uchiyama, "Autonomous three-dimensional formation flight for a swarm of unmanned aerial vehicles," *Journal of Guidance, Control, and Dynamics*, vol. 34, no. 6, pp. 1899–1908, Dec. 2011.
[16] M. Defoort, T. Floquet, A. Kokosy, and W. Perruquetti, "Sliding-mode formation control for cooperative autonomous mobile robots," *IEEE Transactions on Industrial Electronics*, vol. 55, no. 11, pp. 3944–3953, Nov. 2008.
[17] J. Lawton, R. Beard, and B. Young, "A decentralized approach to formation maneuvers," *IEEE Transactions on Robotics and Automation*, vol. 19, no. 6, pp. 933–941, Dec. 2003.
[18] P. Ogren, M. Egerstedt, and X. Hu, "A control lyapunov function approach to multiagent coordination," *IEEE Transactions on Robotics and Automation*, vol. 18, no. 5, pp. 847–851, Oct. 2002.
[19] W. Yu, G. Chen, W. Ren, J. Kurths, and W. Zheng, "Distributed higher order consensus protocols in multiagent dynamical systems," *IEEE Transactions on Circuits and Systems*, vol. 58, no. 8, pp. 1924–1932, Feb. 2011.
[20] S. J. Ho, C. Y. Chen, and B. S. Chen, "Analysis for the robust H_∞ synchronization of nonlinear stochastic coupling networks through Poisson processes and core coupling design," *IEEE Transactions on Control of Network Systems*, vol. 4, no. 2, pp. 223–235, June 2017.
[21] Z. Li, Z. Duan, G. Chen, and L. Huang, "Consensus of multiagent systems and synchronization of complex networks: A unified viewpoint," *IEEE Transactions on Circuits and Systems I: Regular Chapters*, vol. 57, no. 1, pp. 213–224, June 2009.
[22] Z. Li, Z. Duan, and L. Huang, "Leader-follower consensus of multiagent systems," in *American Control Conference*, St. Louis, MO, June 2009, pp. 3256–3261.
[23] C. Ramazan, I. Serhat, and S. Serhat, "Statistical Wiener process model for vibration signals in accelerated aging processes of electric motors," *Journal of Vibroengineering*, vol. 16, no. 2, pp. 1392–8716, Mar. 2014.
[24] R. C. L. F. Oliveira, A. N. Vargas, J. B. R. do Val, and P. L. D. Peres, "Mode-independent H2-control of a DC motor modeled as a Markov jump linear system," *IEEE Transactions on Control System Technology*, vol. 22, no. 5, pp. 1915–1919, Sep. 2014.
[25] B. Øksendal and A. Sulem, *Applied Stochastic Control of Jump Diffusions*, 2nd ed., Berlin: Springer, 2007.

[26] F. Hanson, *Applied Stochastic Processes and Control for Jump-Diffusions: Modeling, Analysis and Computation*, 2nd ed., Philadelphia, PA: SIAM, 2007.
[27] B. S. Chen, B. K. Lee, and L. B. Guo, "Optimal tracking design for stochastic fuzzy systems," *IEEE Transactions on Fuzzy Systems*, vol. 11, no. 6, pp. 796–813, Dec. 2003.
[28] K. Watanabe, "Stochastic fuzzy control. I. Theoretical derivation," in *IEEE International Conference on Fuzzy Systems*, Yokohama, Japan, Mar. 1995, pp. 547–554.
[29] K. Watanabe, K. Izumi, and F. Han, "Stochastic fuzzy servo control using multiple linear dynamic models," in *IEEE International Conference on Knowledge-Based Intelligent Electronic Systems*, Adelaide, SA, Australia, Apr. 1998, pp. 474–482.
[30] C. S. Tseng, B. S. Chen, and H. J. Uang, "Fuzzy tracking control design for nonlinear dynamic systems via T-S fuzzy model," *IEEE Transactions on Fuzzy Systems*, vol. 9, no. 3, pp. 381–392, June 2001.
[31] T. Takagi and M. Sugeno, "Fuzzy identification of systems and its applications to modeling and control," *IEEE Transactions on Systems, Man, and Cybernetics*, vol. 15, no. 1, pp. 116–132, Feb. 1985.
[32] B. Kosko, "Fuzzy systems as universal approximators," *IEEE Transactions on Computers*, vol. 43, no. 11, pp. 1329–1333, Nov. 1994.
[33] J. J. Xiong and E. H. Zheng, "Position and attitude tracking control for a quadrotor UAV," *ISA Transactions*, vol. 53, no. 3, pp. 725–731, May 2014.
[34] Z. Lin, W. Ding, G. Yan, C. Yu, and A. Giua, "Leader–follower formation via complex Laplacian," *Automatica*, vol. 49, no. 6, pp. 1900–1906, June 2013.
[35] S. Boyd, L. E. Ghaoui, E. Feron, and V. Balakrishnan, *Linear Matrix Inequalities in System and Control Theory*, Philadelphia, PA: SIAM, 1994.

3 Stochastic Robust H_∞ Team Formation Tracking Design of Multi-Quadrotor UAV Networked Control System under Time-Varying Delay and Random Fluctuation in Smart City

3.1 INTRODUCTION

UNMANNED aerial vehicles (UAVs) have received a growing interest no matter in academic research or industrial field in recent years. Unlike the fixed-wing UAV, the quadrotor UAV uses multiple rotors to generate vertical thrust, which can force the quadrotor UAV body hovering in the air or flying in the vertical motion like helicopter [1]. Thus, due to its high maneuverability and payload capacity, the quadrotor-UAV has given rise to interest nowadays. Physically, the quadrotor UAV is an under-actuated system with six degrees of freedom which contains three attitudes and three positions but only four independent control inputs. Although the under-actuated system can reduce the manufacturing difficulty of the system, the strong nonlinear coupling and static instability will make the quadrotor UAV control problem very difficult and challenging [2–4].

Recently, one of the emerging areas of quadrotor UAVs is their involvement in future smart cities. Since the quadrotor UAV missions become more and more complex and challenging, various missions need to be completed by the team formation flight, such as agriculture application, battlefield reconnaissance, and multi-target attacking [5,6]. Nowadays, there are several formation control strategies, such as virtual leader structure (V-S) approach [7], leader-follower (L-F) approach [8], and behavior-based method [9]. The idea of L-F approach is that one of the quadrotor UAVs is selected as the leader to represent the team behavior and is asked to track

DOI: 10.1201/9781003545965-3

the reference path, while other quadrotor UAVs are the followers which are required to track the trajectory of the leader quadrotor UAV with a team formation.

Despite the fact that L-F approach is the most earlier developed and commonly used method, the main drawback is that the entire team formation depends on only one agent (leader). Thus, if the leader quadrotor UAV is crashed during the guidance process, the whole quadrotor UAV team formation cannot be maintained [10]. Compared with the L-F approach, V-S approach can avoid aforementioned problems and becomes more robust. The concept of the V-S approach is that the virtual leader quadrotor UAV is at the central of the desired quadrotor UAV team formation shape to represent the movement of the team formation. In this situation, the virtual leader generates the team formation reference path [11], and then each quadrotor UAV in the team formation has its own trajectory which is asked to track the reference path of the virtual leader quadrotor UAV with a certain shape. Hence, there are no interactions between the agents in the V-S approach [12]. The main cause of robustness is that the virtual leader quadrotor UAV will not suffer from the perturbation and can substantially reduce the possibility of team formation failure in reality. Several experimental works of V-S control strategy of team formation can be found in References [13,14].

In recent years, advances in wireless communication technologies have facilitated multi-agent control over wireless communication networks [15,16]. Most of team formation control schemes within the networked control system (NCS) so far are based on the time-triggered communication, which makes full use of the sampled data at every time instant. Under the time-triggered scheme of NCS, different transmission signals will occupy certain channel resources at the same bandwidth. However, in a smart city [17], due to the limited wireless communication bandwidth and the massive number of agents, it is necessary and important to consider the energy-waste and resource-saving problems. To overcome these power-saving problems, the event-triggered mechanism is proposed to replace the conventional time-triggered mechanism in networked control system (NCS). In the event-triggered mechanism, the event (energy constraint) will be executed and the controlled command will be transmitted only when the event-triggered condition is met, that is, the controller will be updated only when it is necessary. As a result, event-triggered mechanism is regarded as an efficient way to reduce the load of the wireless communication network. An event-triggered state-feedback approach is proposed in Reference [18] and the event-triggered output-feedback control is developed by studies in Reference [19]. In Reference [20], the authors proposed the general event-triggered controller with a fixed constant threshold from the mean square error perspective which is the simplest triggering condition in all kinds of event-triggered schemes. There are other event-triggered schemes such as the triggering function related to the Lyapunov function from the perspective of energy [21–26].

In the previous studies, most of the research about quadrotor UAV NCS have considered the external disturbance or noise as deterministic signals for the convenience of control design [27,28]. However, in the real smart city, the quadrotor UAV NCS will suffer from parametric fluctuations due to the rotors, rigid bodies, electrical circuits and communication channels, or sensors. For example, there are continuous fluctuations from rotors or channel variations and discontinuous sudden voltage jumps from the sensors or electrical circuits [29–33]. Through Itô-Lévy integral [34,35], these intrinsic fluctuations should be modeled by random processes, including continuous

Wiener process and discontinuous Poisson process in the multi-quadrotor UAV team tracking problem. Besides, there exist some external disturbances in multi-quadrotor UAV system during flight process, such as the unpredictable aerodynamic perturbations or wireless interferences on the wireless network. Thus, the multi-quadrotor UAV system in the smart city should be modeled as a stochastic nonlinear system from the practical point of view.

In this chapter, an event-triggered mechanism and a robust state feedback controller are designed simultaneously to ensure that the controlled multi-quadrotor UAV networked control system could gradually track on their desired team path with the desired attitude despite intrinsic random fluctuation in the quadrotor UAV networked control system and external disturbance from the environment. Also, we can save the energy resources and release the load of communication bandwidth in the wireless network with the consideration of the time-varying delay in transmission channel. To deal with the stochastic event-triggered robust team tracking control problem with external disturbance, the H_∞ event-triggered robust team formation tracking control design is proposed to efficiently attenuate the effect of random fluctuation and external disturbance on the team formation tracking performance of the multiple quadrotor UAV networked control system. Then, by using Itô-Lévy formula, the event-triggered nonlinear multi-quadrotor UAV robust optimal H_∞ team formation tracking design problem is transformed to an equivalent Hamilton-Jacobi inequality (HJI)-constraint optimization problem. However, since HJI is a partial differential inequality, it is difficult to be solved analytically and numerically [36,37]. As a result, Takagi-Sugeno (T-S) fuzzy technique [38,39] is adopted to efficiently approximate the nonlinear multi-quadrotor UAV system to simplify the design procedure. T-S fuzzy model can interpolate several locally linearized systems by fuzzy bases to approximate the nonlinear multi-UAV system. Therefore, the HJI-constrained optimization problem for the H_∞ event-triggered robust formation team tracking control can be transformed to a linear matrix inequality (LMI)-constrained optimization problem that can be easily solved with the help of MATLAB LMI toolbox. In the simulation, a square-shape team formation tracking task for four UAVs stochastic networked control system is provided to validate the effectiveness of proposed H_∞ event-triggered robust formation team tracking control.

The main purposes of this chapter are as follows:

(i) To completely describe the more realistic multi-quadrotor UAV team formation tracking system and complicated wireless communication network in future smart city, the stochastic nonlinear event-triggered multi-quadrotor UAV networked control system is firstly formulated with the consideration of time-varying delay, external disturbance, intrinsic continuous Wiener diffusion fluctuations and discontinuous Poisson jump fluctuation.

(ii) Without using the conventional time-triggered control for multi-quadrotor UAV networked control system in previous studies, the event-triggered mechanism is utilized for the multi-quadrotor UAV team formation tracking control design to relieve the communication load and save the energy consumption for the quadrotor UAV networked control system in future smart city. Then, an augmented and shifted nonlinear multi-quadrotor UAV tracking system is developed to transform the complex multi-quadrotor UAV team formation tracking control design to an equivalent stabilization control design of the augmented system to significantly simplify the complex design procedure.

(iii) By constructing the error tracking dynamic and using T-S fuzzy model, the robust optimal H_∞ stochastic event-triggered multi-UAV team formation tracking control could be designed by solving an LMI-constrained optimization scheme to efficiently achieve the dynamic virtual structure team formation tracking. As a result, the robust optimal H_∞ stochastic event-triggered multi-UAV team formation tracking control design can be simply implemented for more practical applications.

NOTATION

A^T: the transpose of matrix A; $A \geq 0 (A > 0)$: symmetric positive semi-definite (symmetric positive definite) matrix A; I_n: the n-dimensional identity matrix; $\|\mathbf{x}\|_2$: the Euclidean norm for the given vector $\mathbf{x} \in \mathbb{R}^n$; C^2: the class of functions $V(\mathbf{x})$ with twice continuous derivatives with respect to $\mathbf{x}$; $f_{\mathbf{x}}$:the gradient column vector of continuously differentiable function $\mathbf{f}(\mathbf{x})$ (i.e., $\frac{\partial f(\mathbf{x})}{\partial \mathbf{x}}$); $f_{\mathbf{xx}}$: the Hessian matrix with elements of second partial derivatives of twice continuously differentiable function $\mathbf{f}(\mathbf{x})$, (i.e., $\frac{\partial^2 f(\mathbf{x})}{\partial \mathbf{x}^2}$); $\mathbf{E}$: the expectation operator; $\mathcal{L}_{\mathcal{F}}^{\,2}(\mathbb{R}^+, \mathbb{R}^l)$: the space of nonanticipative stochastic processes $y(t) \in \mathbb{R}^l$ with respect to an increasing σ-algebras $\mathcal{F}_t (t \geq 0)$ satisfying $\|\mathbf{y}(t)\|_{\mathcal{L}^2(\mathbb{R}^+;\mathbb{R}^l)} \triangleq \mathbf{E}\left\{\int_0^\infty \mathbf{y}^T(t)\mathbf{y}(t)dt\right\}^{\frac{1}{2}} < \infty$; The matrix $\begin{bmatrix} A & B \\ B^T & C \end{bmatrix}$ is represented as $\begin{bmatrix} A & B \\ * & C \end{bmatrix}$ for the simplicity.

3.2 SYSTEM DESCRIPTION AND PRELIMINARIES

3.2.1 Stochastic System Model of the Quadrotor UAV NCS

In order to deal with the attitude and position tracking control in the quadrotor UAV NCS, we consider a translation subsystem with respect to the inertial frame and the attitude subsystem with respect to the body frame, respectively. By considering the velocity and acceleration of the translation subsystem, and the angular velocity and acceleration of the attitude subsystem, the quadrotor dynamic equations can be described as a dynamic system of 12 components, including the translation dynamic of quadrotor-related UAV to its position motion and the attitude dynamic associated with its angular motion. By the Euler method and the conception above, the dynamic system of the quadrotor UAV in Figure 3.1 can be described as follows [45]:

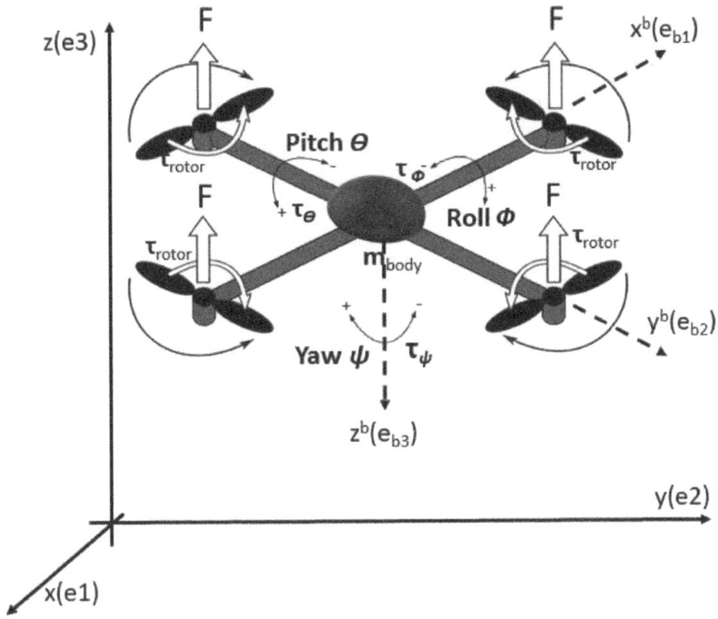

FIGURE 3.1 The *i*th quadrotor UAV dynamic system with the attitude and translation subsystem

$$
\begin{aligned}
&\dot{x}_1(t) = x_2(t) \\
&\dot{x}_2(t) = -\frac{f_x}{m}x_2(t) + (\cos\phi_1(t)\sin\theta_1(t)\cos\psi_1(t) \\
&\qquad + \sin\phi_1(\mathrm{t})\sin\psi_1(\mathrm{t}))\frac{\mathrm{F(t)}}{\mathrm{m}} + \mathrm{v_x(t)} \\
&\dot{y}_1(t) = y_2(t) \\
&\dot{y}_2(t) = -\frac{f_y}{m}y_2(t) + (\cos\phi_1(t)\sin\theta_1(t)\sin\psi_1(t) \\
&\qquad - \sin\phi_1(\mathrm{t})\cos\psi_1(\mathrm{t}))\frac{\mathrm{F(t)}}{\mathrm{m}} + \mathrm{v_y(t)} \\
&\dot{z}_1(t) = z_2(t) \\
&\dot{z}_2(t) = -\frac{f_x}{m}z_2(t) - g + (\cos\phi_1(t)\cos\theta_1(t))\frac{F(t)}{m} + v_z(t) \\
&\dot{\phi}_1(t) = \phi_2(t) \\
&\dot{\phi}_2(t) = \frac{J_\theta - J_\psi}{J_\phi}\theta_2(t)\psi_2(t) - \frac{f_\phi}{J_\phi}\phi_2(t) + \frac{1}{J_\phi}\tau_\phi(t) + v_\phi(t)
\end{aligned}
\tag{3.1}
$$

$$\dot{\theta}_1(t) = \theta_2(t)$$
$$\dot{\theta}_2(t) = \frac{J_\psi - J_\phi}{J_\theta}\phi_2(t)\psi_2(t) - \frac{f_\theta}{J_\theta}\theta_2(t) + \frac{1}{J_\theta}\tau_\theta(t) + v_\theta(t)$$
$$\dot{\psi}_1(t) = \psi_2(t)$$
$$\dot{\psi}_2(t) = \frac{J_\phi - J_\theta}{J_\psi}\phi_2(t)\theta_2(t) - \frac{f_\psi}{J_\psi}\psi_2(t) + \frac{1}{J_\psi}\tau_\psi(t) + v_\psi(t)$$

where the attitude vector is denoted by $\Theta(t) = [\phi_1(t), \theta_1(t), \psi_1(t)]^T$ in the body frame associated with the unit vector basis (e_{b1}, e_{b2}, e_{b3}). The parameters $\phi_1(t), \theta_1(t), \psi_1(t)$ are Euler angles of rotation fixed to the body of a quadrotor UAV. These three angles respectively represent roll angle ($-\frac{\pi}{2} < \phi_1(t) < \frac{\pi}{2}$), pitch angle ($-\frac{\pi}{2} < \theta_1(t) < \frac{\pi}{2}$), and yaw angle ($-\pi < \psi_1(t) < \pi$) related to the orientation of the quadrotor to be controlled by changing the rotational speed of four rotors. The position vector $\Xi(t) = [x_1(t), y_1(t), z_1(t)]^T$ represents the position of mass center of the quadrotor in the inertial frame associated with the unit vector basis (e_1, e_2, e_3). Specifically, $x_1(t), y_1(t), z_1(t) \in \mathbb{R}^1$ are the position states of UAV, which are on the Cartesian coordinate with respect to the inertial frame. $v_x(t)$, $v_y(t)$, and $v_z(t)$ are the external disturbances of the quadrotor UAV in the three translation dynamics. $v_\phi(t)$, $v_\theta(t)$, and $v_\psi(t)$ are the external disturbances of the quadrotor UAV caused by the unexpected rotation force in raw, pitch, and yaw dynamics, respectively. f_x, f_y, $f_z \in \mathbb{R}^+$ are the coefficients of the translation drag forces and f_ϕ, f_θ, $f_\psi \in \mathbb{R}^+$ represent the aerodynamic friction coefficient of the quadrotor. The total thrust $F(t) \in \mathbb{R}^1$ and the rotational forces $\tau_\phi(t), \tau_\theta(t), \tau_\psi(t) \in \mathbb{R}^1$ are produced by four rotors of the quadrotor UAV. $m \in \mathbb{R}^+$ is the total mass of the quadrotor UAV and g denotes the acceleration of gravity. $J_\phi, J_\theta, J_\psi \in \mathbb{R}^+$ are the moments of inertia of the quadrotor UAV.

In the realistic situation, the quadrotor UAV system will unavoidably suffer from parametric fluctuations due to the rotors, rigid bodies, electrical circuits or sensors. To mimic the real quadrotor UAV system, the dynamical model of the ith UAV in (3.1) should be modified by continuous and discontinuous intrinsic random fluctuations as the following nonlinear stochastic system:

$$\begin{aligned} dx_i(t) = (f_i(x_i(t)) + g_i(x_i(t))u_i(t) + v_i(t))dt + \sigma(x_i(t))dW_i(t) \\ + \Gamma(x_i(t))dN_i(t) \end{aligned} \tag{3.2}$$

with

$$f_i(x_i(t)) = [x_2^i(t), -\frac{f_x^i}{m^i}x_2^i, y_2{}^i(t), -\frac{f_y^i}{m^i}y_2^i, z_2^i(t), \text{-g-}\frac{\mathrm{f}_\mathrm{z}^\mathrm{i}}{\mathrm{m}^\mathrm{i}}\mathrm{z}_2^\mathrm{i}(\mathrm{t}), \phi_2{}^\mathrm{i}(\mathrm{t}),$$
$$\frac{\mathrm{J}_\theta^\mathrm{i}\text{-}\mathrm{J}_\psi^\mathrm{i}}{\mathrm{J}_\phi^\mathrm{i}}\theta_2^\mathrm{i}(\mathrm{t})\psi_2^\mathrm{i}(\mathrm{t})\text{-}\frac{\mathrm{f}_\phi^\mathrm{i}}{\mathrm{J}_\phi^\mathrm{i}}\phi_2^\mathrm{i}(\mathrm{t}), \theta_2^\mathrm{i}(\mathrm{t}), \frac{\mathrm{J}_\psi^\mathrm{i}\text{-}\mathrm{J}_\phi^\mathrm{i}}{\mathrm{J}_\theta^\mathrm{i}}\phi_2^\mathrm{i}(\mathrm{t})\,\psi_2^\mathrm{i}(\mathrm{t})\text{-}\frac{\mathrm{f}_\theta^\mathrm{i}}{\mathrm{J}_\theta^\mathrm{i}}\theta_2^\mathrm{i}(\mathrm{t}), \psi_2^\mathrm{i}(\mathrm{t}),$$
$$\frac{\mathrm{J}_\phi^\mathrm{i}\text{-}\mathrm{J}_\theta^\mathrm{i}}{\mathrm{J}_\psi^\mathrm{i}}\phi_2^\mathrm{i}(\mathrm{t})\theta_2^\mathrm{i}(\mathrm{t})\text{-}\frac{\mathrm{f}_\psi^\mathrm{i}}{\mathrm{J}_\psi^\mathrm{i}}\psi_2{}^\mathrm{i}(\mathrm{t})]^\mathrm{T}$$

$$g_i(x_i(t)) = \begin{bmatrix} 0 & 0 & 0 & 0 \\ \frac{1}{m^i}(\cos\phi_1^i(t)\sin\theta_1^i(t) \times\cos\psi_1^i(t)+\sin\phi_1^i(t)\times\sin\psi_1^i(t)) & 0 & 0 & 0 \\ 0 & 0 & 0 & 0 \\ \frac{1}{m^i}(\cos\phi_1^i(t)\sin\theta_1^i(t)\times\sin\psi_1^i(t)-\sin\phi_1^i(t)\times\cos\psi_1^i(t)) & 0 & 0 & 0 \\ 0 & 0 & 0 & 0 \\ \frac{1}{m^i}(\cos\phi_1^i(t)\cos\theta_1^i(t)) & 0 & 0 & 0 \\ 0 & 0 & 0 & 0 \\ 0 & \frac{1}{J_\phi^i} & 0 & 0 \\ 0 & 0 & 0 & 0 \\ 0 & 0 & \frac{1}{J_\theta^i} & 0 \\ 0 & 0 & 0 & 0 \\ 0 & 0 & 0 & \frac{1}{J_\psi^i} \end{bmatrix}$$

$$\begin{aligned} x_i(t) &= [x_1^i(t), x_2^i(t), y_1^i(t), y_2^i(t), z_1{}^i(t), z_2^i(t), \phi_1^i(t), \\ &\quad \phi_2{}^i(t), \theta_1^i(t), \theta_2^i(t), \psi_1^i(t), \psi_2{}^i(t)]^T \\ u_i(t) &= [F^i(t), \tau_\phi^i(t), \tau_\theta^i(t), \tau_\psi{}^i(t)]^T \\ v_i(t) &= [0, v_x^i(t), 0, v_y^i(t), 0, v_z^i(t), 0, v_\phi^i(t), 0, v_\theta^i(t), 0, v_\Psi^i(t)]^T \end{aligned}$$

for $i = 1, 2, \cdots, N$, where $x_i(t)$ denotes the state variable of the ith quadrotor UAV and $u_i(t)$ denotes the control input, which is produced by the four rotors of the ith UAV. $v_i(t) \in \mathcal{L}_F^2(\mathbb{R}^{12})$ denotes the finite energy external disturbance in the ith quadrotor UAV networked control system, which is produced by the unpredictable interference from the wireless network and the neighboring quadrotor UAVs. $W_i(t) \in \mathbb{R}^1$ is the Wiener process, which is continuous but non-differentiable, and $\sigma_i(x_i(t))dW_i(t)$ denotes the effect of continuous stochastic intrinsic fluctuation caused by the modeling uncertainty of the ith UAV. $N_i(t) \in \mathbb{R}^1$ denotes the Poisson counting process with jump intensity $\lambda_i > 0$, and $\Gamma_i(x_i(t))dN_i(t)$ denotes the effect of discontinuous abruptly random fluctuation caused by the sudden incident at time instant t such as the packet loss in network transmission or deformations of faults of mechanical elements during the flight. It is assumed that $W_i(t)$ and $N_i(t)$ are independent. The above two processes are defined on the complete filtration probability space $(\Omega, \mathcal{F}, \mathcal{F}_t, \mathcal{P})$, where Ω is the sample space, $\mathcal{F}$ denotes the filtration, and P denotes the probability measure. $\mathcal{F} = \{\mathcal{F}_t : t \geq 0\}$ and $\mathcal{F}_t$ is an σ -algebra filtration generated by the Wiener process $W_i(s)$ and the Poisson counting processes $N_i(s)$, for $s < t$.

Remark 3.1 Some important properties of Wiener process and Poisson counting process are given as follows [34,35]: (1) $E\{W_i(t)\} = E\{dW_i(t)\} = 0$. (2) $E\{dW_i(t)^T dW_j(t)\} = 0$, for $i \neq j$. $E\{dW_i(t)^T dW_j(t)\} = dt$, for $i = j$. (3) $E\{dN_i(t)\} = \lambda_i dt$, where the finite scalar number $\lambda_i > 0$ is the Poisson jump intensity.

3.2.2 Event-Triggered Team Formation Tracking Networked Control System of the Multi-Quadrotor UAV

We now introduce the event-triggered scheme for the ith quadrotor UAV in the multi-quadrotor UAV networked control system, where the sensor is clock-driven while the data transmission is event-driven. To begin with, we first make some definitions related to the time instant. The set $S_1 = \{0, h, 2h, \cdots, jh\}$ denotes the sampling sequence of the sensor under the time-triggered with sampling period $h \in \mathbb{R}^+$ and some $j \in \mathbb{N}$, that is, they are discrete-time instants of all the sampled data from the sensors. The set $S_2 = \{0, t_1 h, t_2 h, \cdots, t_k h\} \subseteq S_1$, for some $t_k \in \mathbb{N}$ denotes the transmission sequence after the event-triggered mechanism, by which whether the sampled data would be transmitted or not is determined by the event-triggered condition. The transmission condition of the event-triggered mechanism can be described as [21–24]:

$$\begin{aligned} t_{k+1}h = t_k h + \min_{l>0, l\in\mathbb{N}} \{lh \mid E\{[x_i(i_k h) - x_i(t_k h)]^T \\ \times \Phi[x_i(i_k h) - x_i(t_k h)]\} \geq \sigma E\{x_i^T(t_k h)\Phi x_i(t_k h)\}\} \end{aligned} \tag{3.3}$$

where $\sigma \in [0,1)$ is the proper threshold parameter, which represents the level of the event-triggered mechanism. $i_k h = t_k h + lh$ means the next sampled instant which should be decided to transmit or not, for some $l \in \mathbb{N}$. Φ is the positive definite event-triggered matrix which should be designed. $x_i(i_k h) - x_i(t_k h)$ is the threshold error between the states at the current sampling instant and the latest transmitted instant, $t_k h$ is the time instant at sensor successively transmitting data to the controller. Note that the parameters σ, Φ and h of the event-triggered scheme in (3.3) are related to the communication load of networked control system. When the event-triggered condition is satisfied, the new transmission $x_i(t_{k+1}h)$ occurs through the wireless network and it will be stored in the event-triggered mechanism for the next computation and decision (see Figure 3.2). Especially, if $\sigma = 0$, it implies the scheme in (3.3) becomes the conventional time-triggered mechanism.

Remark 3.2 Different from the conventional continuous-time event-triggered mechanism, the event-triggered mechanism in (3.3) evaluates the constraint at each time instants kh, for $k \in \mathbb{N}$, with sample period $h > 0$. Thus, the minimum triggered time interval is greater than one sample period h and the Zeno effect is excluded, that is, there are no infinite triggered instants in any finite time interval.

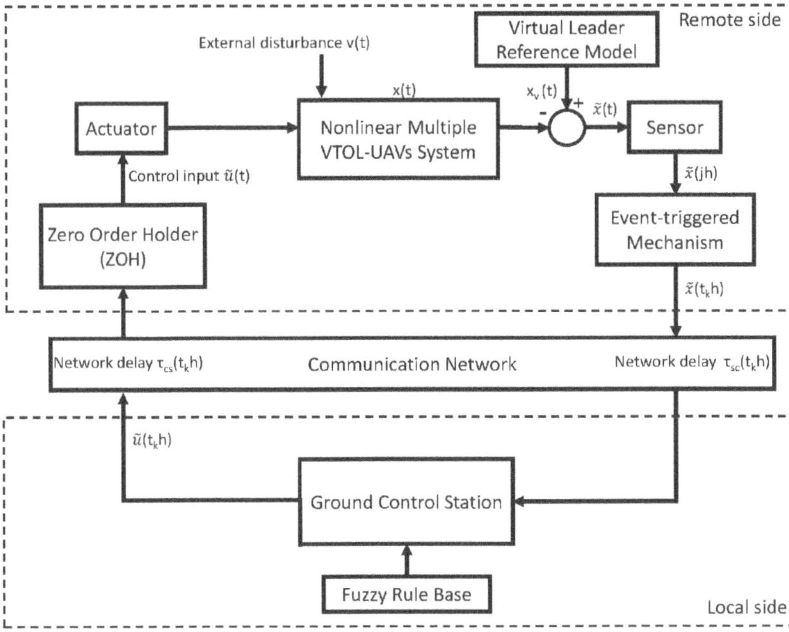

FIGURE 3.2 The event-triggered scheme for the quadrotor UAV networked tracking control system.

Remark 3.3 The function of the ZOH at the actuator stores the last transmitted control signal and keeps the control input of the plant until the next trigger updating has occurred.

In order to analyze the stability of the event-based quadrotor UAV networked team tracking control system, we need to make a detailed timing analysis. As shown in Figure 3.3, the holding interval $\Omega^{(t_k)} \equiv [t_k h + \tau_{t_k}, t_{k+1}h + \tau_{t_{(k+1)}})$ can be partitioned into several subsets, that is, $\Omega^{(t_k)} = \bigcup_{l=0}^{t_{k+1}-t_k-1} \Omega_l^{(t_k)}$. The subset $\Omega_l^{(t_k)}$ is defined as:

$$\Omega_l^{(t_k)} = (i_k h + \tau_{i_k}, i_k h + h + \tau_{i_{(k+1)}}), \tag{3.4}$$

where $i_k h = t_k h + lh$, for $l = 0,1,2,\ldots,t_{k+1} - t_k - 1$. In order to provide a unified framework for $x_i(i_k h)$, that is, $i_k h = t - (t - i_k h)$, a piecewise delay function $\eta(t)$ is defined as [46]:

$$\eta(t) \equiv t - i_k h, t \in \Omega_l^{(t_k)} \tag{3.5}$$

which denotes the time-varying delay in the control signal. From the scope of the subset $\Omega_l^{(t_k)}$, it is clear that $\eta(t)$ is a linear differentiable function satisfying [41]

$$\dot{\eta}(t) = 1, \ 0 \le \eta(t) \le \bar{\eta} = h + \bar{\tau}, t \in \Omega_l^{(t_k)} \tag{3.6}$$

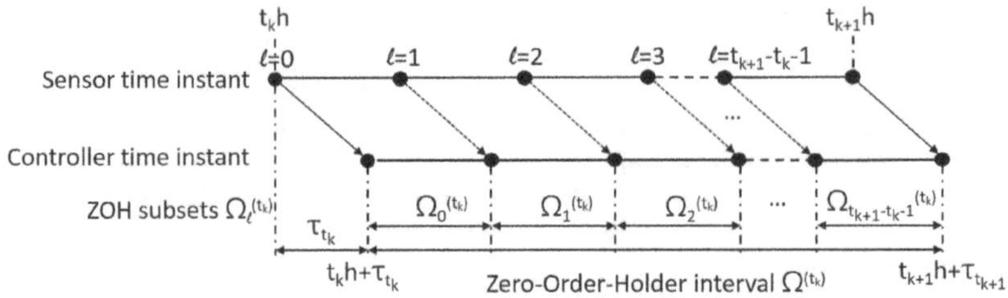

FIGURE 3.3 The zero-order-holder, l is the positive integer related to the time of the current sampling, that is, $i_k h = t_k h + lh$.

where $\bar{\eta}$ is the sum of maximum allowable upper delay bound $\bar{\tau}$ and one sampling period h. As a result, we can obtain a trade-off between the sampling period and the allowable delay upper bound when $\bar{\eta}$ is specified. Then, by considering the network-induced delay, the i th quadrotor UAV event-triggered networked control system can be formulated as the following stochastic nonlinear diffusion jump system:

$$\begin{aligned} dx_i(t) &= (f_i(x_i(t)) + g_i(x_i(t))u_i(t) + v_i(t))dt \\ &\quad +\sigma_i(x_i(t))dW_i(t)+\Gamma_i(x_i(t))dN_i(t), \\ x_i(s) &= \varsigma_i(s), \\ \text{with} \quad u_i(t) &= \hat{u}_i(x_i(t_k h), x_v(t_k h)) \\ t &\in [t_k h + \tau_{t_k}, t_{k+1}h + \tau_{t_{(k+1)}}), \\ \forall i &= 1, 2, \cdots, N, \ k \in \mathbb{N} \cup \{0\} \\ \forall -\bar{\eta} &\le s \le 0 \end{aligned} \tag{3.7}$$

where $x_v(t_k h)$ denotes the reference state to be tracked at $t = t_k h$, $\hat{u}_i(x_i(t_k h), x_v(t_k h))$ is the control input after ZOH for the ith quadrotor UAV in the event-triggered multi-quadrotor UAV team formation tracking networked control system and will be designed in the sequel, and $\varsigma_i(s)$ denotes known continuous function in (3.7) to describe the state value of ith UAV from $-\bar{\eta}$ to 0.

3.3 PROBLEM DESCRIPTION

In this chapter, we employ a virtual leader structure to deal with the multi-quadrotor UAV team formation tracking control. Figure 3.2 shows a framework of the proposed event-triggered scheme for the multi-quadrotor UAV team tracking control system through wireless network. Compared with other types of quadrotor UAV formation such as leader-follower approach, the best advantage of virtual leader formation is that not only the leader can affect the followers, but also the follower information can feedback to the leader, which could make the formation more robust and flexible. The detail of the approach is described in the sequel.

3.3.1 Team Formation Tracking Design of Multi-Quadrotor UAV Virtual Leader Structure

For a connected N-quadrotor UAV team formation networked control system, the formation structure can be constructed by two parts: one is the real members of all N quadrotor UAVs as the followers; the other is the virtual leader quadrotor UAV in the central of the followers' formation shape and its path is defined by a reference model via the software. The formation pattern is settled in such a way that each follower quadrotor UAV will track a trajectory specified by the state of virtual leader. The dynamic of each follower quadrotor UAV with event-triggered team tracking control is represented as (3.7) and $u_i(t)$ is the control input after ZOH to be designed to track the virtual leader, with a team formation shape to maintain the location and attitude of the ith quadrotor UAV in the team. The state of virtual leader is generated by the reference model. The advantage of the virtual structure is that such a team formation shape of quadrotor UAVs could simplify the team tracking design procedure. Therefore, the desired trajectory of virtual leader including position, velocity and attitude (including roll, pitch and yaw angles) of multi-quadrotor UAV team could be generated by the following virtual leader reference model:

$$dx_v(t) = (A_r x_v(t) + r(t))dt \tag{3.8}$$

where $x_v(t) \in \mathbb{R}^{12}$ denotes the reference state including the desired position, velocity, and row, pitch, yaw angles; $r(t) \in \mathbb{R}^{12}$ denotes the bounded reference input vector; and the matrix A_r denotes a specific asymptotically stable matrix.

Remark 3.4 It is obvious that at the steady state of the reference model, $x_v(t) = -A_r^{-1}r(t)$. If we choose $A_r = -I$, the desired trajectory $x_v(t)$ is equivalent to the reference input $r(t)$ at the steady state. If this is the case, $x_v(t)$ will converge to $r(t)$ at the steady state. In this situation $r(t)$ can be considered as the desired target of the virtual leader.

As shown in Figure 3.4, four quadrotor UAVs are assumed to maintain a square formation during the flight, which will be used to illustrate in the simulation example in the sequel. The virtual leader in Figure 3.4 represents the movement of the overall quadrotor UAV team structure and is requested to be the same as the reference path, velocity, and attitude generated by reference model in (3.8). By the above reference information of the virtual leader quadrotor, the follower quadrotor UAVs will be controlled to track the virtual leader to maintain the formation on the desired path, velocity, and attitude despite intrinsic random fluctuation, external disturbance, and arbitrary reference input signal from the robust H_∞ tracking perspective. Since the external disturbances such as weather conditions or channel interferences during the flight tracking process of multi-quadrotor UAV team formation networked control system are unpredictable and the reference input $r(t)$ is arbitrary and unknown to the followers, the following robust H_∞ tracking control strategy is considered to efficiently attenuate the effect of external disturbances $\bar{v}(t) = [r^T(t), v_1^T(t), \ldots, v_N^T(t)]^T$

on the tracking performance of the stochastic event-triggered multi-quadrotor UAV team formation networked control system:

$$J_\infty(\{u_i\}_{i=1}^N) = \sup_{\substack{\bar{v}(t)\in\mathcal{L}^2_{\mathcal{F}_t} \\ (\mathbb{R}^+,\mathbb{R}^{12(N+1)})}} \frac{\begin{array}{l} E[\int_0^{t_f}\sum_{i=1}^N (x_i(t)-x_v(t)-e_{di})^T \\ \times Q_i(x_i(t)-x_v(t)-e_{di})dt \\ -\sum_{i=1}^N (x_i(0)-x_v(0)-e_{di})^T \\ \times P_i(x_i(0)-x_v(0)-e_{di})\end{array}}{E\{\int_0^{t_f}\bar{v}^T(t)\bar{v}(t)dt\}} \quad (3.9),$$

where $Q_i \in R^{12\times 12}$ denotes the positive definite tracking error weighting matrix on the ith quadrotor UAV. $x_i(t)-x_v(t)$ denotes the tracking error between the ith quadrotor UAV and the virtual leader. e_{di} is the desired formation between the ith quadrotor UAV and the virtual leader UAV. If $x_i(t)-x_v(t)-e_{di}\to 0$, then $x_i(t)\to x_v(t)+e_{di}$, that is, the quadrotor UAV team can maintain the formation structure $e_d=[0^T, e_{d1}^T,\ldots,e_{dN}^T]^T$ around the virtual leader $x_v(t)$. $\sum_{i=1}^N (x_i(0)-x_v(0)-e_{di})^T P_i(x_i(0)-x_v(0)-e_{di})$ is the effect of the initial conditions on the team formation tracking of quadrotor UAVs with some positive definite matrices $\{P_i\}_{i=1}^N$ and should be excluded in the H_∞ tracking performance in (3.9), t_f denotes the terminal time of team formation tracking control process. In (3.9), our objective is to design the

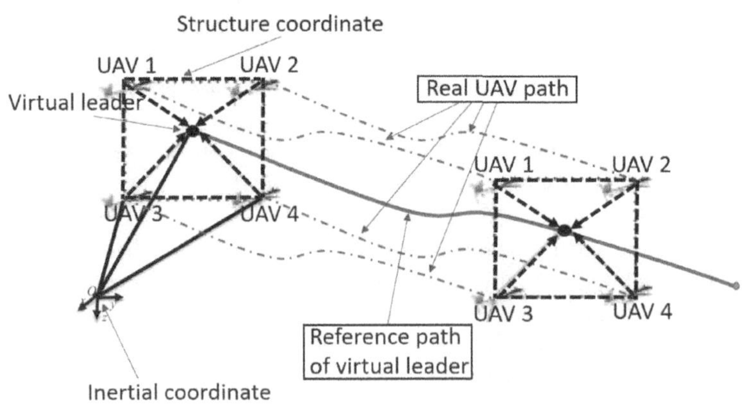

FIGURE 3.4 The virtual structure team formation of the multi-quadrotor UAV networked control system to track the desired trajectory. The example considers four follower-quadrotor UAVs. The black dot is the virtual leader quadrotor UAV at the central of the quadrotor UAV team formation shape, and the dynamic of virtual leader quadrotor UAV is given by the reference model in (3.8).

specific controllers $\{u_i(t)\}_{i=1}^N$ for each quadrotor UAV so that the worst-case effect of external disturbance $\overline{\mathbf{v}}(t)$ on the reference team formation tracking performance could be below a prescribed disturbance attenuation level ρ^2 for the quadrotor UAVs team tracking system from the energy perspective. i.e., we aim to design the specific controllers $\{u_i^*(t)\}_{i=1}^N$ such that $J_\infty(\{u_i^*(t)\}_{i=1}^N) \le \rho^2$.

3.3.2 The Virtual Leader Team Tracking Design of Multi-Quadrotor UAV Event-triggered Networked Control System

In order to investigate the H$_\infty$ team tracking performance of the stochastic multi-quadrotor UAV dynamic system, we first aggregate the states of all follower quadrotor UAVs and the virtual leader and then convert the original team formation tracking system into a team formation tracking error dynamic system to simplify the design procedure. At first, the augmented states and control input by (3.7) and (3.8) are given as:

$$\begin{aligned}\overline{x}(t) &= [x_v^T(t), x_1^T(t) - x_v^T(t), x_2^T(t) - x_v{}^T(t),\\ &\quad \mathrm{x}_3^{\mathrm{T}}(\mathrm{t})\text{-}\mathrm{x}_{\mathrm{v}}^{\mathrm{T}}(\mathrm{t}),\cdots,\mathrm{x}_{\mathrm{N}}{}^{\mathrm{T}}(\mathrm{t})\text{-}\mathrm{x}_{\mathrm{v}}^{\mathrm{T}}(\mathrm{t})]^{\mathrm{T}}\\ \overline{u}(t) &= [u_1^T(t), u_2^T(t),\cdots,u_N^T(t)]^T\end{aligned} \tag{3.10}$$

By the time instant defined in (3.3), the set $S_1 = \{jh \mid J \in N\}$ represents the sampled instants. Let $x_1(jh), x_2(jh),\ldots,\ x_N(jh)$ denote the sampled state of the N controlled quadrotor UAVs and $x_v(jh)$ denote the sampled states of the virtual leader reference. Moreover, these sampled states are augmented as a packet $\overline{x}(jh)$ and then transmitted to the event-triggered mechanism in Figure 3.2 at the ith sampling instant. After the decision from the event-triggered mechanism, the signal $\overline{x}(t_k h)$ is transmitted to the local side through the network. To simplify the discrete state $\overline{x}(t_k h)$, the threshold error between the augmented state at the current instant $i_k h$ and the one at the last transmitted instant $t_k h$ is defined as:

$$e_k(t) \overset{\Delta}{=} \overline{x}(i_k h) - \overline{x}(t_k h), t \in \Omega_l^{(t_k)} \tag{3.11}$$

Then, from (3.5) and (3.11),

$$\overline{x}(t_k h) = \overline{x}(t - \eta(t)) -_k (t), t \in \Omega_l^{(t_k)}, 0 < \eta(t) \le \overline{\eta} \tag{3.12}$$

Equation (3.12) is the feedback state information transmitted to the ground control station through the network. Besides, the nonlinear tracking control input $\overline{u}(t)$ in (3.10) of the event-triggered multi-UAV networked team tracking control system is given as

$$\overline{u}(t) = \overline{K}(\overline{x}(t_k h)) = \overline{K}(\overline{x}(t - \eta(t)) - e_k(t)),\ t \in [t_k h + \tau_{t_k}, t_{k+1}h + \tau_{t_{(k+1)}}) \tag{3.13}$$

where $\overline{K}(\cdot)$ is the nonlinear state-feedback function to be designed.

Remark 3.5 In (3.12), the value of the controller $\overline{K}(\overline{x}(t_k h))$ at $t \in [t_k h + \tau_{t_k}, t_{k+1}h + \tau_{t_{(k+1)}})$ depends on the information $\overline{x}(t_k h)$ at $t = t_k h$. As a result, the time-varying delay is not used in designing the controller. On the other hand, it is worth to point out that the piecewise delay function $\eta(t)$ is introduced to simplify the design. By using the piecewise delay function $\eta(t)$, the controller $\bar{u}(t)$ can be designed as the form $\bar{K}(\bar{x}(t-\eta(t))-e_k(t))$, which is associated with two variables $\bar{x}(t-\eta(t))$ and $e_k(t)$. In this situation, the design conditions are more flexible.

Thus, the event-triggered stochastic quadrotor UAV networked team formation tracking error dynamic system in consideration of the networked-induced delay is derived as follows:

$$\begin{aligned} d\bar{x}(t) = {} & [\bar{f}(\bar{x}(t)) + \bar{g}(\bar{x}(t))\bar{u}(t) + H\bar{v}(t)]dt \\ & + \sum_{i=1}^{N} \left[\bar{\sigma}_i\left(\bar{x}(t)\right)dW_i(t) + \bar{\Gamma}\left(\bar{x}(t)dN_i(t)\right)\right] \end{aligned} \tag{3.14}$$

where $\bar{u}(t)$ is as in (3.13) and

$$\bar{v}(t) = [r(t), v_1(t), \cdots, v_N(t)]^T$$

$$\bar{f}(\bar{x}(t)) = [(A_r x_v(t))^T, f_1(x_1(t))^T - (A_r x_v(t))^T, \cdots, f_N(x_N(t))^T - (A_r x_v(t))^T]^T$$

$$\bar{\sigma}_1(\bar{x}(t)) = [0, \sigma_1^T(x_1(t)), 0, \cdots, 0]^T$$

$$\vdots$$

$$\bar{\sigma}_i(\bar{x}(t)) = [0, 0, \cdots, \sigma_i^T(x_i(t)), \cdots,, 0]^T$$

$$\vdots$$

$$\bar{\sigma}_N(\bar{x}(t)) = [0, 0, \cdots, 0, \sigma_N^T(x_N(t))]^T$$

$$\bar{\Gamma}_1(\bar{x}(t)) = [0, \Gamma_1^T(x_1(t), 0, \cdots, 0]^T$$

$$\vdots$$

$$\bar{\Gamma}_i(\bar{x}(t)) = [0, 0, \cdots, \Gamma_i^T(x_i(t)), \cdots, 0]^T$$

$$\vdots$$

$$\bar{\Gamma}_N(\bar{x}(t)) = [0, 0, \cdots, 0, \Gamma_N^T(x_N(t))]^T$$

$$\bar{g}(\bar{x}(t)) = \begin{bmatrix} 0 & 0 & \cdots & 0 \\ 0 & g_1(x_1(t)) & \cdots & 0 \\ \vdots & \vdots & \ddots & \vdots \\ 0 & 0 & \cdots & g_N(x_N(t)) \end{bmatrix}, \quad H = \begin{bmatrix} I & 0 & \cdots & 0 \\ -I & I & \cdots & 0 \\ \vdots & \vdots & \ddots & \vdots \\ -I & 0 & \cdots & I \end{bmatrix}$$

In order to arrange N quadrotor UAVs in a virtual structure formation, the formation structure of the quadrotor UAV team is specified as $e_d = [0^T, e_{d1}^T, \ldots, e_{dN}^T]^T$ and the shifted error state is defined as:

$$x(t) = \bar{x}(t) - e_d \tag{3.15}$$

where $x(t)$ denotes the desired tracking error state of all quadrotor UAVs in the virtual formation structure of team tracking process.

Then, we get the following shifted nonlinear multi-quadrotor UAV networked team formation control system:

$$\begin{aligned} dx(t) &= [\tilde{f}(x(t)) + \tilde{g}(x(t))\tilde{u}(t) + H\bar{v}(t)]dt \\ &+ \sum_{i=1}^{N}[\tilde{\sigma}_i(x(t))dW_i(t) + \tilde{\Gamma}_i(x(t))dN_i(t)] \end{aligned} \tag{3.16}$$

where

$$\begin{aligned} &\tilde{f}(x(t)) = \bar{f}(x(t)+e_d),\ \tilde{g}(x(t)) = \bar{g}(x(t)+e_d), \\ &\tilde{u}(t) = K(x(t-\eta(t)) - e_k(t)) = \bar{K}(x(t-\eta(t)) - e_k(t) + e_d), \\ &\tilde{\sigma}_i(x(t)) = \bar{\sigma}_i(x(t)+e_d), \\ &\tilde{\Gamma}_i(x(t)) = \bar{\Gamma}_i(x(t)+e_d), e_k(t) = x(i_k h) - x(t_k h),\ t \in \Omega_l^{(t_k)}. \end{aligned}$$

For the simplicity of notation, the transmission condition of the event-triggered mechanism for the multi-quadrotor UAV networked team tracking system in Figure 3.2 can be reformulated as [40]:

$$\begin{aligned} t_{k+1}h = t_k h + \min_{l>0, l\in\mathbb{N}} \{lh \mid E\{e_k(t)^T \Phi e_k(t)\} \\ \geq \sigma \mathrm{E}\{[\mathrm{x}(\mathrm{t}\text{-}\eta(\mathrm{t}))\text{-}\mathrm{e_k}(\mathrm{t})]^{\mathrm{T}} \Phi [\mathrm{x}(\mathrm{t}\text{-}\eta(\mathrm{t}))\text{-}\mathrm{e_k}(\mathrm{t})]\} \end{aligned} \tag{3.17}$$

where $i_k h = t_k h + lh$. The transmission condition depends on the tracking error of the latest transmission instant $x(t_k h) = x(t-\eta(t)) - e_k(t)$, and the variation of the tracking error between the current sampling instant and the latest transmission instant, that is, $x(i_k h) - x(t_k h) = e_k(t)$. Equation (3.17) is similar to the event-triggered mechanism in (3.16) but with one quadrotor UAV state being replaced by the shifted tracking error of the multi-quadrotor UAV team.

Thus, the origin $x\ (t) = 0$ of the nonlinear multi-quadrotor UAV networked team tracking error dynamic system in (3.16) is at the desired steady state (target of the desired team formation) e_d of the augmented nonlinear multi-quadrotor UAV networked team tracking dynamic system in (3.14), that is, the multi-quadrotor UAV team formation tracking problem in (3.14) with the desired tracking target e_d transformed to an equivalent stabilization problem at the origin $x\ (t) = 0$ of the shifted nonlinear multi-UAV networked team tracking error dynamic system in (3.16). This origin shift makes the design procedure of virtual leader formation tracking problem of multi-quadrotor UAV team much simpler.

With the help of the shifted event-triggered multi-UAV networked team tracking system in (3.16) and (3.17), the H_∞ robust team formation tracking strategy in (3.9) can be reformulated as the following equivalent H_∞ stabilization problem of the shifted multi-quadrotor UAV networked team tracking system in (3.16):

$$J_\infty(u(t)) = \sup_{\substack{\bar{\mathbf{v}}(t)\in \\ L_F^2(\mathbb{R}^+;\mathbb{R}^{12(N+1)})}} \frac{E\{\int_0^{t_f}[x(t)\bar{Q}x(t)]dt - V(x(0))\}}{E\{\int_0^{t_f}[\bar{\mathbf{v}}^T(t)\bar{\mathbf{v}}(t)]dt\}} \tag{3.18}$$

where $\bar{Q} = diag\{0, Q_1, \cdots, Q_N\}$, $V(x(0))$ is the positive function of the initial used to represent the effect of the initial condition. In the robust H_∞ event-triggered multi-UAV team formation tracking design, our goal is to design a specific controller $u(t)$ to achieve the H_∞ team formation tracking performance in (3.18) under a prescribed disturbance attenuation level ρ^2, that is, $J_\infty(u(t)) \le \rho^2$ in (3.18).

Remark 3.6 By the augmented multi-quadrotor UAV networked control system in (3.10) and (3.14) and the shifted multi-quadrotor UAV networked control system in (3.15) and (3.16), the complex H_∞ team formation control problem of N quadrotor UAVs in (3.9) to track the desired reference model becomes the robust H_∞ stabilization problem in (3.18) of the shifted multi-quadrotor UAV networked team tracking system in (3.16).

Before the design of event-triggered H_∞ team formation tracking control of multi-quadrotor UAV system, the following lemmas are necessary.

Lemma 3.1 [34]. Consider the Lyapunov function $V(\cdot) \in C^2$ with .. and $V(0) = 0$. For the stochastic nonlinear system in (3.16), the Itô-Lévy formula of $V(x(t))$ is given as follows:

$$\begin{aligned} dV(x(t)) = &[V_x^T [f(x(t)) + g(x(t))\tilde{u}(t) + H\bar{\mathbf{v}}(t)] \\ &+ \sum_{i=1}^{N} \frac{1}{2} \tilde{\sigma}_i^T (\tilde{x}(t)) V_{\tilde{x}\tilde{x}} \tilde{\sigma}_i(\tilde{x}(t)) dt + \sum_{i=1}^{N} [V_x^T \times \tilde{\sigma}_i (\tilde{x}(t)) dW_i(t) \\ &+ \{V(x(t) + \Gamma_i(x(t))) - V(x(t))\} dN_i(t)]] \end{aligned} \tag{3.19}$$

Lemma 3.2 [42,43] For any arbitrary matrix A and B (or vector) with appropriate dimension, the following inequality holds for any $\beta > 0$:

$$A^T B + B^T A \le \beta^{-1} A^T A + \beta B^T B \tag{3.20}$$

Then the main theorem of the event-triggered H_∞ team tracking control design for the nonlinear stochastic multi-quadrotor UAV networked control system is given as follows:

Theorem 3.1 Consider the stochastic nonlinear event-triggered multi-quadrotor UAV networked team tracking system in (3.16) with the event-triggered mechanism in (3.17). If one could specify a nonlinear tracking controller $u(t) = K(x(t - \eta(t)) - e_k(t))$ and the event-triggered weighting matrix $\Phi = \Phi^T > 0$ such that the following HJI has a positive solution $V(x(t)) > 0$ with $V(0) = 0$ and $V(\cdot) \in C^2$

$$\begin{aligned} &x^T(t)\bar{Q}x(t) + V_x^T f(x(t)) + V_x^T g(x(t))u(t) \\ &+ \frac{1}{4\rho^2} V_x^T H H^T V_x + \sum_{i=1}^{N} \frac{1}{2} \tilde{\sigma}_i^T (x(t)) V_{xx} \tilde{\sigma}_i (x(t)) \\ &+ \sum_{i=1}^{N} \lambda_i \{V(x(t) + \Gamma_i(x(t))) - V(x(t))\} \le 0 \end{aligned} \tag{3.21}$$

then the robust event-triggered H_∞ team tracking control performance (3.18) of the multi-quadrotor UAV networked team tracking system under the event-triggered mechanism in (3.17) is guaranteed for a prescribed attenuation level ρ^2 for the arbitrary external disturbance $\bar{\mathbf{v}}(t) \in \mathcal{L}^2_{\mathcal{F}}(\mathbb{R}^+;\mathbb{R}^{12(N+1)})$.

Proof. Please refer to Appendix 3A.

3.4 ROBUST H$_\infty$ TEAM TRACKING CONTROL DESIGN OF STOCHASTIC EVENT-TRIGGERED MULTI-QUADROTOR UAV NETWORKED CONTROL SYSTEM VIA T-S FUZZY MODEL

In general, in order to deal with the robust H_∞ event-triggered team tracking control problem for stochastic nonlinear multi-quadrotor UAV networked control system, we need to solve the nonlinear partial differential HJI in (3.21). However, HJI in (3.21) is difficult to be solved analytically and directly. Therefore, a T-S fuzzy model is introduced to approximate the nonlinear stochastic system of quadrotor UAV by interpolating several local linear stochastic systems through fuzzy bases. Under the assumption that the state variables are accessible and bounded, the mth rule of the T-S fuzzy model for the stochastic nonlinear event-triggered multi-quadrotor UAV shifted team tracking dynamic system in (3.16) can be described as [39]

$$\begin{aligned}
&\textbf{Plant Rule } \boldsymbol{m}\textbf{:}\\
&\text{If } \varpi_{i,1}(t) \text{ is } G_{m,1},\cdots, \text{ and } \varpi_{i,g}(t) \text{ is } G_{m,g},\\
&\textbf{Then}\\
&d\tilde{x}_i(t) = (A_{im}\tilde{x}_i(t) + B_{im}\tilde{u}_i(t) + v_i(t)\\
&-r(t))dt + C_{im}\tilde{x}_i(t)dW_i(t) + D_{im}\tilde{x}_i(t)dN_i(t)\\
&\text{for } m = 1,2,...,L, \text{and } i = 1,2,...,N
\end{aligned} \tag{3.22}$$

where $\tilde{x}_i(t) = x_i(t) - x_v(t) - e_{di}$ and L is the number of the fuzzy rules. A_{im}, B_{im}, C_{im}, and D_{im} are the local linearization matrices with appropriate dimensions; $\tilde{u}_i(t)$ is the ith control input in $\tilde{u}(t)$; $\varpi_{i,1}(t), \varpi_{i,2}(t), \cdots, \varpi_{i,g}(t)$ are the premise variables related to the state of the ith quadrotor UAV in the virtual structure; $G_{m,1}, G_{m,2}, \cdots, G_{m,g}$ are the fuzzy sets; and g is the number of the premise variables of the each quadrotor UAV. Then we define the grade of membership function of the mth rule $\mu_m(\varpi_i(t))$ as:

$$\mu_m(\varpi_i(t)) = \prod_{s=1}^{g} G_{m,s}(\varpi_{i,s}(t)) \geq 0 \text{ for } m = 1,2,\cdots,L \tag{3.23}$$

where $G_{m,s}(\varpi_{i,s}(t))$ is the membership grade of $\varpi_{i,s}(t)$ in $G_{m,s}$. $\varpi_i(t) = [\varpi_{i,1}(t), \cdots, \varpi_{i,g}(t)]$. It is obvious that $\sum_{m=1}^{L} \mu_m(\varpi_i(t)) \geq 0$ and the mth interpolation functions of the ith quadrotor UAV can be inferred as follows:

$$\bar{h}_m(\varpi_i(t)) = \frac{\mu_m(\varpi_i(t))}{\sum_{m=1}^{L} \mu_m(\varpi_i(t))} \geq 0 \tag{3.24}$$

which satisfies the following interpolation property

$$\sum_{m=1}^{L}\bar{h}_m(\varpi_i(t))=1 \tag{3.25}$$

Based on T-S fuzzy models (3.22)–(3.25), the overall nonlinear event-triggered multi-quadrotor UAV team tracking system in (3.16) can be represented by:

$$\begin{aligned} dx(t)&=\sum_{m=1}^{L}h_m(\varpi(t))[\bar{A}_m x(t)+\bar{B}_m\tilde{u}(t)+H\bar{v}(t)]dt\\ &+\sum_{i=1}^{N}\{\bar{C}_{im}x(t)dW_i(t)+\bar{D}_{im}x(t)dN_i(t)\},k\in\mathbb{N},\\ &t\in[t_k h+\tau_{t_k},t_{k+1}h+\tau_{t_{(k+1)}}) \end{aligned} \tag{3.26}$$

where

$$\begin{aligned} \bar{A}_m&=diag\{A_r,A_{1m},A_{2m},\cdots,A_{Nm}\}\\ \bar{B}_m&=diag\{0,B_{1m},B_{2m},\cdots,B_{Nm}\}\\ \bar{C}_{1m}&=diag\{0,C_{1m},0,\cdots,0\}\\ &\vdots\\ \bar{C}_{im}&=diag\{0,0,\cdots,C_{im},\cdots,0\}\\ &\vdots\\ \bar{C}_{iN}&=diag\{0,0,\cdots,C_{Nm}\}\\ \bar{D}_{1m}&=diag\{0,D_{1m},0,\cdots,0\}\\ &\vdots\\ \bar{D}_{im}&=diag\{0,0,\cdots,D_{im},\cdots,0\}\\ &\vdots\\ \bar{D}_{Nm}&=diag\{0,0,,\cdots,D_{Nm}\}\\ h_m(\varpi(t))&=diag\{I_{12},\bar{h}_m(\varpi_1(t))I_{12},\cdots,\bar{h}_m(\varpi_N(t))I_{12}\} \end{aligned}$$

By the similar way, the fuzzy model of the event-based controller for the ith UAV in the multi-quadrotor UAV team tracking system can be described as follows:

$$\begin{aligned} &\textbf{Control Rule n}:\\ &\textit{If } \varpi_{i,1}(t_k h) \textit{ is } G_{n,1},\ldots, \textit{and } \varpi_{i,g}(t_k h) \textit{ is } G_{n,g}\\ &\textbf{Then}\\ &\tilde{u}_i(t)=\bar{K}_{in}[x_i(t_k h)-x_v(t_k h)-e_{di}],k\in\mathbb{N},\\ &t\in[t_k h+\tau_{t_k},t_{k+1}h+\tau_{t_{(k+1)}})\\ &\textit{for } n=1,2,\ldots,L, \textit{ and } i=1,2,\ldots,N \end{aligned} \tag{3.27}$$

By combining the definitions of (3.5) and (3.11), the control signal of multi-quadrotor UAV team tracking system in (3.16) is generated after the zero-order holder:

$$\begin{aligned}\tilde{u}(t) &= K(x(t-\eta(t))-e_k(t)) \\ &= \sum_{n=1}^{L} h_n(\varpi(\bar{t}_k))[\bar{K}_n(x(t-\eta(t))-e_k(t))], \\ & k\in\mathbb{N}, t\in[t_k h+\tau_{t_k}, t_{k+1}h+\tau_{t_{(k+1)}})\end{aligned} \tag{3.28}$$

where $\bar{K}_n = diag\{0, \bar{K}_{1n}, \cdots, \bar{K}_{Nn}\}$ denotes the fuzzy control matrices to be designed, $h_n(\varpi(\bar{t}_k)) = diag\{I_{12}, \bar{h}_n(\varpi_1(\bar{t}_k)), \cdots, \bar{h}_n(\varpi_N(\bar{t}_k))\}$ is the augmented interpolation function and $\bar{t}_k = t_k h$ is the delay time instant at which ground control station receives the state information.

Remark 3.7 Due to the networked induced delay, the ground control station receives the delayed state information and the fuzzy controller depends on the delayed state information. This fact implies that the firing mechanisms of the plant and controller are asynchronous [47].

Then by combining the T-S fuzzy model of the multi-quadrotor UAV team tracking system and the fuzzy controller above, the overall stochastic nonlinear event-triggered multi-quadrotor UAV shifted team tracking dynamic system in (3.16) can be rewritten as follows:

$$\begin{aligned}dx(t) &= \sum_{m=1}^{L} h_m(\varpi(t)) \sum_{n=1}^{L} h_n(\varpi(\bar{t}_k))\{[\bar{A}_m x(t) + \bar{B}_m \bar{K}_n(x(t-\eta(t))-e_k(t)) \\ &+ H\bar{v}(t)]dt + \sum_{i=1}^{N}[\bar{C}_{im}x(t)dW_i(t) + \bar{D}_{im}x(t)dN_i(t)]\}, k\in\mathbb{N}, \\ & t\in[t_k h+\tau_{t_k}, t_{k+1}h+\tau_{t_{(k+1)}})\end{aligned} \tag{3.29}$$

Remark 3.8 In this design, the fuzzy approximation error is state dependent and very complex. It can be merged into the random fluctuations and its effect can be efficiently attenuated by the proposed robust H_∞ team formation tracking control.

3.5 H∞ FUZZY EVENT-TRIGGERED TEAM TRACKING CONTROL DESIGN OF STOCHASTIC MULTI-QUADROTOR UAV NETWORKED CONTROL SYSTEM WITH TIME-VARYING DELAY

After using the T-S fuzzy model in (3.29) to approximate stochastic nonlinear event-triggered multi-quadrotor UAV team tracking system in (3.16), we are now going to transform the complicated HJIs in (3.21) into a set of LMIs. To begin with, we select the Lyapunov function for the event-triggered multi-quadrotor UAV shifted team tracking error dynamic system in (3.16) as $V(t) = \tilde{x}^T(t)P\tilde{x}(t)$, where P is a positive definite symmetric matrix $P = P^T > 0$. Then, we have the following theorem:

Theorem 3.2 In the stochastic event-triggered multi-UAV shifted team tracking error dynamic system (3.29), for the given scalars $1 > \sigma \geq 0$, $\rho > 0, \bar{\eta} > 0$, and the event-triggered scheme in (3.17), if we can find some matrices $W = diag\{W_0, W_1, \cdots, W_N\}$ with $W_j > 0$, for $j = 0, \cdots, N$, $\bar{Y}_n = diag\{0, \bar{Y}_{1n}, \ldots \bar{Y}_{Nn}\}$, for $n = 1, \cdots, L$, $\Phi = \Phi^T > 0, \alpha > 0$ with appropriate dimension as the solutions of the following bilinear matrix inequalities (BMIs):

$$\begin{bmatrix} \Xi_{11}^m & \begin{matrix}\bar{B}_m \bar{Y}_n - \alpha I \\ +(\bar{\eta}+2)W\end{matrix} & -\bar{B}_m \bar{Y}_n & 2W \\ * & \Xi_{22} & -\sigma W \Phi W & -\bar{\eta} W \\ * & * & (\sigma - 1) W \Phi W & 0 \\ * & * & * & -\alpha I \end{bmatrix} \leq 0, \tag{3.30}$$

$$for\ m, n = 1, 2, \ldots, L$$

where $\Xi_{11}^m = W\bar{Q}W + W\bar{A}_m^T + \bar{A}_m W + \frac{1}{\rho^2} HH^T + \sum_{i=1}^{N} [W\bar{C}_{im}^T W^{-1} \bar{C}_{im} W + \lambda_i (W\bar{D}_{im}^T + \bar{D}_{im} W + W\bar{D}_{im}^T W^{-1}\ \bar{D}_{im} W)] - 4W + \alpha I$, $\Xi_{22} = -2\bar{\eta} W + \alpha I\ + \sigma W \phi W$, then the fuzzy controller gains can be constructed as $\bar{K}_n = \bar{Y}_n W^{-1}, n = 1, 2, \ldots, L$ and the robust H$_\infty$ team tracking control performance in (3.18) of the multi-quadrotor UAV system (3.16) under the discrete event-triggered scheme in (3.17) can be guaranteed under a prescribed attenuation level ρ^2 for an arbitrary external disturbance $\bar{\mathbf{v}}(t) \in \mathcal{L}_{\mathcal{F}}^2(\mathbb{R}^+; \mathbb{R}^{12(N+1)})$.

Proof. Please refer to Appendix 3B.

Remark 3.9 Due to the fact $\alpha > 0$, $\Phi > 0$, and $1 > \sigma \geq 0$, the diagonal terms (3.3) and (3.4) in (3.30) must be negative definite. Moreover, there are design variables $-4W < 0$ and $-2\bar{\eta} W < 0$ in (3.1) block and (3.2) block, respectively. Thus, it is possible to solve (3.30) in Theorem 3.2 by using Matlab LMI TOOLBOX and the proposed two-step design procedure in the sequel.

In fact, the inequalities in (3.30) are BMIs which are still complicated to be solved, and therefore we employ a two-step procedure to solve the event-triggered H$_\infty$ robust team formation tracking control problem. The reason is that the diagonal components in (3.30) have some unsolvable coupled-variables such as $\sigma W \Phi W$ which can not be dealt by the Schur complement. Since all the diagonal terms in (3.30) should be negative definite if (3.30) holds, we first solve the term $\Xi_{11}^m \leq 0$ in (3.30) to obtain the variable W and α, and then by substituting W into the BMI in (3.30) as an LMI, the solvable LMI for $\bar{Y}_n$ and Φ can be obtained. The two-step procedure for solving BMIs in (3.30) is given in the following:

Step 1: The inequality (3.30) implies that $\Xi_{11}^m \leq 0$, that is,

$$\begin{aligned} & W\bar{Q}W + W\bar{A}_m^T + \bar{A}_m W + \frac{1}{\rho^2} HH^T + \sum_{i=1}^{N} [W\bar{C}_{im}^T W^{-1} \bar{C}_{im} W \\ & + \lambda_i (W\bar{D}_{im}^T + \bar{D}_{im} W + W\bar{D}_{im}^T W^{-1} \bar{D}_{im}\ W) - 4W + \alpha I \leq 0 \end{aligned} \tag{3.31}$$

By Schur complement, (3.31) is equivalent to the following LMI:

$$\begin{bmatrix} \Xi_{1,1}^{m,n} & \Xi_{1,2}^{m,n} \\ * & \Xi_{2,2}^{m,n} \end{bmatrix} \leq 0 \tag{3.32}$$

where $\Xi_{1,1}^{m,n} = W\bar{A}_m^T + \bar{A}_m W + \sum_{i=1}^{N} \lambda_i (W\bar{D}_{im}^T + \bar{D}_{im} W) + \alpha I - 4W$, $\Xi_{1,2}^{m,n} = [H, W\bar{Q}^{\frac{1}{2}}, W\bar{C}_{1m}^T, \cdots, W\bar{C}_{Nm}^T, W\bar{D}_{1m}^T, \cdots, W\bar{D}_{Nm}^T]$, $\Xi_{2,2}{}^{m,n} = diag\{-\rho^2, -I, -W, \cdots, -W, \cdots, -W, -\frac{1}{\lambda_1} W, \cdots, -\frac{1}{\lambda_N} W\}$.

By solving the LMI in (3.32) for W and α, it will be used in the rest of the design procedure.

Step 2: Substituting W and α obtained from LMIs in (3.32) into the BMI in (3.30), (3.30) becomes LMIs and the design variables $\bar{Y}_n$, Φ can be obtained by solving the LMIs in (3.30). Moreover, the corresponding fuzzy control gains can be constructed as $\bar{K}_n = \bar{Y}_n W^{-1}$.

Thus, the optimal H_∞ team formation tracking control problem of the multi-UAV networked control system can be formulated as the following LMIs-constrained optimization problem to achieve the optimal H_∞ event-triggered team tracking performance:

$$\rho_0{}^2 = \min_{W, \{\bar{Y}_n\}_{n=1}^{L}, \Phi} \rho^2$$
$$\textit{subject to } W = W^T > 0, \Phi > 0, \text{ (3.30) and (3.32).} \tag{3.33}$$

By solving the EVP in (3.33), we can achieve the optimal H_∞ team formation tracking performance in (3.18) to optimally suppress the effect of external disturbance and continuous and discontinuous intrinsic random fluctuation and efficiently save the energy and communication load of the networked device under the event-triggered weighting matrix Φ at the same time.

Remark 3.10 For the proposed LMIs-constrained optimization problem in (3.33), the design variables are α, W, $\{\bar{Y}_n\}_{n=1}^{L}$, and Φ. According to the dimensions of these design variables, the computation complexity in (3.33) is $O((n(n+1))^{2.75} L^{1.5})$ where n is the dimension of variable W, L is the number of fuzzy if–then rules [42].

Remark 3.11 Since the upper bound of induced delay $\bar{\eta} > 0$ and threshold parameter of event-triggered mechanism σ are considered in the LMI constraints in (3.30), the feasibility of (3.30) depends on these two prescribed parameters. Mathematically, the LMIs will be more feasible if these parameters are decreased. In this situation, however, the designed fuzzy sample controller becomes much conservative. For example, if $\sigma = 0$, the designed fuzzy sample controller becomes the conventional time-triggered controller and it will increase the energy consumption and bandwidth load in the wireless networked control system. On the other hand, with the consideration of small $\bar{\eta}$ in (3.30), the designed fuzzy sample controller can only be applied to the networked control system with great channel quality. As a result, the trade-off of these two parameters in designing the controller should be carefully treated by the designer.

3.6 SIMULATION RESULTS

In this section, we provide a simulation example to validate the effectiveness of the proposed robust event-triggered H$_\infty$ virtual structure formation team tracking control strategy for the four-quadrotor UAV networked control system, as shown in Figure 3.4, with stochastic jump diffusions, time-varying delays, and external disturbances. The stochastic shifted team tracking error dynamic model of the four quadrotor UAVs is shown in (3.16) with the event-triggered mechanism in (3.17), and the parameters of four quadrotor UAVs in the networked control system are given as follows [44]: $m = 2kg$ is the mass of each quadrotor UAV. The coefficients of the translation drag forces $f_x = f_y = f_z = 0.01Ns/m$ and aerodynamic friction coefficients $f_\phi = f_\theta = f_\psi = 0.012$Ns/m; $g = 9.8m/s^2$ denotes the acceleration of gravity. $J_\phi = J_\theta = J_\psi = 0.013Ns^2/rad$ are the moments of inertia of each UAV. In this simulation example, the virtual structure of quadrotor UAV team is set as a square-shape structure. As a result, we respectively set the initial positions of four quadrotor UAVs at (+0.2, +0.2, 0), (–0.2, +0.2, 0), (–0.2, –0.2, 0), (+0.2, –0.2, 0) on the ground. Obviously, the virtual leader reference is initially at the origin.

To construct the T-S fuzzy model of each quadrotor UAV, the premise variables of each quadrotor UAV are defined as $\varpi_{i,1} = \phi_1^i$, $\varpi_{i,2} = \theta_1^i$ for $i = 1,2,3,4$ in (3.22) and (3.27), and they are used to interpolate the nonlinear quadrotor UAV system by a set of local linear UAV systems. For each premise variable, the operation points are chosen as $-\pi/9, 0, \pi/9$ and the grade of membership function is selected as trapezoidal function. Based on the above setting, each quadrotor UAV has 9 fuzzy rules, and there are in total 6,561 fuzzy rules for the whole multi-quadrotor UAV networked team tracking control system.

By using MATLAB system identification toolbox, the dynamic model of the qth rule within 6,561 rules of the fuzzy-based nonlinear stochastic shifted multi-quadrotor UAV event-triggered team tracking networked control system is written as follows:

$$
\begin{aligned}
d\tilde{x}(t) &= [A_q\tilde{x}(t) + B_qK_q(\tilde{x}(t-\eta(t) - e_k(t)) + H\bar{v}(t)]dt \\
&\quad + \sum_{i=1}^{N}[\bar{C}_i\tilde{x}(t)dW_i(t) + \bar{D}_i\tilde{x}(t)dN_i(t)]
\end{aligned}
\tag{3.34}
$$

where A_q and B_q are matrices generated by MATLAB system identification toolbox and $\{\bar{C}_i, \bar{D}_i\}_{i=1}^4$ are matrices of random process:

$$
\begin{aligned}
\bar{C}_1 &= diag\{0, C, 0, 0, 0\}, \bar{D}_1 = diag\{0, D, 0, 0, 0\} \\
\bar{C}_2 &= diag\{0, 0, C, 0, 0\}, \bar{D}_2 = diag\{0, 0, D, 0, 0\} \\
\bar{C}_3 &= diag\{0, 0, 0, C, 0\}, \bar{D}_3 = diag\{0, 0, 0, D, 0\} \\
\bar{C}_4 &= diag\{0, 0, 0, 0, C\}, \bar{D}_4 = diag\{0, 0, 0, 0, D\}
\end{aligned}
$$

with $C = diag\{0, 0.2, 0, 0.2, 0, 0.2, 0, 0.2, 0, 0.2, 0, 0.2\}$ and $D = diag\{0, 0.2, 0, 0.2, 0, 0.2, 0, 2, 0, 2, 0, 2\}$. Besides, the Poisson jump intensities are set as $\lambda_i = 0.1\}_{i=1}^4$ and the external disturbances are set to be Gaussian noises with zero mean and unit variance. If the event-triggered threshold parameter $\sigma = 0.1$ and the sampling period $h = 0.01$s are set, then from (3.6) and [41], $\bar{\eta} = 0.0869$, which means the maximum allowable upper delay bound $\bar{\tau} = 0.0769$, that is, the time-varying delay from the sensor to the controller is set within $0 \le \tau_{t_k} \le 0.0769$.

(a) H_∞ event-triggered team formation tracking control scheme with h = 0.01, σ = 0.1.
(b) H_∞ periodic time-triggered team formation tracking control scheme with h = 0.01.

In this simulation, we tend to make the quadrotor UAV team tracking on a desired spiral upward path and the initial position of the quadrotor UAV team is set as the square formation. The virtual leader reference input vector in (3.8) is given as follows:

$$r(t) = [x_d(t), \dot{x}_d(t), y_d(t), \dot{y}_d(t), z_d(t), \dot{z}_d(t),$$
$$\phi_d(t), \dot{\phi}_d(t), \theta_d(t), \dot{\theta}_d(t), \psi_d(t), \dot{\psi}_d(t)]$$

with the reference position: $x_d(t) = sin(0.5t)$, $y_d(t) = \cos(0.5t)$, $z_d(t) = 0.8t$, respectively. Besides, the reference attitude of ϕ_d, θ_d, and ψ_d are given as follows:

$$\phi_d = \frac{1}{2}\arcsin(d_x \sin(\psi_d) - d_y \cos(\psi_d)), \ \psi_d(t) = 0,$$
$$\theta_d = \frac{1}{2}\arcsin\left(\frac{d_x \cos(\psi_d) + d_y \sin(\psi_d)}{\cos(\phi_d)}\right).$$

where $d_x = 10(x - x_d) + 5(x - x_d), d_y = 10(y - y_d) + 5(y - y_d)$. Moreover, the asymptotically stable matrix is chosen as $A_r = -I$ in (3.8). The reason why we define such complicated reference attitude is that we tend to simulate the real behavior of the multiple quadrotor UAVs during the team flight, that is, when quadrotor UAV leads to the changes in the x position and y position, it needs to change the roll or pitch attitude to cause the translation motion. However, the yaw attitude is set to zero in this case.

The H$_\infty$ team formation tracking control strategy in (3.18) is used to design the event-triggered multiple quadrotor UAV networked control system. Since the tracking of position and attitude of each quadrotor UAV are more significant than the tracking of velocity and angle velocity, the following weighting matrices of UAV are given as:

$$Q_i = diag\{1, 0.001, 1, 0.001, 1, 0.001, 1, 0.001, 1, 0.001, 1, 0.001\}$$
$$for\ i = 1, 2, 3, 4.$$

By solving the LMIs-constrained optimization problem in (3.33), we have the optimal disturbance attenuation level $\rho^* = 3.2105$ and the corresponding fuzzy controller in (3.28). The result in Figure 3.5a shows the actual 3-D quadrotor UAV

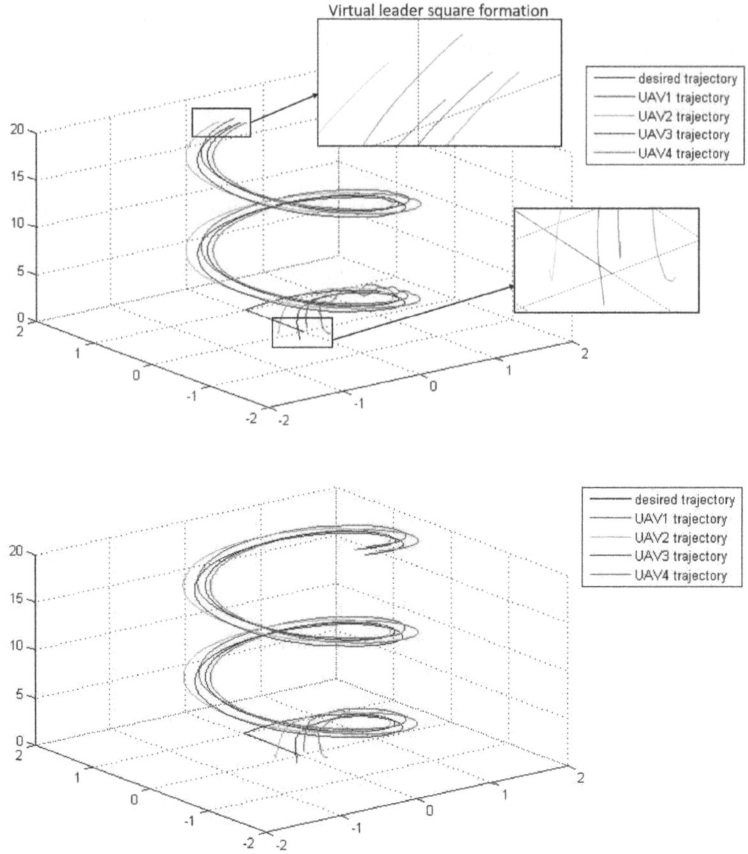

FIGURE 3.5 The 3D formation tracking flight trajectory of the four-quadrotor UAV team. (a) the trajectory is under event-triggered scheme $\sigma = 0.1$. (b) The trajectory is under periodic time-triggered scheme

team formation flight trajectory under the proposed H$_\infty$ event-triggered team tracking scheme. In the result of Figure 3.5a, the four quadrotor UAVs can well track on the trajectory of the virtual leader while maintaining the square team formation structure despite external disturbances and intrinsic continuous Wiener fluctuation and discontinuous Poisson fluctuation. Compared with the result in Figure 3.5b with the traditional periodically time-triggered scheme with $\sigma = 0$, we can see that the trajectory in Figure 3.5a oscillates more substantially during the transient stage. It means that the team formation tracking performance in event-triggered scheme will be slightly degraded due to the saving of communication loading and energy waste. However, the multi-quadrotor UAV team can finally finish the team tracking task of spiral upward maneuvering flight in a square shape formation under such a severe situation. The quadrotor UAV team tracking trajectories in Figures. 3.6 and 3.7 are separated into 12 components of each quadrotor UAV state. The simulation result in Figure 3.6 includes the positions $x, y,$ and z and the velocities $v_x, v_y,$ and v_z of the four quadrotor UAVs, while simulation result in Figure 3.7 includes the angles

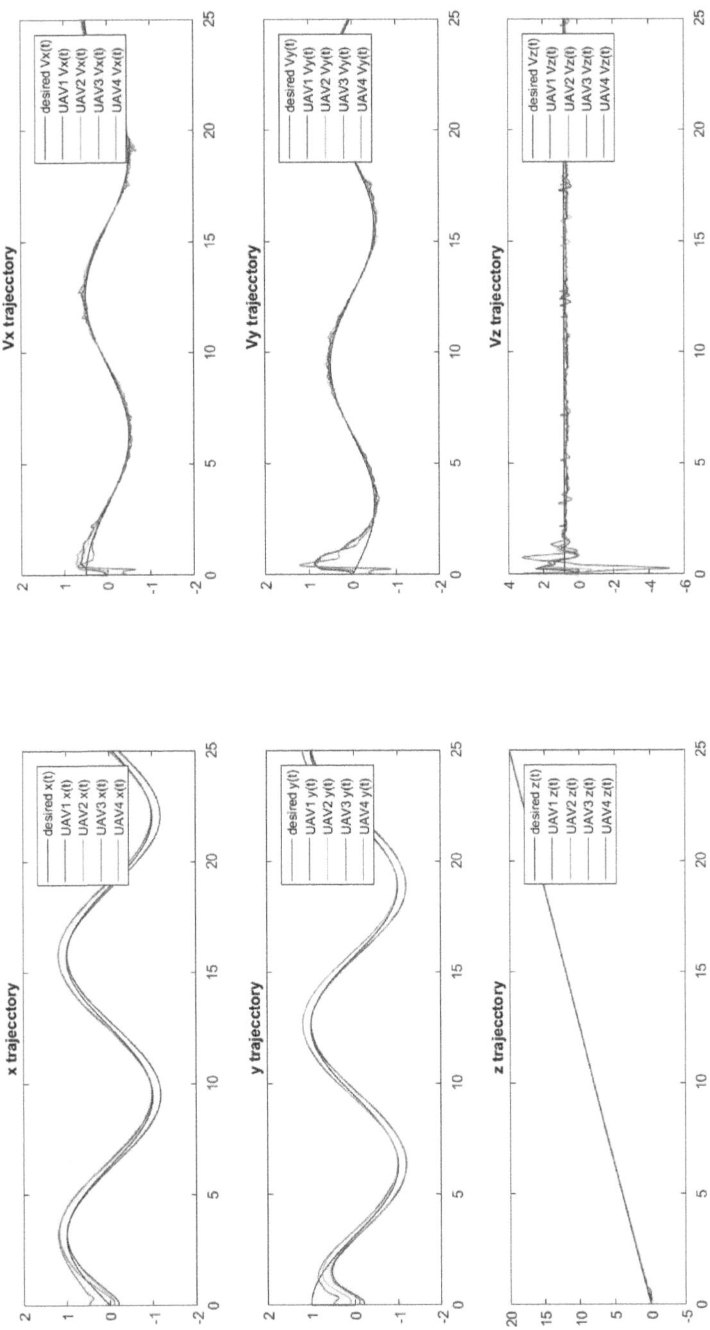

FIGURE 3.6 The position and the velocity tracking trajectory of the virtual structure quadrotor UAV formation team. The effect of random fluctuation due to Wiener process and Poisson process in Figure 3.8 and external disturbance are effectively attenuated by the proposed H_∞ team formation tracking control strategy.

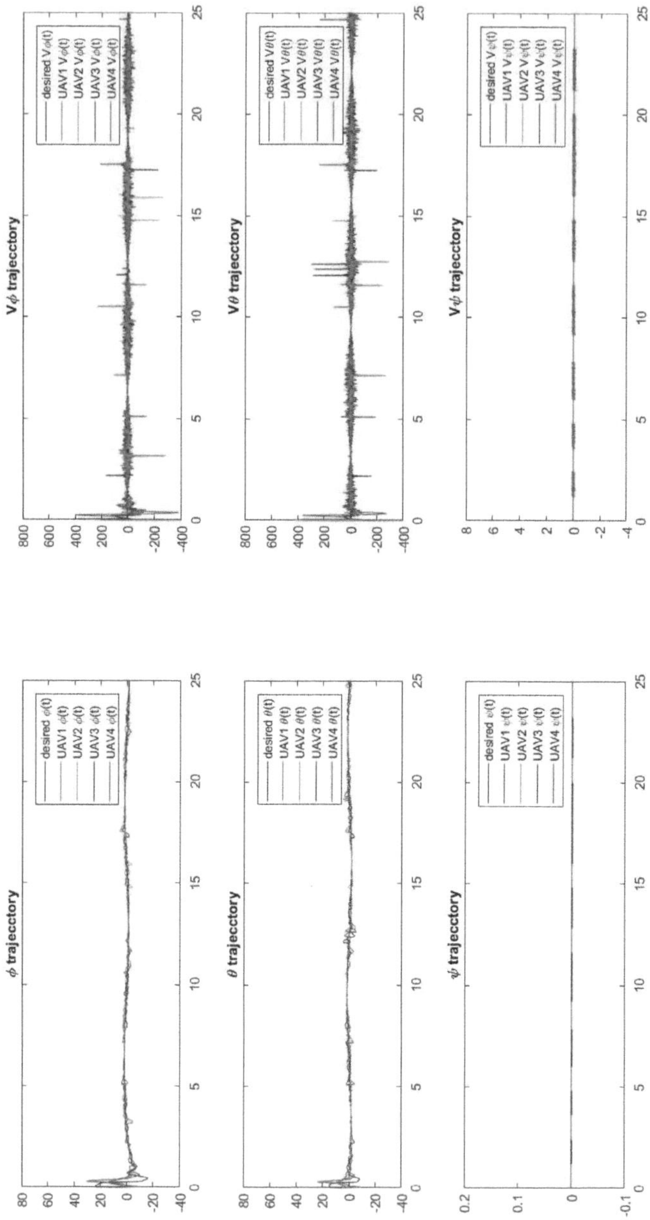

FIGURE 3.7 The angular and angular velocity tracking trajectory of the virtual structure multi-quadrotor UAV formation team. The effect of the intrinsic fluctuations (Wiener process and Poisson process in Figure 3.8) and time-varying delay can be observed in the simulation, especially on V_ψ.

(a) Wiener process of the four quadrotor UAVs.
(b) Poisson counting process of the four quadrotor UAVs.

ϕ, θ, and ψ and the angular velocities v_ϕ, v_θ, and v_ψ of the four quadrotor UAVs. We can see the effect of time-varying delay in Figure 3.7 especially on v_ψ. The continuous Wiener process and the discontinuous Poisson counting process of the four quadrotor UAVs are described in Figure 3.8a and Figure 3.8b, respectively.

Figure 3.9 shows the released instants of the stochastic quadrotor UAV team formation tracking system under the event-triggered scheme in (3.17), that is, if the

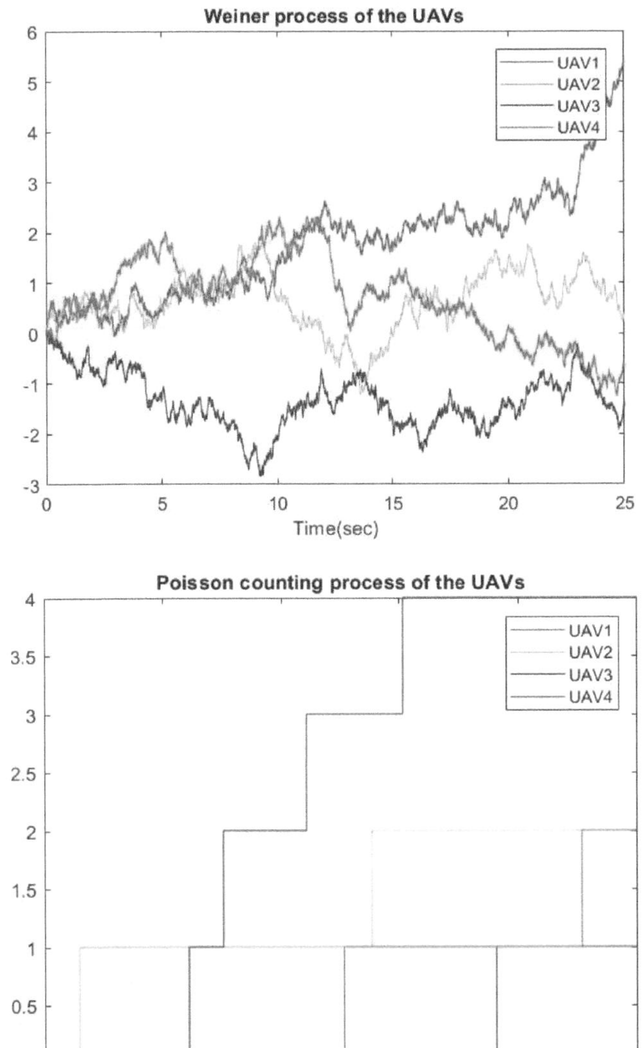

FIGURE 3.8 The stochastic random process of the quadrotor UAV formation team tracking system, including continuous Wiener process in part (a) and the discontinuous Poisson process in part (b).

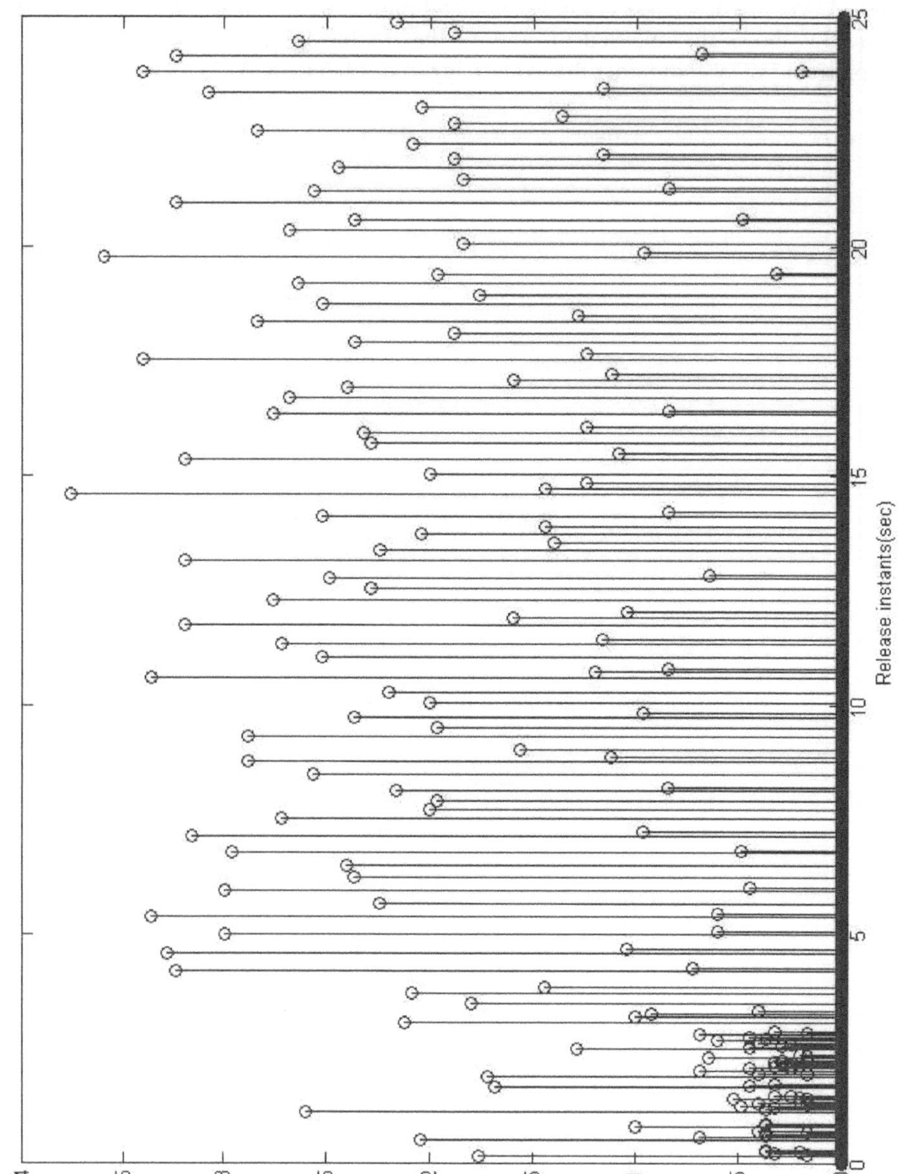

FIGURE 3.9 The released instants and the released intervals of the proposed event-triggered scheme for the multi-quadrotor UAV team formation tracking system. The maximum released interval is 0.38 s. Due to the transient response at the beginning, the control signal needs to be released more times than the sequels.

event-triggered condition is met, sensor will release a signal $(t_k h)$ to the controller to generate a control signal $u\ (t_k h)$ to actuator, otherwise the actuator would use the last control signal and sensor would not release signal. Released interval is defined as the interval between two released sequential instants and we can see that the maximum released interval is $(t_k h)$ in our case. Compared with the periodically time-triggered scheme, the stochastic multi-quadrotor UAV team tracking strategy we proposed can save 67% of energy and communication loading. It means that our method only costs 33% of the consumption of energy and communication resource. In summary, the proposed robust $0.38s$ event-triggered team formation tracking control strategy will make the formation tracking performance of multi-quadrotor UAV team within a desired disturbance attenuation level but it can save lots of energy and communication loading which become more and more important in the future application for the multi-quadrotor UAV networked control system in smart city.

3.7 CONCLUSION

In this chapter, the stochastic H H_∞ virtual structure team tracking control strategy embedded with event-triggered mechanism is proposed for the team formation design problem of multi-quadrotor UAV networked control system in future smart city. To describe the uncertainties, interference, and delay in networked control system, the intrinsic random fluctuation, time-varying delay, and unknown external disturbances from the environment are considered in the proposed nonlinear stochastic multi-quadrotor UAV networked team formation control system. To make the multiple quadrotor UAVs track a fixed virtual structure formation, a shifted team tracking dynamic system is constructed to simplify the design procedure. Therefore, the event-triggered team formation tracking control problem can be transformed to an HJI-constrained optimization problem. However, for practical application, it is still difficult to be solved analytically and numerically. Thus, T-S fuzzy method is applied such that the HJI-constrained optimization problem can be replaced by the LMI-constraint optimization problem and the H_∞ event-triggered team formation tracking problem of the multi-quadrotor UAV system can be solved efficiently with the help of MATLAB LMI toolbox. In the simulation result, the four-quadrotor-UAV networked control system successfully achieve the square-shape formation tracking with a robust H_∞ team formation tracking performance and the communication load and energy consumption are greatly reduced than the traditional periodically time-triggered scheme. In the future, along with the increasing of mission complexity, the number of quadrotor UAVs in the team formation system may increase substantially. In this case, a large amount quadrotor UAVs in the network team tracking control system can not be controlled through one controller due to the large computational complexity. Thus, the centralized control is not workable for a large quadrotor UAV networked team tracking control system. Moreover, under the framework of networked control design, the quadrotor UAVs team formation tracking control system not only suffers from channel interference but also receives malicious signal from the attacker. Hence, the future works need to focus on the decentralized control design of quadrotor

UAV networked team formation tracking system and fault-tolerant control design of quadrotor UAV networked team formation tracking control system under malicious attack. These team formation control design problems will be discussed in the following chapters.

APPENDIX 3A: PROOF OF THEOREM 3. 1

Defining a positive function $V(x(t)) \geq 0$ with $V(0)=0,\ V(\cdot) \in C^2$, we have

$$E\int_0^{t_f} V(x(t)) + V(x_0) - \lim_{t\to t_f} V(x(t))\} = 0 \tag{3.35}$$

Then, by using Itô-Lévy formula in Lemma 3.1 with the fact that $E[dW_i(t)] = 0$ and $E[dN_i(t)] = \lambda_i dt$, for $i = 1,\cdots,N$, (3.35) can be written as:

$$\begin{aligned} &E\{V(x_0) - \lim_{t\to t_f} V(x(t)) + \int_0^{t_f} [V_x^T f(x(t)) + V_x^T g(x(t))\tilde{u}(t) + V_x{}^T H\bar{v}(t) \\ &+\sum_{i=1}^{N} \frac{1}{2}\tilde{\sigma}_i{}^T(x(t))V_{xx}\tilde{\sigma}_i(x(t)) + \sum_{i=1}^{N}\lambda_i[V(x(t)+\Gamma_i(x(t))) - V(x(t))]]dt\} = 0 \end{aligned} \tag{3.36}$$

By (3.35) and (3.36), the numerator of robust H$_\infty$ tracking performance in (3.18) can be written as:

$$\begin{aligned} E\{\int_0^{t_f} x^T(t)\bar{Q}x(t)\} &= E\int_0^{t_f} x^T(t)\bar{Q}x(t)dt + \int_0^{t_f} V(x(t))\} \\ &+ E\{V(ex_0)\} - E\{\lim_{t\to t_f} V(x(t))\} \leq E\{V(x_0) \\ &+ E\int_0^{t_f} x^T(t)\bar{Q}x(t) + V_x^T nf(x(t)) + V_x^T g(x(t))\tilde{u}(t) \\ &+ V_x^T H\bar{v}(t) + \sum_{i=1}^{N}\frac{1}{2}\tilde{\sigma}_i^T(x(t))V_{xx}\tilde{\sigma}_i(x(t)) \\ &+ \sum_{i=1}^{N}\lambda_i[V(x(t)+\Gamma_i(x(t))) - V(x(t))]dt\} \end{aligned} \tag{3.37}$$

By applying Lemma 3.2, the term $V_x^T H\bar{v}(t)$ in (3.37) can be relaxed as

$$\begin{aligned} V_x^T H\bar{v}(t) &= \frac{1}{2}V_x{}^T H\bar{v}(t) + \frac{1}{2}\bar{v}^T(t)H^T V_x \leq \frac{1}{4\rho^2}V_x^T HH^T V_x \\ &+ \rho^2\bar{v}^T(t)\bar{v}(t) \end{aligned} \tag{3.38}$$

By substituting (3.38) into (3.37), the following inequality holds:

$$
\begin{aligned}
&E\{\int_0^{t_f} x^T(t)\bar{Q}x(t)\} \le E\{V(x_0)+\int_0^{t_f} x^T(t)\bar{Q}x(t)+V_x^T f(x(t)) \\
&+V_x^T g(x(t))\tilde{u}(t)+\frac{1}{4\rho^2}V_x^T HH^T V_x+\rho^2\bar{v}^T(t)\bar{v}(t) \\
&+\sum_{i=1}^{N}\frac{1}{2}\tilde{\sigma}_i^T(x(t))V_{xx}\tilde{\sigma}_i(x(t)) \\
&+\sum_{i=1}^{N}\lambda_i[V(x(t)+\Gamma_i(x(t)))-V(x(t)]\}]dt\}
\end{aligned}
\tag{3.39}
$$

As a result, if the HJI in (3.21) is satisfied, that is,

$$
\begin{aligned}
&x^T(t)\bar{Q}x(t)+V_x^T f(x(t))+V_x^T g(x(t))u(t) \\
&+\frac{1}{4\rho^2}V_x^T HH^T V_x+\sum_{i=1}^{N}\frac{1}{2}\tilde{\sigma}_i^T(x(t))V_{xx}\tilde{\sigma}_i(x(t)) \\
&+\sum_{i=1}^{N}\lambda_i\{V(x(t)+\Gamma_i(x(t)))-V(x(t))\}\le 0
\end{aligned}
\tag{3.40}
$$

then we can obtain the following inequality:

$$
E\{\int_0^{t_f} x^T(t)\bar{Q}x(t)dt\} \le E\{V(x_0)+\rho^2\int_0^{t_f}\bar{v}^T(t)\bar{v}(t)dt\}
\tag{3.41}
$$

that is, the robust H$_\infty$ team formation tracking performance in (3.18) is satisfied with $J_\infty(u(t))\le\rho^2$.

Q.E.D.

APPENDIX 3B: PROOF OF THEOREM 3. 2

In Theorem 3.1, the stochastic nonlinear event-triggered multi-quadrotor UAV shifted dynamic system (3.16) will achieve the H$_\infty$ team formation tracking performance if the constrained HJI in (3.21) is satisfied. By the T-S fuzzy model in (3.22) with the Lyapunov function $V(x(t))=x^T(t)Px(t)$ and positive matrix P, we can rewrite the terms in (3.21) as follows:

$$
V_x^T f(x(t))=\sum_{m=1}^{L}h_m(\varpi(t))x^T(t)\left[\bar{A}_m^T P+P\bar{A}_m\right]x(t)
\tag{3.42}
$$

$$
\begin{aligned}
V_x^T g(x(t))K(x(t-\eta(t))) = \sum_{m=1}^{L} h_m(\varpi(t)) \sum_{n=1}^{L} nh_n(\varpi(\bar{t}_k))\{x^T(t)P\bar{B}_m\bar{K}_n \\
\times x(t-\eta(t)) + x^T(t-\eta(t))\bar{K}_n^T\bar{B}_m^T Px(t) \\
- x^T(t)P\bar{B}_m\bar{K}_n e_k(t) - e_k^T(t)\bar{K}_n^T\bar{B}_m^T Px(t)\}
\end{aligned} \tag{3.43}
$$

$$
\begin{aligned}
&\frac{1}{2}\sigma_i^T\left(x(t)\right)V_{xx}\sigma_i\left(x(t)\right) \\
&= (\sum_{m=1}^{L} h_m(\varpi(t))\bar{C}_{im}x(t))^T P(\sum_{k=1}^{L} h_k(\varpi(t))\bar{C}_{im}x(t)) \\
&\leq \sum_{m=1}^{L} h_m(\varpi(t))x^T(t)\left[\bar{C}_{im}^T P\bar{C}_{im}\right]x(t)
\end{aligned} \tag{3.44}
$$

$$
\begin{aligned}
&\lambda_i\{V(x(t)+\Gamma_i\left(x(t)\right)) - V(x(t))\} \\
&= \lambda[(x(t) + \sum_{m=1}^{L} h_m(\varpi(t))\bar{D}_{im}x(t))^T P(x(t) \\
&+ \sum_{k=1}^{L} h_k(\varpi(t))\bar{D}_{ik}x(t)) - x^T(t)Px(t)] \\
&\leq \sum_{m=1}^{L} h_m(\varpi(t))x^T(t)\lambda_i[\bar{D}_{im}^T P + P\bar{D}_{im} + \bar{D}_{im}^T P\bar{D}_{im}]x(t), \\
&\forall\ i = 1,\cdots,N
\end{aligned} \tag{3.45}
$$

By the Newton-Leibniz formula [38] for the event-triggered team formation shifted tracking dynamic system in (3.26), we have:

$$
x(t) - x(t-\eta(t)) = \int_{t-\eta(t)}^{t} dx(s) \tag{3.46}
$$

$$
\begin{aligned}
\alpha(\int_{t-\eta(t)}^{t} dx(s))^T PP \int_{t-\eta(t)}^{t} dx(s) = \alpha(x(t) - x(t-\eta(t))^T \\
PP(x(t) - x(t-\eta(t))
\end{aligned} \tag{3.47}
$$

where $\alpha > 0$ is the scalar design variable.

From the event-triggered mechanism in (3.17), we note that the control signal between the two transmitted sequential instants must satisfy the following inequality:

$$
\begin{aligned}
E\{[x(t-\eta(t)) - e_k(t)]^T \sigma\Phi[x(t-\eta(t)) - e_k(t)] - e_k(t)^T \Phi e_k(t)\} > 0 \\
for\ t \in [t_k h + \tau_{t_k}, t_{j+1}h + \tau_{t_{(k+1)}}), \mathrm{k} \in \mathbb{N}
\end{aligned} \tag{3.48}
$$

By (3.46), (3.47) and inequality (3.48), (3.39) can be written as:

$$
\begin{aligned}
&E\{\int_0^{t_f} x^T(t)\bar{Q}x(t)dt\} \le E\{V(x(0)) - V(x(t_f)) + \int_0^{t_f}[\sum_{k=0}^{T}\sum_{m=1}^{L}\sum_{n=1}^{L}\chi_k(t) \\
&\times h_m(\varpi(t))h_n(\varpi(\bar{t}_k))x^T(t)(\bar{Q} + \bar{A}_m^T P + P\bar{A}_m + \frac{1}{\rho^2}PHH^T P \\
&+\sum_{i=1}^{N}\{\bar{C}_{im}^T P\bar{C}_{im} + \lambda_i[\bar{D}_{im}^T P + P\bar{D}_{im} + \bar{D}_{im}^T P\bar{D}_{im}]\})x(t) \\
&+x^T(t)P\bar{B}_m\bar{K}_n x(t-\eta(t)) \\
&+x^T(t-\eta(t))\bar{K}_n^T\bar{B}_m^T Px(t) - x^T(t)P\bar{B}_m\bar{K}_n e_k(t) \\
&-e_k^T(t)\bar{K}_n^T\bar{B}_m^T Px(t)] + 2(-2x(t) + \bar{\eta}x(t-\eta(t)))^T \\
&\times P(x(t) - x(t-\eta(t)) - \int_{t-\eta(t)}^{t} x(s)) \\
&+\alpha(x(t) - x(t-\eta(t))^T PP(x(t) - x(t-\eta(t)) \\
&-\alpha(\int_{t-\eta(t)}^{t} x(s))^T PP\int_{t-\eta(t)}^{t} x(s) \\
&+[x(t-\eta(t)) - e_k(t)]^T\sigma\Phi[x(t-\eta(t)) - e_k(t)] \\
&-e_k(t)^T\Phi e_k(t)]dt + E\{\int_0^{t_f}\rho^2\bar{v}^T(t)\bar{v}(t)dt\} \\
&= E\{\int_0^{t_f}\sum_{k=0}^{T}[\sum_{m=1}^{L}\sum_{n=1}^{L}\chi_j(t)h_m(\varpi(t))h_n(\varpi(\bar{t}_k))\xi_k^T(t) \\
&\times\Xi_{m,n}\xi_k(t)]dt\} + E\{\rho^2\int_0^{t_f}\bar{v}^T(t)\bar{v}(t)dt + V(x(0))\}
\end{aligned}
\tag{3.49}
$$

with

$$
\xi_k(t) = [x^T(t), x^T(t-\eta(t)), e_k^T(t), \int_{t-\eta(t)}^{t} x^T(s)]^T
$$

$$
\Xi_{m,n} = \begin{bmatrix}
\Xi_{11}^m & \begin{matrix} P\bar{B}_m\bar{K}_n - \alpha PP \\ +(\bar{\eta}+2)P \end{matrix} & -P\bar{B}_m\bar{K}_n & 2P \\
* & \Xi_{22} & -\sigma\Phi & -\bar{\eta}P \\
* & * & (\sigma-1)\Phi & 0 \\
* & * & * & -\alpha PP
\end{bmatrix}
\tag{3.50}
$$

$$
\chi_j(t) = \begin{cases} 1, & t \in [t_j h + \tau_{t_j}, t_{j+1}h + \tau_{t_{(j+1)}}) \\ 0, & otherwise \end{cases}
$$

where

$$\begin{aligned}\Xi_{11}^{m} &= \bar{Q} + \bar{A}_m^T P + P\bar{A}_m + \frac{1}{\rho^2} PHH^T P \\ &+ \sum_{i=1}^{N} \{\bar{C}_{im}^T P\bar{C}_{im} + \lambda_i[\bar{D}_{im}^T P + P\bar{D}_{im} + \bar{D}_{im}^T P\bar{D}_{im}]\} - 4P + \alpha PP, \\ \Xi_{22} &= -2\bar{\eta}P + \sigma\Phi + \alpha PP.\end{aligned}$$

From the matrix structure of $\Xi_{m,n}$, $\Xi_{m,n}$ is independent from index k. Thus, if the following matrix inequalities hold:

$$\Xi_{m,n} \leq 0 \;\; for \; m,n = 1,\cdots,L \tag{3.51}$$

then we immediately have the following inequality:

$$E\{\int_0^{t_f} x^T(t)Qx(t)dt\} \leq E\{V(x_0) + \rho^2 \int_0^{t_f} \bar{v}^T(t)\bar{v}(t)dt\} \tag{3.52}$$

which shows the event-triggered multi-quadrotor UAV system will achieve the robust H$_\infty$ team formation tracking performance in (3.18) under a prescribed level ρ^2.

However, due to the cross-couple terms in $\Xi_{m,n}$, the matrix inequality $\Xi_{m,n} \leq 0$ can not be solved by current convex techniques. Thus, the following relaxing condition is developed to deal with this problem. Define the positive matrix P as $P = diag\{P_0, P_1, \cdots, P_N\}$ with $P_i > 0$, for $i = 0,1,\cdots, N$. Then, by pre-multiplying and post-multiplying $\bar{W} = \text{diag}\{W,W,W,W\}$ with $W = P^{-1}$ to (3.51), we have

$$\begin{aligned}&\bar{\Xi}_{m,n} \leq 0 \\ &for \; m,n = 1,\cdots,L\end{aligned} \tag{3.53}$$

where

$$\bar{\Xi}_{m,n} = \begin{bmatrix} \bar{\Xi}_{11}^{m} & \begin{matrix}\bar{B}_m\bar{Y}_n - \alpha I \\ +(\bar{\eta}+2)W\end{matrix} & -\bar{B}_m\bar{Y}_n & 2W \\ * & \bar{\Xi}_{22} & -\sigma W\Phi W & -\bar{\eta}W \\ * & * & (\sigma-1)W\Phi W & 0 \\ * & * & * & -\alpha I \end{bmatrix} \tag{3.54}$$

$$\begin{aligned}\bar{\Xi}_{11}^{m} &= W\bar{Q}W + W\bar{A}_m^T + \bar{A}_m W + \frac{1}{\rho^2}HH^T + \sum_{i=1}^{N}\{W\bar{C}_{im}^T W^{-1}\bar{C}_{im}W + \lambda_i[W\bar{D}_{im}^T \\ &+ \bar{D}_{im}W + W\bar{D}_{im}^T W^{-1}\bar{D}_{im}W]\} - 4W + \alpha I, \;\; \bar{\Xi}_{22} = -2\bar{\eta}W + \sigma W\Phi W + \alpha I, \\ \bar{Y}_{in} &= \bar{K}_{in}W_i \text{ and } \bar{Y}_n = diag\{0, \bar{Y}_{1n}, \bar{Y}_{2n}, \ldots, \bar{Y}_{Nn}\} = \bar{K}_n W. = \bar{K}_n W\end{aligned}$$

As a results, the matrix inequalities in (3.51) are transformed to the more relaxed BMIs in (3.53), that is, the robust H$_\infty$ team formation tracking performance in (3.18) is achieved under a prescribed level ρ^2 if the BMIs in (3.30) hold. The proof is done.

REFERENCES

[1] P. Castillo, A. Dzul, and R. Lozano, "Real-time stabilization and tracking of a four-rotor mini rotorcraft," *IEEE Transactions on Control Systems Technology*, vol. 12, no. 4, pp. 510–516, July 2004.

[2] F. Santoso, M. A. Garratt, and S. G. Anavatti, "Hybrid PD-fuzzy and PD controllers for trajectory tracking of a quadrotor unmanned aerial vehicle: Autopilot designs and real-time flight tests," *IEEE Transactions on Systems, Man, and Cybernetics: Systems*, to be published. http://doi.org/10.1109/TSMC.2019.2906320.

[3] X. L. Lin, C. F. Wu, and B. S. Chen, "Robust H_∞ adaptive fuzzy tracking control for MIMO nonlinear stochastic Poisson jump diffusion systems," *IEEE Transactions on Cybernetics*, vol. 49, no. 8, Aug. 2019.

[4] A. Abbaspour, K. K. Yen, P. Forouzannezhad, and A. Sargolzaei, "A neural adaptive approach for active fault-tolerant control design in UAV," *IEEE Transactions on Systems, Man, and Cybernetics: Systems*, to be published. http://doi.org/10.1109/TSMC.2018.2850701.

[5] R. Xue and G. Cai, "Formation flight control of multi-UAV system with communication constraints," *Journal of Aerospace Technology and Management*, vol. 8, no. 2, pp. 203–210, May 2016.

[6] A. Giyenko and Y. I. Cho, "Intelligent UAV in smart cities using IoT," in *Proceedings of the 16th International Conference on Control, Automation and Systems (iCCAS)*, Gyeongju, Korea (South), Oct. 2016, pp. 207–210.

[7] J. Lawton, R. Beard, and B. Young, "A decentralized approach to formation maneuvers," *IEEE Transactions on Robotics and Automation*, vol. 19, no. 6, pp. 933–941, Dec. 2003.

[8] B. S. Chen, C. P. Wang, and M. Y. Lee, "Stochastic robust team tracking control of multi-UAV networked control system under Wiener and Poisson random fluctuations," *IEEE Transactions on Cybernetics*, to be published. http://doi.org/10.1109/TCYB.2019.2960104.

[9] P. Ogren, M. Egerstedt, and X. Hu, "A control Lyapunov function approach to multiagent coordination," *IEEE Transactions on Robotics and Automation*, vol. 18, no. 5, pp. 847–851, Oct. 2002.

[10] D. Scharf, F. Hadaegh, and S. Ploen, "A survey of spacecraft formation flying guidance and control. Part II: Control," in *Proceeding of the 2004. American Control Conference*, Boston, MA, USA, 2004, pp. 2976–2985.

[11] Q. Zhang, et al. "Distributed control of coordinated path tracking for networked nonholonomic mobile vehicles," *IEEE Transactions on Industrial Informatics*, vol. 9, no. 1, pp. 472–484, Feb. 2013.

[12] O. Saif, I. Fantoni, and A. Zavala-Rio, "Real-time flocking of multiple-quadrotor system of systems," in *2015 10th System of Systems Engineering Conference (soSE)*, San Antonio, TX, USA, May 2015, pp. 286–291.

[13] A. Kushleyev, D. Mellinger, C. Powers, and V. Kumar, "Towards a swarm of agile micro quadrotors," *Autonomous Robots*, vol. 35, pp. 287–300, Nov. 2013.

[14] A. Schollig, F. Augugliaro, S. Lupashin, and R. D'Andrea, "Synchronizing the motion of a quadrocopter to music," in *Proceeding of IEEE International Conference on Robotics and Automation, 2010*, Anchorage, Alaska, May 2010, pp. 3355–3360.

[15] J. S. Song and X. H. Chang, "H_∞ controller design of networked control systems with a new quantization structure," *Applied Mathematics and Computation*, vol. 376, p. 125070, July 2020.

[16] C. C. Ren, N. Rong, and S. P. He, "Finite-time positiveness and distributed control of Lipschitz nonlinear multi-agent systems," *Journal of the Franklin Institute*, vol. 356, no. 15, pp. 8080–8092, Oct. 2019.

[17] I. Jawhar, N. Mohamed, and J. Al-Jaroodi, "Networking architectures and protocols for smart city systems," *Journal of Internet Services and Applications*, vol. 9, no. 26, Dec. 2018.

[18] J. Lunze and D. Lehmann, "A state-feedback approach to event-based control," *Automatica*, vol. 46, no. 1, pp. 211–215, Jan. 2010.

[19] D. Lehmann and J. Lunze, "Event-based output-feedback control," in *19th Mediterranean Conference on Control and Automation*, Corfu, Greece, 2011, pp. 982–987.

[20] D. V. Dimarogonas and E. Frazzoli, "Distributed event-triggered control strategies for multi-agent systems," in *2009 47th Annual Allerton Conference on Communication, Control, and Computing*, Allerton, Sep. 2009.

[21] X. Wang and M. D. Lemmon, "Event-triggering in distributed networked control systems," *IEEE Transactions on Automatic Control*, vol. 56, no. 3, pp. 586–601, Mar. 2011.

[22] D. Yue, E. Tian, and Q. Han, "A delay system method to design of event-triggered control of networked control systems," in *IEEE Conference on Decision and Control and European Control Conference*, Orlando, FL, USA, Dec. 2011.

[23] H. C. Yan, J. Y. Sun, X. S. Zhan, and F. W. Yang, "Event-triggered H_∞ state estimation of 2-DOF quarter-car suspension systems with nonhomogeneous markov switching," *IEEE Transactions on Systems, Man, and Cybernetics: Systems*, to be published. http://doi.org/10.1109/TSMC.2018.2852688.

[24] H. C. Yan, C. Y. Hu, H. Zhang, H. R. Karimi, X. W. Jiang, and M. Liu, "H_∞ output tracking control for networked control systems with adaptively adjusted event-triggered scheme," *IEEE Transactions on Systems, Man, and Cybernetics: Systems*, vol. 49, no. 10, pp. 2050–2058, Oct. 2019.

[25] H. Zhang, Z. P. Wang, H. C. Yan, F. W. Yang, and X. Zhou, "Adaptive event-triggered transmission scheme and H_∞ filtering co-design over a filtering network with switching topology," *IEEE Transactions on Cybernetics*, vol. 49, no. 12, pp. 4296–4307, Sep. 2018.

[26] Z. M. Lee, X. H. Chang, and J. H. Park, "Quantized static output feedback fuzzy tracking control for discrete-time nonlinear networked control systems with asynchronous event-triggered constraints," *IEEE Transactions on Systems, Man, and Cybernetics: Systems*, to be published. http://doi.org/10.1109/TSMC.2019.2931530.

[27] Z. Zhou, H. Wang, and Z. Hu, "Event-based time varying formation control for multiple quadrotor UAVs with Markovian switching topologies," *Complexity*, vol. 2018, ID: 8124861, pp. 1–15, Apr. 2018.

[28] T. Dierks and S. Jagannathan, "Output feedback control of a quadrotor UAV using neural networks," *IEEE Transactions on Neural Networks*, vol. 21, no. 1, pp. 50–66, Jan. 2010.

[29] H. C. Yan, Y. X. Tian, H. Y. Li, H. Zhang, and Z. C. Li, "Input–output finite-time mean square stabilization of nonlinear semi-Markovian jump system," *Automatica*, vol. 104, pp. 82–89, June 2019.

[30] W. H. Zhang, L. H. Lee, and B. S. Chen, *Stochastic H_2/H_∞ Control: A Nash Game Approach*, London: CRC Press, Taylor & Francis, 2017.

[31] R. C. L. F. Oliveira, A. N. Vargas, J. B. R. do Val, and P. L. D. Peres, "Mode-independent H2-control of a DC motor modeled as a Markov jump linear system," *IEEE Transactions on Control System Technology*, vol. 22, no. 5, pp. 1915–1919, Sep. 2014.

[32] X. C. Chang, Y. Liu, and M. Q. Shen, "Resilient control design for lateral motion regulation of intelligent vehicle," *IEEE/ASME Transactions on Mechatronics*, vol. 24, no. 6, pp. 2488–2497, Dec. 2019.

[33] C. C. Ren and S. P. He, "Finite-time stabilization for positive Markovian jumping neural networks," *Applied Mathematics and Computation*, vol. 365, p. 124631, Jan. 2020.
[34] B. Øksendal and A. Sulem, *Applied Stochastic Control of Jump Diffusions*, 2nd ed., Berlin: Springer, 2007.
[35] F. Hanson, *Applied Stochastic Processes and Control for Jump-Diffusions: Modeling, Analysis and Computation*, 2nd ed., Philadelphia, PA: SIAM, 2007.
[36] S. P. He, H. Y. Fang, M. G. Zhang, F. Liu, X. L. Luan, and Z. D. Ding, "Online policy iterative-based H_∞ optimization algorithm for a class of nonlinear systems," *Information Sciences*, vol. 495, pp. 1–13, Aug. 2019.
[37] S. P. He, H. Y. Fang, M. G. Zhang, F. Liu, and Z. T. Ding, "Adaptive optimal control for a class of nonlinear systems: The online policy iteration approach," *IEEE Transactions on Neural Networks and Learning Systems*, vol. 31, no. 2, pp. 549–558, Feb. 2020.
[38] K. Watanabe, "Stochastic fuzzy control. I. Theoretical derivation," in *Proceeding of IEEE International Conference on Fuzzy Systems*, Yokohama, Japan, Mar. 1995, pp. 547–554.
[39] K. Tanaka and Hua O. Wang, *Fuzzy Control Systems Design and Analysis: A Linear Matrix Inequality Approach*, John Wiley & Sons, New York, 2001.
[40] C. Peng and T. C. Yang, "Event-triggered communication and control co-design for networked control systems," *Automatica*, vol. 49, no. 5, pp. 1326–1332, May 2013.
[41] D. Yue, E. Tian, and Q. Han, "A delay system method for designing eventtriggered controllers of networked control systems," *IEEE Transactions on Automatic Control*, vol. 58, no. 2, pp. 475–481, June 2012.
[42] S. P. Boyd, L. E. Ghaoui, E. Feron, and V. Balakrishnan, *Linear Matrix Inequalities in System and Control Theory*, Philadelphia, PA: SIAM, 1994.
[43] C. S. Tseng, B. S. Chen, and H. J. Uang, "Fuzzy tracking control design for nonlinear dynamic systems via T-S fuzzy model," *IEEE Transactions on Fuzzy Systems*, vol. 9, no. 3, pp. 381–392, June 2001.
[44] J. Xiong and E. Zheng, "Position and attitude tracking control for a quadrotor UAV," *ISA Transactions*, vol. 53, no. 3, pp. 725–731, May 2014.
[45] T. Luukkonen, *Modelling and Control of Quadcopter, Independent Research Project in Applied Mathematics*, Espoo: Aalto University, 2011.
[46] Z. C. Li, H. C. Yan, H. Zhang, X. S. Shen, and C. Z. Huang, "Improved inequality-based functions approach for stability analysis of time delay system," *Automatica*, vol. 108, pp. 1–7, Oct. 2019.
[47] X. C. Jia, D. W. Zhang, X. H. Hao, and N. N. Zheng, "Fuzzy H_∞ tracking control for nonlinear networked control systems in T–S fuzzy model," *IEEE Transactions on Systems, Man, and Cybernetics-Part B: Cybernetics*, vol. 39, no. 4, pp. 1073–1079, Aug. 2009.

4 Decentralized H_∞ PID Team Formation Tracking Control of Large-Scale Quadrotor UAVs under External Disturbance and Vortex Coupling

4.1 INTRODUCTION

Recently, quadrotor unmanned aerial vehicles (UAVs) have attracted a growing interest in academic and industrial field due to wide and high maneuverability [1,2]. Since the quadrotor UAV is a highly nonlinear dynamic system with six degrees of freedom and four actuators (i.e., the underactuated characteristics), it is not easy to control quadrotor UAV for a desired trajectory especially by a conventional PID control because conventional PID control designs are always based on a transfer function at an operation point and therefore need a set of PID controllers for different flying operation points of nonlinear quadrotor UAV. Nonlinear adaptive control [3,4] and nonlinear robust control [5–8] have been developed for UAV. However, a single quadrotor UAV has many limitations on applications, for example, battery power and loading capacity. Unlike a single quadrotor UAV, a team formation of multiple UAVs can achieve more challenging and complex tasks [9]. Consequently, a team formation of multiple quadrotor UAVs is becoming practical and popular in many application areas such as environmental monitoring, exploration of unsecured regions, detection, and interception, target searching and rescuing, battlefield surveillance [9].

Since the missions of quadrotor UAV are more challenging and complex, several formation control strategies have been developed based on leader-follower method [10], virtual leader method [11], and so on. Team formation of quadrotor UAVs based on leader-follower method is that one of quadrotor UAVs in the team plays the role of leader to represent the behavior trajectory of the team and is controlled to track a desired reference path, while other quadrotor UAVs are asked as followers to track the trajectory of leader quadrotor UAV with a formation shape [10]. Although leader-follower method was developed more earlier and commonly used in the team formation of UAVs, there exist some drawbacks. One is that the team formation is always dependent on the leader UAV. Thus, if the leader UAV is failed or disintegrated, the whole team formation of quadrotor UAVs can not be maintained or will collapse.

DOI: 10.1201/9781003545965-4

When compared with leader-follower method, the virtual leader method is more robust and can avoid the above failed or collapsed problems, that is, virtual leader is a reference model at the central of the desired shape to generate a reference trajectory with a desired formation shape [11,12].

Since the virtual leader is a reference model and will not suffer from perturbation, it can efficiently reduce the probability of team formation failure in reality. Recently, the formation control strategy design of large-scale quadrotor UAVs has been widely investigated. The robust H_∞ team formation network control is designed in Reference [11] in consideration of internal fluctuation in quadrotor UAV system. Also, to save the communication resource in large-scale quadrotor UAV network, the robust event-triggered team formation tracking design of quadrotor UAV network control system is developed in Reference [12]. Further, to reduce the interconnected effect in UAV network control system, the decentralized H_∞ tracking control of a large-scale team formation quadrotor UAV network control system is proposed to passively attenuate the interconnected effect [13]. Therein, to relieve the design difficulty of solving a set of nonlinear Hamilton-Jacobi inequalities (HJIs) for decentralized H_∞ team formation tracking control of large-scale quadrotor UAVs, the Takagi-Sugeno (T-S) fuzzy system is employed to interpolate several local linearized systems to approximate nonlinear quadrotor UAV. In this case, the nonlinear HJIs for the H_∞ team formation tracking control design can be reformulated as a set of coupled linear matrix inequalities (LMIs), which can be easily solved by the current convex optimization techniques [11–13]. On the other hand, based on a specified potential function associated with network topology of UAV network, a decentralized multi-subgroup formation algorithm is developed to achieve the desired formation control and collision avoidance [14].

In the aforementioned articles, it is worth to point out that the computational complexity of proposed team formation control strategies of large-scale UAVs in References [11–14] is still large. More specifically, a large computational time is needed to calculate the fuzzy control law $u(t)=\sum_{j=1}^{L} h_j(z(t))K_j(x(t)-r(t))$ for the centralized fuzzy team formation control in References [11,12] and $u_i(t)=\sum_{j=1}^{L} h_{i,j}(z_{i,j}(t))K_{i,j}(x_i(t)-r_i(t))$, for $i=1,\ldots,N$ for the decentralized control of team formation tracking of N quadrotor UAVs [13,15], respectively. More efforts are still needed for these fuzzy team formation tracking controls of large-scale quadrotor UAVs for practical applications. On the other hand, to achieve an accurate formation control, the potential function in Reference [14] for the controller synthesis is still with high nonlinearity, which will lead to the difficulty of implementation in practical applications. As a result, to remedy the complex T-S fuzzy team formation tracking control design or other nonlinear control strategy design for large-scale quadrotor UAVs, the PID team formation tracking control is probably the most simple method to overcome the complex and highly nonlinear team formation tracking control design problem.

Conventional PID controls have been successfully applied to the design of autopilots as real-time control and online navigation system for quadrotor UAVs due to their simple structure and easy implementation with an adequate performance [16–20]. Recently, an adaptive PID controller is proposed for fault-tolerant control of a quadrotor helicopter system in Reference [16], a robust PID control with the deadbeat response is proposed in Reference [19], and a PID control through nonlinear tan^{-1} function (i.e., the nonlinear PID) has been employed for a quadrotor UAV in Reference [20]. However, traditional PID control designs are always based on the linearized system model [21]. In this situation, a different controller is employed for a different linearized model at a different operation points of a nonlinear dynamic system. Moreover, parameter adjustment or tuning is still needed to obtain the corresponding PID controller gain in the design procedure. Furthermore, once the trajectory of nonlinear system is far away from the operation point of linearized system, the PID controller gain should be updated to enhance the tracking performance [22]. Hence, more efforts are still needed to apply PID controller to the team formation tracking control of large-scale and highly nonlinear quadrotor UAVs under external disturbance and vortex coupling.

In this chapter, a robust decentralized H_∞ PID team formation tracking control design is proposed for large-scale quadrotor UAVs with a desired time-varying formation shape under wind disturbance and coupling of trailing vortex from neighboring quadrotor UAVs. The proposed decentralized H_∞ PID team formation tracking control design for each quadrotor UAV can efficiently attenuate the external disturbance and decouple the coupling effect of trailing vortex from the neighboring quadrotor UAVs. Therefore, it can significantly reduce the control design complexity as well as the computational complexity of practical team formation tracking of large-scale quadrotor UAVs.

At the beginning, the proposed robust decentralized H_∞ PID team formation tracking control design of large-scale quadrotor UAVs can be transformed to a set of independent nonlinear partial differential HJIs for each quadrotor UAV in the team formation. Then, PID controller in each quadrotor UAV needs to be specified to satisfy the corresponding HJI. Further, the optimal decentralized H_∞ PID team formation tracking control design needs to solve an HJI-constrained optimization problem for PID control parameters of each quadrotor UAV in the team formation. However, at present, it is not easy to specify PID controller parameters to solve the corresponding HJI-constrained optimization problem for each quadrotor UAV to achieve the optimal decentralized H_∞ PID team formation tracking control design of large-scale quadrotor UAVs. To overcome the difficulty of solving nonlinear partial differential HJI, the T-S fuzzy interpolation scheme is employed to interpolate multiple local linearized stochastic systems to approximate the nonlinear stochastic system of quadrotor UAV [23]. Therefore, each HJI-constrained optimization problem of the optimal decentralized H_∞ PID team formation tracking control design of each quadrotor UAV in the team formation can be transformed to a corresponding Riccati-like inequalities (RLIs)-constrained optimization problem. Consequently, the HJI-constrained optimization problem of the optimal decentralized H_∞ robust PID team formation tracking control design of large-scale quadrotor UAVs can be transformed to an independent LMIs-constrained optimization problem for each quadrotor UAV in the team formation.

Since the decentralized H_∞ robust PID team formation control can be independently designed, the number of quadrotor UAVs in the team formation can be increased to a very large scale. A simulation example of time-varying team formation task for 25-quadrotor UAV system is provided to validate the performance of proposed decentralized H_∞ team formation PID control strategy in comparison with the conventional decentralized H_∞ T-S fuzzy team formation tracking control strategy of large-scale UAVs in Reference [15].

The main purposes of this chapter are described as follows:

1. The previous PID control designs of quadrotor UAV are always dependent on the linearized model of quadrotor UAV at one operation point and there are little PID control designs for the team formation tracking control of large-scale nonlinear quadrotor UAVs. In this chapter, we consider the robust decentralized H_∞ PID team formation tracking control design problem of large-scale nonlinear stochastic quadrotor UAVs with external disturbance and coupling of trailing vortex from neighboring UAVs. The proposed robust decentralized H_∞ PID team formation tracking control of large-scale quadrotor UAVs can significantly simplify the design complexity and save much computation time of PID controller than the conventional H_∞ fuzzy team formation control of large-scale quadrotor UAVs in References [13–15]. Therefore, the proposed robust PID team formaition control scheme in this chapter has a great potential to practical application for large-scale time-varying team formation control of UAVs.
2. Based on T-S fuzzy interpolation scheme to interpolate multiple local linear stochastic systems to approximate nonlinear stochastic UAV, the proposed optimal decentralized H_∞ PID team formation tracking control design problem of quadrotor UAVs can be transformed to an independent LMIs-constrained optimization problem for each quadrotor UAV in the team formation. Therefore, the number of quadrotor UAVs in the team formation can be easily extended to a very large scale without increasing the design difficulty.
3. When compared with other conventional PID control designs for the linearized quadrotor UAV system, the uncertain wind disturbance, intrinsic random Wiener fluctuation, and wind coupling of trailing vortex from neighborhood UAVs can be efficiently attenuated by the proposed optimal decentralized H_∞ PID team formation tracking control strategy of large-scale nonlinear stochastic UAVs. Further, unlike the conventional complex tuning process of PID control parameters, the proposed optimal H_∞ decentralized team formation control parameters can be found in a single round of convex optimization algorithm with the help of LMI toolbox in MATLAB.
4. Unlike the conventional PID controllers in quadrotor UAVs, which need to change their controllers' parameters based on the linearized model at different operation points according to different flying conditions like taking-off, hovering, and landing, the proposed robust H_∞ PID team formation controller is specified fixedly for all flight conditions of nonlinear dynamic quadrotor UAVs and therefore easy to implement for more practical applications.

The remainder of the chapter is organized as follows. In Section 4.2, the large-scale quadrotor UAV systems are introduced and the problem formulation of the decentralized robust stochastic H_∞ PID control for the team formation quadrotor UAVs is presented. In Section 4.3, the design condition is transformed into an equivalent HJI problem for each quadrotor UAV. In Section 4.4, with the help of T-S fuzzy method, the HJI problem of each quadrotor UAV is transformed into a set of bilinear matrix inequalities (BMIs). Furthermore, with the help of the novel variable transformation method, a set of independent BMIs are transformed into a set of independent LMIs for each quadrotor UAV and are solved by using the LMI toolbox in MATLAB. In Section 4.5, the decentralized robust stochastic H_∞ PID team formation tracking control method for 25 quadrotor UAVs in comparison with the traditional decentralized H_∞ T-S fuzzy team formation tracking control method is given to illustrate the effectiveness of the proposed method. Finally, the conclusion is made in Section 4.6.

Notation: A^T denotes the transpose of matrix A; $A \geq 0$ $(A > 0)$ denotes symmetric semi-positive definite matrix (symmetric positive definite matrix); I_n denotes the n-dimensional identity matrix; $E\{\cdot\}$ is the expectation operator; $\| x \|_2$ denotes the Euclidean norm for the given vector $x \in \mathbb{R}^n$; $x(t) \in \mathcal{L}_F^2[0,\infty)$ if $E\{\int_0^\infty x^T(t)x(t)dt\} < \infty$, that is, $x(t)$ is the finite mean energy; C^2 denotes the class of function $f(x)$ which is twice continuously differentiable with respect to x; $\frac{\partial f(x)}{\partial x}$ denotes the gradient column vector of differentiable function $f(x) \in \mathbb{R}$; and $\frac{\partial^2 f(x)}{\partial x^2}$ is the Hessian matrix of twice-differentiable function $f(x) \in \mathbb{R}$.

4.2 PRELIMINARY AND PROBLEM FORMULATION

4.2.1 Stochastic Dynamic Model Physical Plant of Quadrotor UAVs

In this chapter, we simultaneously consider the three position variables with the corresponding three velocities of the quadrotor UAV and three attitude states with the corresponding three angular velocities of the quadrotor UAV as the state variables in the state space dynamic model of the quadrotor UAV. To model the quadrotor UAV dynamic, two reference frames including inertial frame and body frame are introduced, and the free body diagram of the quadrotror UAV is shown in Figure 4.1. The position of the mass center of the quadrotor UAV is represented as coordinates (x, y, z) with respect to an inertial frame associated with the unit vector basis (e_x, e_y, e_z). On the other hand, the attitude of the quadrotor UAV is denoted by the three Euler angles (ϕ, θ, ψ) which define the orientation vector of the quadrotor UAV with respect to a body frame associated with the unit vector basis (e_{bx}, e_{by}, e_{bz}). Besides, ϕ is a roll angle $(-\frac{\pi}{2} < \phi < \frac{\pi}{2})$, θ is a pitch angle $(-\frac{\pi}{2} < \theta < \frac{\pi}{2})$, and ψ is a yaw angle $(-\pi < \psi < \pi)$. With the Newton-Euler method, the ith quadrotor UAV dynamic model of the large-scale quadrotor UAV networked system can be described as the following state space model [12]:

$$\begin{bmatrix} x_{i,2}(t) \\ y_{i,2}(t) \\ z_{i,2}(t) \\ \phi_{i,2}(t) \\ \theta_{i,2}(t) \\ \psi_{i,2}(t) \end{bmatrix} \tag{4.1}$$

$$\begin{bmatrix} \dot{x}_{i,1}(t) \\ \dot{y}_{i,1}(t) \\ \dot{z}_{i,1}(t) \\ \dot{\phi}_{i,1}(t) \\ \dot{\theta}_{i,1}(t) \\ \dot{\psi}_{i,1}(t) \\ \dot{x}_{i,2}(t) \\ \dot{y}_{i,2}(t) \\ \dot{z}_{i,2}(t) \\ \dot{\phi}_{i,2}(t) \\ \dot{\theta}_{i,2}(t) \\ \dot{\psi}_{i,2}(t) \end{bmatrix} = \begin{bmatrix} f(x_{i,2}(t),\phi_{i,1}(t),\theta_{i,1}(t), \\ \psi_{i,1}(t),F_i(t),v_{i,x}(t)) \\ f(y_{i,2}(t),\phi_{i,1}(t),\theta_{i,1}(t), \\ \psi_{i,1}(t),F_i(t),v_{i,y}(t)) \\ -\frac{d_{i,z}}{m_i}z_{i,2}-g \\ +\cos\phi_{i,1}(t)\cos\theta_{i,1}(t)\frac{F_i(t)}{m_i}+v_{i,z}(t) \\ \frac{J_{i,\theta}-J_{i,\psi}}{J_{i,\phi}}\theta_{i,2}(t)\psi_{i,2}(t) \\ -\frac{d_{i,\phi}}{J_{i,\phi}}\phi_{i,2}(t)+\frac{1}{J_{i,\phi}}\tau_{i,\phi}(t)+v_{i,\phi}(t) \\ \frac{J_{i,\psi}-J_{i,\phi}}{J_{i,\theta}}\psi_{i,2}(t)\phi_{i,2}(t) \\ -\frac{d_{i,\theta}}{J_{i,\theta}}\theta_{i,2}(t)+\frac{1}{J_{i,\theta}}\tau_{i,\theta}(t)+v_{i,\theta}(t) \\ \frac{J_{i,\phi}-J_{i,\theta}}{J_{i,\psi}}\phi_{i,2}(t)\theta_{i,2}(t) \\ -\frac{d_{i,\psi}}{J_{i,\psi}}\psi_{i,2}(t)+\frac{1}{J_{i,\psi}}\tau_{i,\psi}(t)+v_{i,\psi}(t) \end{bmatrix} \tag{4.1}$$

with

$$\begin{aligned} &f(x_{i,2}(t),\phi_{i,1}(t),\theta_{i,1}(t),\psi_{i,1}(t),F_i(t),v_{i,x}(t)) \\ &=-\frac{d_{i,x}}{m_i}x_{i,2}(t)+(\cos\phi_{i,1}(t)\sin\theta_{i,1}(t)\cos\psi_{i,1}(t) \\ &+\sin\phi_{i,1}(t)\sin\psi_{i,1}(t))\frac{F_i(t)}{m_i}+v_{i,x}(t) \\ &f(y_{i,2}(t),\phi_{i,1}(t),\theta_{i,1}(t),\psi_{i,1}(t),F_i(t),v_{i,y}(t)) \\ &=-\frac{d_{i,y}}{m_i}y_{i,2}(t)+(\cos\phi_{i,1}(t)\sin\theta_{i,1}(t)\sin\psi_{i,1}(t) \\ &-\sin\phi_{i,1}(t)\cos\psi_{i,1}(t))\frac{F_i(t)}{m_i}+v_{i,y}(t)\ for\ i=1,2,...,N \end{aligned}$$

where $x_{i,1}(t),\ y_{i,1}(t),\ z_{i,1}(t)\in\mathbb{R}$ denote the three position states of the ith quadrotor UAV in the inertial frame, $x_{i,2}(t),\ y_{i,2}(t),\ z_{i,2}(t)\in\mathbb{R}$ denote the three velocity states of the ith quadrotor UAV in the inertial frame, $\phi_{i,1}(t),\theta_{i,1}(t),\psi_{i,1}(t)\in\mathbb{R}$ denote the three attitude states of the ith quadrotor UAV in the body frame, $\phi_{i,2}(t),\theta_{i,2}(t),\psi_{i,2}(t)\in\mathbb{R}$ denote the angular velocity states of the ith quadrotor UAV in the body frame, $F_i(t)\in\mathbb{R}$ denotes the total thrust of the ith quadrotor

UAV, $\tau_{i,\phi}(t), \tau_{i,\theta}(t), \tau_{i,\psi}(t) \in \mathbb{R}$ denote the rotation torques of the ith quadrotor UAV, $v_{i,x}(t)$, $v_{i,y}(t)$, $v_{i,z}(t)$, $v_{i,\phi}(t)$, $v_{i,\theta}(t)$, and $v_{i,\psi}(t)$ represent the external disturbances of the quadrotor UAV, $m_i \in \mathbb{R}^+$ denotes the total mass of the quadrotor UAV, $g \in \mathbb{R}^+$ denotes the gravitational acceleration, $J_{i,\phi}$, $J_{i,\theta}$, $J_{i,\psi} \in \mathbb{R}^+$ denote the moments of the inertia of $\phi_i(t)$, $\theta_i(t)$, and $\psi_i(t)$, respectively, and $d_{i,x}$, $d_{i,y}$, $d_{i,z}$, $d_{i,\phi}$, $d_{i,\theta}$, $d_{i,\psi} \in \mathbb{R}^+$ represent aerodynamic damping coefficients of the quadrotor UAV.

After introducing a single dynamic system of quadrotor UAV, the large-scale team formation for the dynamic system of quadrotor UAVs is denoted as a set $S = \{S_1, S_2, \ldots, S_N\}$, which is composed by N quadrotor UAVs. From the practical point of view, each quadrotor UAV in the team formation is affected by the external disturbances, coupling effects, internal continuous random fluctuations, and internal discontinuous random fluctuations. To make the dynamic system of quadrotor UAVs in the team formation more practical, the Wiener process and Poisson counting process are considered to formulate the internal continuous and discontinuous random fluctuations due to parameters fluctuations in each quadrotor UAV system. Therefore, the ith nonlinear stochastic quadrotor UAV dynamic system S_i in (4.1) can be reformulated as follows [24,25]:

$$\begin{aligned} S_i : dX_i(t) = (&f_i(X_i(t)) + g_i(X_i(t))u_i(t) \\ &+ \sum\nolimits_{j \in N_i} f_{i,j}(X_i(t))X_j(t - \tau_{i,j}(t)) \\ &+ v_i(t))dt + \sigma_i(X_i(t))dw_i(t) + E_i(X_i(t))dp_i(t) \\ \text{for } i = 1, \ldots, N \end{aligned} \tag{4.2}$$

with

$$\begin{aligned} X_i(t) &= \left[X_{i,P}^T(t), X_{i,V}^T(t)\right]^T \\ X_{i,P}(t) &= [x_{i,1}(t), y_{i,1}(t), z_{i,1}(t), \phi_{i,1}(t), \theta_{i,1}(t), \psi_{i,1}(t)]^T \\ X_{i,V}(t) &= [x_{i,2}(t), y_{i,2}(t), z_{i,2}(t), \phi_{i,2}(t), \theta_{i,2}(t), \psi_{i,2}(t)]^T \\ u_i(t) &= \left[F_i(t), \tau_{i,\phi}(t), \tau_{i,\theta}(t), \tau_{i,\psi}(t)\right]^T \\ v_i(t) &= [0_{1\times 6}, v_{i,x}(t), v_{i,y}(t), v_{i,z}(t), v_{i,\phi}(t), v_{i,\theta}(t), v_{i,\psi}(t)]^T \end{aligned}$$

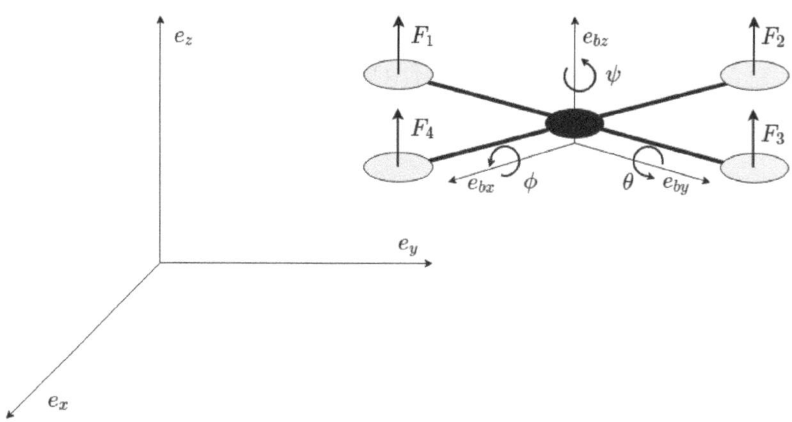

FIGURE 4.1 The free body diagram of the quadrotor UA

$$f_i(X_i(t)) = \begin{bmatrix} x_{i,2}(t) \\ y_{i,2}(t) \\ z_{i,2}(t) \\ \phi_{i,2}(t) \\ \theta_{i,2}(t) \\ \psi_{i,2}(t) \\ -\dfrac{d_{i,x}}{m_i} x_{i,2}(t) \\ -\dfrac{d_{i,y}}{m_i} y_{i,2}(t) \\ \dfrac{J_{i,\theta} - J_{i,\psi}}{J_{i,\phi}} \theta_{i,2}(t)\psi_{i,2}(t) - \dfrac{d_{i,\phi}}{J_{i,\phi}} \phi_{i,2}(t) \\ \dfrac{J_{i,\psi} - J_{i,\phi}}{J_{i,\theta}} \psi_{i,2}(t)\phi_{i,2}(t) - \dfrac{d_{i,\theta}}{J_{i,\theta}} \theta_{i,2}(t) \\ \dfrac{J_{i,\phi} - J_{i,\theta}}{J_{i,\psi}} \phi_{i,2}(t)\theta_{i,2}(t) - \dfrac{d_{i,\psi}}{J_{i,\psi}} \psi_{i,2}(t) \end{bmatrix}$$

$$g_i(X_i(t)) = [0_{6\times 4}^T, g_{i,1}^T(X_P(t))]^T$$

where

$$g_{i,1}(X_{i,P}(t)) = \begin{bmatrix} \begin{matrix}(\cos\phi_{i,1}(t)\sin\theta_{i,1}(t) \\ \times\cos\psi_{i,1}(t) \\ +\sin\phi_{i,1}(t) \\ \times\sin\psi_{i,1}(t))\dfrac{1}{m_i}\end{matrix} & 0 & 0 & 0 \\ \begin{matrix}(\cos\phi_{i,1}(t)\sin\theta_{i,1}(t) \\ \times\sin\psi_{i.1}(t) \\ -\sin\phi_{i,1}(t) \\ \cos\psi_{i,1}(t))\dfrac{1}{m_i}\end{matrix} & 0 & 0 & 0 \\ \cos\phi_{i,1}(t)\cos\theta_{i,1}(t)\dfrac{1}{m_i} & 0 & 0 & 0 \\ 0 & \dfrac{1}{J_{i,\phi}} & 0 & 0 \\ 0 & 0 & \dfrac{1}{J_{i,\theta}} & 0 \\ 0 & 0 & 0 & \dfrac{1}{J_{i,\psi}} \end{bmatrix}$$

$X_i(t)=\left[X_{i,P}^T(t),X_{i,V}^T(t)\right]^T \in \mathbb{R}^{12}$ represents the ith dynamic system state of the quadrotor UAV, $u_i(t)\in \mathbb{R}^4$ represents the control input of the ith quadrotor UAV, $X_j(t-\tau_{i,j}(t)) \in \mathbb{R}^{12}$ represents the trailing vortex coupling term from the jth UAV to the ith UAV with a time-varying delay $\tau_{i,j}(t)\in \mathbb{R}^+$, $\sum_{j\in N_i} f_{i,j}(X_i(t))X_j(t-\tau_{i,j}(t))\in \mathbb{R}^{12}$ represents the interconnected couplings as the trailing vortex effect caused by the neighborhood quadrotor UAVs of the ith UAV, N_i represents a neighborhood set of quadrotor UAVs nearing the ith UAV, $v_i(t)\in \mathcal{L}_F^2[0;\infty)$ denotes the external disturbance of the ith quadrotor UAV, $w_i(t)\in \mathbb{R}$ denotes the Wiener process of the ith quadrotor UAV which is continuous but non-differentiable, $\sigma_i(X(t))dw_i(t)\in \mathbb{R}^{12}$ denotes the effect of the continuous stochastic intrinsic fluctuation caused in the ith quadrotor UAV, $p_i(t)\in \mathbb{R}$ denotes the Poisson counting process with jump intensity λ_i, and $E_i(X_i(t))dp_i(t) \in \mathbb{R}^{12}$ is the discontinuous intrinsic fluctuation caused in the ith quadrotor UAV.

Remark 4.1 The Wiener processes $\{w_i(t)\}$ and Poisson counting processes $\{p_i(t)\}$ have the following properties [8,11]:

$$(I)\ E\{dw_i(t)\}=0\ (II) E\{dw_i(t)dw_i(t)\}=dt\ (III)\ E\{dp_i(t)\}=\lambda_i dt$$
$$(IV) E\{dw_i(t)dp_i(t)\}=0$$

4.2.2 Problem Formulation of PID Team Formation Control of Large-Scale UAVs

In order to let N quadrotor UAVs maintain a specific time-varying formation shape during the flight, the virtual leader method is used in this chapter. The team formation shape of N quadrotor UAVs is constructed by one virtual leader and N quadrotor UAVs and then each quadrotor UAV will track the virtual leader's trajectory with a specific time-varying formation shape to achieve the formation task. The reference model of the desired trajectory for N quadrotor UAVs of the team formation can be defined as follows:

$$\begin{aligned} &dX_{i,r}(t)=[A_{i,r}X_{i,r}(t)+C_{i,r}(r(t)+d_i(t))]dt \\ &\text{for } i=1,...,N \end{aligned} \tag{4.3}$$

where $X_{i,r}(t)=[X_{i,r1}^T(t)\ \ X_{i,r2}^T(t)]^T \in \mathbb{R}^{12}$ denotes the desired trajectory of the ith quadrotor UAV to be tracked. $X_{i,r1}(t)\in \mathbb{R}^6$ denotes the three positions and attitudes of the ith quadrotor UAV and $X_{i,r2}(t)\in \mathbb{R}^6$ denotes the three velocities and angular velocities of the ith quadrotor UAV. $r(t)\in \mathbb{R}^{12}$ is the trajectory of the virtual leader to be generated as the desired flight trajectory of the team formation of N quadrotor UAVs and $d_i(t)\in \mathbb{R}^{12}$ denotes the time-varying offset from the virtual leader to the ith quadrotor UAV. $A_{i,r} \in \mathbb{R}^{12\times 12}$ is an asymptotically stable and nonsingular matrix. $C_{i,r} \in \mathbb{R}^{12\times 12}$ is the reference input matrix.

Remark 4.2 The ith reference model in (4.3) is to generate the reference flight trajectory $X_{i,r}(t)$ of the ith quadrotor UAV. The system matrix $A_{i,r}$ is to be specified as an asymptotically matrix. If we set $C_{i,r}=-A_{i,r}$. Then, at the steady state, the trajectory of the ith reference model in (4.3) is $X_{i,r}(t)=(r(t)+d_i(t))$. Therefore, if we can control N quadrotor UAVs to track the corresponding N trajectories of reference models $X_{i,r}(t)$ $i=1,...,N$ in (4.3), then the N quadrotor UAVs will form a time-varying formation shape $(d_1(t),...,d_i(t),...,d_N(t))$ around the trajectory of the virtual leader $r(t)$ at the steady state. This phenomenon will be confirmed by the simulation example in the sequel.

Because the proportional-integral-derivative (PID) controllers have been used in the most automatic process control applications in industry, the PID controller is employed to control each quadrotor UAV for the team formation in this chapter. The conventional PID control for the ith quadrotor UAV is shown as follows [18]:

$$\begin{aligned} u_i(t) &= K_{i,P}\, e_i(t) + K_{i,I}\int_0^t e_i(\tau)d\tau + K_{i,D}\frac{de_i(t)}{dt} \\ \text{for } i &= 1,...,N \end{aligned} \tag{4.4}$$

where $K_{i,P}\in\mathbb{R}^{4\times6}$, $K_{i,I}\in\mathbb{R}^{4\times6}$, $K_{i,D}\in\mathbb{R}^{4\times6}$ are the controller gains with respect to the PID controller, $e_i(t)=(X_{i,P}(t)-X_{i,r1}(t))\in\mathbb{R}^6$ denotes the tracking error of the three positions and three attitudes of the ith quadrotor UAV, $\frac{de_i(t)}{dt}=X_{i,V}(t)-X_{i,r2}(t)$ denotes the tracking error of the three velocities and three angular velocities of the ith quadrotor UAV, and $\int_0^t e_i(\tau)d\tau$ denotes the integral of the tracking error of the three positions and three attitudes of the ith quadrotor UAV.

In the realistic situation, the effect of external disturbances $v_i(t)$ and trailing vortex $\sum_{j\in N_i} f_{i,j}(X_i(t))X_j(t-\tau_{i,j}(t))$ is inevitable during the team formation tracking process of quadrotor UAVs and it may deteriorate the team formation tracking performance. Besides, the reference input $r(t)$ and the team formation offset $d_i(t)$ in (4.3) of the ith quadrotor UAV are unpredictable for the stochastic quadrotor UAV system in (4.2). Thus, the decentralized H_∞ PID team formation tracking strategy is proposed for each UAV in the team formation to efficiently attenuate the effect of the external disturbances, trailing vortex and reference input on the team formation tracking performance. Then, the decentralized robust stochastic H_∞ PID team formation tracking control strategy of the i th quadrotor UAV is proposed as follows:

$$\begin{aligned} &J_{i,\infty}(u_i(t)) \\ &= \frac{E\{\int_0^{t_f}[(X_i(t)-X_{i,r}(t))^T Q_i(X_i(t)-X_{i,r}(t))]dt - V_i(X_i(0),X_{i,r}(0))\}}{\begin{aligned}E\{\int_0^{t_f}[v_i^T(t)v_i(t)+(r(t)+d_i(t))^T(r(t)+d_i(t)) \\ +\sum_{j\in N_i} X_j^T(t-\tau_{i,j}(t))X_j(t-\tau_{ij}(t))]dt\}\end{aligned}} \le \rho_i \\ &\text{for } i=1,...,N,\ \forall v_i(t),(r(t)+d_i(t)),X_j(t-\tau_{i,j}(t))\in\mathcal{L}_F^2[0;\infty) \end{aligned} \tag{4.5}$$

where $Q_i=diag\{Q_{i,P}\ Q_{i,V}\}\ \ge 0$ with the weighting matrices $Q_{i,P}\ge 0$ and $Q_{i,V}\ge 0$ which denote the weighting matrix on the tracking error of the $(X_{i,P}(t)-X_{i,r1}(t))$

and $(X_{i,V}(t)-X_{i,r2}(t))$, respectively, and $V_i(X_i(0),X_{i,r}(0))$ denotes the effect of the initial condition, which is to be deducted in the design procedure.

By using the decentralized robust stochastic H_∞ PID team formation tracking control strategy to boost the team formation performance of in the ith quadrotor UAV, the aim is to design the specific PID tracking controller $u_i(t)$ to attenuate the worst-case effect of external disturbance, reference input, and trailing vortex on the team formation tracking performance below a prescribed level ρ_i in (4.5) for each quadrotor UAV in the team.

However, the conventional PID control methods are limited to the linear dynamic systems or simple nonlinear dynamic systems so the methods should be improved to solve the complex decentralized H_∞ team formation control problem of highly nonlinear dynamic quadrotor UAV system. In this chapter, we employ the method of coordinate transformation so that the conventional PID control problems can be systematically analyzed and conveniently designed for the decentralized H_∞ team formation tracking control of large-scale UAVs. The PID controller $u_i(t)$ in (4.4) is reformulated as follows:

$$\begin{aligned} u_i(t) &= K_{i,P}\, e_i(t) + K_{i,I}\int_0^t e_i(\tau)d\tau + K_{i,D}\frac{de_i(t)}{dt} \\ &= [K_{i,P}\; K_{i,D}\; K_{i,I}]\begin{bmatrix} e_i(t) \\ \dfrac{de_i(t)}{dt} \\ \zeta_i(t) \end{bmatrix} \\ &= [K_{i,P}\; K_{i,D}\; K_{i,I}]\begin{bmatrix} T_1[X_i(t)-X_{i,r}(t)] \\ T_2[X_i(t)-X_{i,r}(t)] \\ \zeta_i(t) \end{bmatrix} \\ &= K_i T\bar{X}_i(t) \text{ for } i=1,\ldots,N \end{aligned} \tag{4.6}$$

with

$$\begin{aligned} K_i &= [K_{i,P}\; K_{i,D}\; K_{i,I}] \\ T_1 &= [I_6\; 0_{6\times6}], T_2 = [0_{6\times6}\; I_6] \\ T &= \begin{bmatrix} T_1 & 0_{6\times12} & 0_{6\times6} \\ T_2 & 0_{6\times12} & 0_{6\times6} \\ 0_{6\times12} & 0_{6\times12} & I_6 \end{bmatrix} \\ \bar{X}(t) &= \begin{bmatrix} X_i(t)-X_{i,r}(t) \\ X_{i,r}(t) \\ \zeta_i(t) \end{bmatrix} \end{aligned}$$

where $\zeta_i(t)=\int_0^t e_i(\tau)d\tau$

4.3 DECENTRALIZED ROBUST STOCHASTIC H_∞ PID TRACKING CONTROL FOR THE NONLINEAR LARGE-SCALE TEAM FORMATION QUADROTOR UAVS

In this section, the decentralized robust stochastic H_∞ tracking control design for the team formation of the ith quadrotor UAV is addressed. Let us augment the ith stochastic quadrotor UAV system in (4.2), the reference system in (4.3), and the PID controller in (4.6), then we get the augmented system of the ith quadrotor UAV as follows:

$$\begin{aligned}&d\bar{X}_i(t)=[\bar{F}_i(\bar{X}_i(t))+\bar{G}(\bar{X}_i(t),K_i)+\bar{D}_i(\bar{X}_i(t))\bar{v}_i(t)]dt\\&+\bar{\sigma}_i(\bar{X}_i(t))dw_i(t)+\bar{E}_i(\bar{X}_i(t))dp_i(t)\\&\text{for } i=1,\ldots,N\end{aligned} \tag{4.7}$$

where $\bar{X}_i(t)=[X_i^T(t)-X_{i,r}^T(t),\ X_{i,r}^T(t),\ \zeta_i^T(t)]^T$.

The system matrices are given as:

$$\bar{F}_i(\bar{X}_i(t))=\begin{bmatrix}f_i(X_i(t))-A_{i,r}X_{i,r}(t)\\A_{i,r}X_{i,r}(t)\\{[I_6\ 0_{6\times6}][X_i(t)-X_{i,r}(t)]}\end{bmatrix}$$

$$\bar{G}(\bar{X}_i(t),K_i)=\begin{bmatrix}g_i(X_i(t))K_iT\bar{X}_i(t)\\0_{18\times1}\end{bmatrix}$$

$$\bar{D}_i(\bar{X}_i(t))=\begin{bmatrix}I_{12\times12}&-C_{i,r}&f_{i,1}(X_i(t))&\cdots&f_{i,N}(X_i(t))\\0_{12\times12}&C_{i,r}&0_{12\times12}&\cdots&0_{12\times12}\\0_{6\times12}&0_{6\times12}&0_{6\times12}&\cdots&0_{6\times12}\end{bmatrix}$$

$$\begin{aligned}&\bar{v}_i(t)=[v_i^T(t)\quad (r_i(t)+d_i(t))^T\\&X_1^T(t-\tau_{i,1})\ \ \ldots\ \ X_N^T(t-\tau_{i,N})]^T\\&\bar{\sigma}_i(\bar{X}_i(t))=[\sigma_i^T(X_i(t))\ 0_{18\times1}^T]^T\\&\bar{E}_i(\bar{X}_i(t))=[E_i^T(X_i(t))\ 0_{18\times1}^T]^T\end{aligned}$$

Based on the augmented system in (4.7), the decentralized robust stochastic H_∞ PID team formation tracking control strategy of the ith quadrotor UAV can be reformulated as:

$$J_{i,\infty}(K_{i,P},K_{i,D},K_{i,I})=\frac{E\{\int_0^{t_f}[(\bar{X}_i^T(t)\bar{Q}_i\bar{X}_i(t)]dt-\bar{V}_i(\bar{X}_i(0))\}}{E\{\int_0^{t_f}[\bar{v}_i^T(t)\bar{v}_i(t)]dt\}}\le\rho_i \tag{4.8}$$

for $i=1,\ldots,N$

where $\bar{Q}_i = diag\{Q_i, 0_{18\times18}\}$ and $\bar{V}_i(\bar{X}_i(0)) = V_i(X_i(0), X_{i,r}(0), \zeta_i(0))$. Clearly, with the help of the augmented system of the i th UAV in (4.7), the robust decentralized H_∞ PID team formation tracking control design problem in (4.5) is reformulated as the decentralized H_∞ PID stabilization control problem of the augmented system for each quadrotor UAV in (4.8) to simplify the design procedure. However, the conventional Lyapunov stabilization design method can not be directly used in the stochastic nonlinear augmented system in (4.7) because the property of the Wiener process and Poisson counting process in the nonlinear ith quadrotor UAV system is indifferentiable every time. Fortunately, by using the Itô-Lévy formula in Reference [29], the decentralized robust stochastic H_∞ PID control problem for the team formation of quadrotor UAVs can be transformed into a set of independent HJIs. Before we deduce the HJIs for the decentralized H_∞ PID control in Theorem 4.1, the Itô-Lévy formula and technical inequality are described as follows:

Lemma 4.1 Defining the Lyapunov function $\bar{V}_i(\bar{X}_i(t)) \in C^2$ with $\bar{V}_i(0) = 0$ and $\bar{V}_i(\bar{X}_i(t)) > 0$, the Itô-Lévy formula of $\bar{V}_i(\bar{X}_i(t))$ for the stochastic nonlinear augmented system of the ith quadrotor UAV in (4.7) is given as follows [29]:

$$
\begin{aligned}
&d\bar{V}_i(\bar{X}_i(t)) = [\bar{V}_{i,\bar{X}}^T(\bar{X}_i(t))(\bar{F}_i(\bar{X}_i(t)) + \bar{G}(\bar{X}_i(t), K_i)) + \bar{V}_{i,\bar{X}}^T(\bar{X}_i(t))\bar{D}_i(\bar{X}_i(t))\bar{v}_i(t) \\
&+\frac{1}{2}\bar{\sigma}_i^T(\bar{X}_i(t))\bar{V}_{i,\bar{X}\bar{X}}(\bar{X}_i(t))\bar{\sigma}_i(\bar{X}_i(t))]dt + \bar{V}_{i,\bar{X}}^T(\bar{X}_i(t))\bar{\sigma}_i(\bar{X}_i(t))dw_i(t) \\
&+[\bar{V}_i(\bar{X}_i(t) + \bar{E}_i(\bar{X}_i(t))) - \bar{V}_i(\bar{X}_i(t))]dp_i(t) \\
&\text{for } i = 1, \dots, N
\end{aligned} \tag{4.9}
$$

where $\bar{V}_{i,\bar{X}}(\bar{X}_i(t)) = [\partial \bar{V}_i(\bar{X}_i(t))]/[\partial \bar{X}_i(t)]$ and $\bar{V}_{i,\bar{X}\bar{X}}(\bar{X}_i(t)) = [\partial^2 \bar{V}_i(\bar{X}_i(t))]/[\partial \bar{X}_i^2(t)]$.

Lemma 4.2 For any two matrices G and H with appropriate dimensions, the following matrix inequality holds [30]:

$$
G^T H + H^T G \le G^T Z^{-1} G + H^T Z H \tag{4.10}
$$

where Z is any positive definite symmetric matrix.

With the help of above lemmas, the decentralized robust stochastic H_∞ PID reference tracking control design problem for the team formation of quadrotor UAVs is transformed to an equivalent set of independent HJIs in the following theorem.

Theorem 4.1 In the decentralized stochastic nonlinear H_∞ PID team formation tracking control of the ith quadrotor UAV in (4.7), if there exists a PID control gain $K_i = [K_{i,P}, K_{i,D}, K_{i,I}]$ in (4.6) such that the following independent HJI_i has a positive solution $\bar{V}_i(\bar{X}_i(t)) > 0$ with $\bar{V}_i(0) = 0$ for each UAV

$$
\begin{aligned}
HJI_i = E\{&\bar{X}_i^T(t)\bar{Q}_i\bar{X}_i(t) + \bar{V}_{i,\bar{X}}^T(\bar{X}_i(t))(\bar{F}_i(\bar{X}_i(t)) + \bar{G}(\bar{X}_i(t), K_i)) \\
&+\frac{1}{4\rho_i}\bar{V}_{i,\bar{X}}^T(\bar{X}_i(t))\bar{D}_i(\bar{X}_i(t)) \times \bar{D}_i^T(\bar{X}_i(t))\bar{V}_{i,\bar{X}}(\bar{X}_i(t)) \\
&+\frac{1}{2}\bar{\sigma}_i^T(\bar{X}_i(t))\bar{V}_{i,\bar{X}\bar{X}}(\bar{X}_i(t))\bar{\sigma}_i(\bar{X}_i(t)) \\
&+\lambda_i(\bar{V}_i(\bar{X}_i(t) + \bar{E}_i(\bar{X}_i(t))) - \bar{V}_i(\bar{X}_i(t)))\} \le 0 \\
\text{for } i = 1, \dots, N&
\end{aligned} \tag{4.11}
$$

then (i) the decentralized robust stochastic H_∞ PID team formation tracking performance of quadrotor UAVs in (4.8) is guaranteed with a prescribed disturbance level ρ_i for all possible $\bar{v}_i(t) \in \mathcal{L}^2[0;\infty)$ from the perspective of the energy; and (ii) the mean square asymptotical team formation tracking ability is also achieved, that is, $X_i(t) \to X_{i,r}(t)$ as $t_f \to \infty$ if $\bar{v}_i(t) \in \mathcal{L}_F^2[0;\infty)$.

Proof. Please refer to Appendix 4A.

Since the HJI_i in (4.11) for the decentralized H_∞ PID team formation tracking control of the each quadrotor UAV does not contain the information of the other quadrotor UAVs in the formation team, the $\{\mathrm{HJI}_i\}_{i=1}^N$ in (4.11) can be solved independently to obtain PID control parameters of each quadrotor UAV and therefore, the number N of quadrotor UAVs in the team formation can be increased to a very large scale. Furthermore, the optimal decentralized H_∞ PID team formation reference tracking control design of N quadrotor UAVs can also be formulated as the following N independent HJI_i constrained optimization problems.

$$
\begin{aligned}
\rho_i^* &= \min_{\bar{V}_i(\bar{X}_i(t)),K_{i,P},K_{i,D},K_{i,I}} \rho_i \\
& \quad s.t.\ HJI_i\ in\ (4.11) \\
\text{for } i &= 1,\ldots,N
\end{aligned}
\tag{4.12}
$$

Due to nonlinear terms $\bar{F}_i(\bar{X}_i(t)$, $\bar{G}(\bar{X}_i(t),K_i)$, $\bar{D}_i(\bar{X}_i(t))$, at present, there exists no analytical or numerical method to efficiently solve N HJI_i in (4.11) or N HJI_i-constrained optimization problems in (4.12) for the optimal decentralized H_∞ PID team formation tracking control of N quadrotor UAVs.

4.4 DECENTRALIZED H_∞ ROBUST PID CONTROL DESIGN FOR THE TEAM FORMATION OF LARGE-SCALE QUADROTOR UAVS VIA T-S FUZZY APPROACH

In the previous section, the decentralized H_∞ PID team formation tracking control strategy of N stochastic nonlinear quadrotor UAVs in (4.8) is transformed into the N equivalent HJI_i problems in (4.11). However, it is still very difficult to find an analytical or a numerical method to solve N nonlinear partial differential HJIs in (4.11) or N HJI_i-constrained optimization problems in (4.12) for the decentralized H_∞ PID team formation tracking control design of N stochastic nonlinear quadrotor UAVs. Hence, the T-S fuzzy interpolation method is employed to interpolate several local linear stochastic systems to approximate the ith nonlinear stochastic quadrotor UAV system in (4.2). First, the kth rule of the T-S fuzzy model for the i th stochastic nonlinear quadrotor UAV dynamic system in (4.2) can be described as follows [26,27]:

Plant Rule k:

$$\text{If } \hat{p}_{i,1}(t) \text{ is } \hat{F}_{k,1},\ldots, and\ \hat{p}_{i,g}(t) \text{ is } \hat{F}_{k,g},$$

Then,

$$dX_{i,k}(t) = [A_{i,k}X_i(t) + B_{i,k}u_i(t) + v_i(t)\sum_{j\in N_i} + F_{i,j,k}X_j(t-\tau_{i,j}(t))]dt$$
$$+ H_{i,k}X_i(t)dw_i(t) + M_{i,k}X_i(t)dp_i(t) \tag{4.13}$$
$$\text{for } i = 1,...,N,\ k = 1,...,L$$

where L is the number of the fuzzy rules, $A_{i,k} \in \mathbb{R}^{12\times 12}, B_{i,k} \in \mathbb{R}^{12\times 4}, H_{i,k}, M_{i,k} \in \mathbb{R}^{12\times 12}$ are the local matrices of $f_i(X_i(t))$, $g_i(X_i(t))$, $\sigma_i(X_i(t))$, and $E(X_i(t))$ respectively. $(F_{i,1,k}, .. F_{i,i-1,k}, F_{i,i+1,k}, ..., F_{i,N,k})$ are the local matrices of the effectively trailing vortex from other quadrotor UAVs. $\hat{p}_{i,l}(t)$ for $l = 1,...,g$ are the premise variables about the ith quadrotor UAV state, and $\hat{F}_{k,1}, ..., \hat{F}_{k,g}$ are the fuzzy sets and g is the number of the premise variables of each UAV. The kth membership function $\nu_k(\hat{p}_i(t))$ is defined as follows:

$$\nu_k(\hat{p}_i(t)) = \Pi_{l=1}^{g}\hat{F}_{k,l}(\hat{p}_{i,l}(t)) \geq 0, \text{ for } i = 1,...,N,\ k = 1,...,L \tag{4.14}$$

where $\hat{F}_{k,l}(\hat{p}_{i,l}(t))$ is the membership grade of $\hat{p}_{i,l}(t)$ in $\hat{F}_{k,l}$. $\hat{p}_i(t) = [\hat{p}_{i,1}(t),...,\hat{p}_{i,g}(t)]^T$, and $\sum_{k=1}^{L}\nu_k(\hat{p}_i(t)) > 0$. The interpolation functions of the i th quadrotor UAV system can be inferred as follows:

$$h_k(\hat{p}_i(t)) = \frac{\nu_k(\hat{p}_i(t))}{\sum_{k=1}^{L}\nu_k(\hat{p}_i(t))} \geq 0$$

for $k = 1,2,...,L$, and it is clear that

$$\sum_{k=1}^{L} h_k(\hat{p}_i(t)) = 1 \tag{4.15}$$

By the defuzzification process in Reference [27], the nonlinear system of the ith quadrotor UAV can be represented:

$$\begin{aligned} dX_i(t) = \sum_{k=1}^{L} h_k(\hat{p}_i(t))[(A_{i,k}X_i(t) + B_{i,k}u_i(t) + v_i(t) \\ + \sum_{j\in N_i} F_{i,j,k}X_j(t-\tau_{i,j}(t)))dt \\ + H_{i,k}X_i(t)dw_i(t) + M_{i,k}X_i(t)dp_i(t)] \end{aligned} \tag{4.16}$$

That is, the nonlinear stochastic system of the ith UAV in (4.2) can be represented by interpolating L local linearized stochastic systems through L fuzzy interpolation functions $\{h_k(\hat{p}_i(t))\}_{k=1}^{L}$.

By substituting the T-S fuzzy model (4.16) of the ith UAV in (4.2) into the augmented system in (4.7), we have:

$$\begin{aligned} d\bar{X}_i(t) = \sum_{k=1}^{L} h_k(\hat{p}_i(t))[((\bar{A}_{i,k} + \bar{B}_{i,k}K_iT)\bar{X}_i(t) \\ + \bar{D}_{i,k}\bar{v}_i(t))dt + \bar{H}_{i,k}\bar{X}_i(t)dw_i(t) + \bar{M}_{i,k}\bar{X}_i(t)dp_i(t)] \end{aligned} \tag{4.17}$$

With

$$\bar{A}_{i,k} = \begin{bmatrix} A_{i,k} & A_{i,k} - A_{i,r} & 0_{12\times 6} \\ 0_{12\times 12} & A_{i,r} & 0_{12\times 6} \\ [I_6 \ 0_{6\times 6}] & 0_{6\times 12} & 0_{6\times 6} \end{bmatrix}, \ \bar{B}_{i,k} = \begin{bmatrix} B_{i,k} \\ 0_{12\times 4} \\ 0_{6\times 4} \end{bmatrix},$$

$$\bar{D}_{i,k} = \begin{bmatrix} I_{12} & -C_{i,r} & F_{i,1,k} & \cdots & F_{i,N,k} \\ 0_{12\times 12} & C_{i,r} & 0_{12\times 12} & \cdots & 0_{12\times 12} \\ 0_{6\times 12} & 0_{6\times 12} & 0_{6\times 12} & \cdots & 0_{6\times 12} \end{bmatrix}, \ \bar{H}_{i,k} = \begin{bmatrix} H_{i,k} & H_{i,k} & 0_{12\times 6} \\ 0_{12\times 12} & 0_{12\times 12} & 0_{12\times 6} \\ 0_{6\times 12} & 0_{6\times 12} & 0_{6\times 6} \end{bmatrix},$$

$$\bar{M}_{i,k} = \begin{bmatrix} M_{i,k} & M_{i,k} & 0_{12\times 6} \\ 0_{12\times 12} & 0_{12\times 12} & 0_{12\times 6} \\ 0_{6\times 12} & 0_{6\times 12} & 0_{6\times 6} \end{bmatrix}$$

With the help of the T-S fuzzy interpolation method, the complex HJI$_i$-constrained optimization in (4.12) for the decentralized robust H_∞ PID team formation control of N quadrotor UAV systems will be transformed into an equivalent LMIs-constrained optimization problem for each quadrotor UAV, which can be easily solved by using the LMI toolbox in MATLAB. Before the discussion of the main theorem, a quadratic Lyapunov function is selected for HJI$_i$ in (4.11) to simplify the design procedure of the decentralized H_∞ PID control of team formation of N quadrotor UAVs.

To begin with, the Lyapunov function of the ith augmented quadrotor UAV system of the team formation in (4.17) is defined as follows:

$$\bar{V}_i(\bar{X}_i(t)) = \bar{X}_i^T(t)\bar{P}_i\bar{X}_i(t), \text{ for i=1,...,N} \tag{4.18}$$

where $\bar{P}_i \in \mathbb{R}^{30\times 30}$ is a symmetric positive-definite matrix.

Theorem 4.2 For the decentralized H_∞ PID team formation tracking control of the ith quadrotor UAV stochastic system in (4.17), if we can specify the PID control gains $K_i = [K_{i,P}, K_{i,D}, K_{i,I}]$ in (4.6) and symmetric positive-definite matrix $\bar{P}_i$ such that the following independent Riccati-like inequalities (RLIs) hold:

$$\begin{aligned} &\bar{Q}_i + (\bar{A}_{i,k} + \bar{B}_{i,k}K_iT)^T\bar{P}_i + \bar{P}(\bar{A}_{i,k} + \bar{B}_{i,k}K_iT) \\ &+\frac{1}{\rho_i}\bar{P}_i\bar{D}_{i,k}\bar{D}_{i,k}^T\bar{P}_i + \bar{H}_{i,k}^T\bar{P}_i\bar{H}_{i,k} \\ &+\lambda_i(\bar{M}_{i,k}^T\bar{P}_i\bar{M}_{i,k} + \bar{M}_{i,k}^T\bar{P}_i + \bar{P}_i\bar{M}_{i,k}) \le 0 \\ &\text{for } i = 1,\ldots,N, \ k = 1,2,\ldots,L \end{aligned} \tag{4.19}$$

then (i) the decentralized robust H_∞ PID team formation tracking control strategy of N quadrotor UAVs in (4.8) is achieved with a prescribed disturbance level $\rho_i > 0$ for all possible $\bar{v}_i(t) \in \mathcal{L}^2[0;\infty)_F$ from the perspective of the energy; and (ii) the mean square asymptotical team formation tracking ability is also achieved, that is, $X_i(t) \to X_{i,r}(t)$ as $t_f \to \infty$ if $\bar{v}_i(t) \in \mathcal{L}_F^2[0;\infty)$.

Proof. Please refer to Appendix 4B.

Remark 4.3 If PID controller of ith quadrotor UAV in (4.6) can satisfy with RLIi, $k = 1,...,L$, in (4.19), that is, N local fuzzy linearized systems are taken care by robust H_∞ PID team formation tracking controller, then the PID controller in (4.6) can guarantee the H_∞ team formation tracking performance in (4.5) for L local fuzzy linearized systems of ith nonlinear stochastic quadrotor UAV. Further, N RLIs are decoupled for N quadrotor UAVs, that is, they can be solved independently without the information of other quadrotor UAVs.

If the sufficient condition in (4.19) is satisfied, then the decentralized robust stochastic H_∞ PID team formation tracking control performance for the team formation of N quadrotor UAVs in (4.8) is guaranteed with a prescribed disturbance level ρ_i. However, the design parameters $\bar{P}_i$ and K_i are coupled and bilinear matrix inequalities in (4.19) can not be solved by the current convex optimization method. To deal with the problem in (4.19), the following transformation method is adopted.

By pre-multiplying and post-multiplying $\bar{P}_i^{-1} = \bar{W}_i$, the RLIs in (4.19) are equivalent to

$$\begin{aligned}&\bar{W}_i\bar{Q}_i\bar{W}_i + \bar{W}_i(\bar{A}_{i,k} + \bar{B}_{i,k}K_iT)^T + (\bar{A}_{i,k} + \bar{B}_{i,k}K_iT)\bar{W}_i + \frac{1}{\rho_i}\bar{D}_{i,k}\bar{D}_{i,k}^T \\ &+\bar{W}_i\bar{H}_{i,k}^T\bar{W}_i^{-1}\bar{H}_{i,k}\bar{W}_i + \lambda_i(\bar{W}_i\bar{M}_{i,k}^T\bar{W}_i^{-1}\bar{M}_{i,k}\bar{W}_i + \bar{W}_i\bar{M}_{i,k}^T + \bar{M}_{i,k}\bar{W}_i) \le 0 \qquad (4.20)\\ &for\ i = 1,...,N,\ k = 1,...,L\end{aligned}$$

Furthermore, set $\bar{W}_i = diag\{W_{i,1}, W_{i,2}, W_{i,3}\}$ with $W_{i,1} = diag\{W_{i,1,1}, W_{i,1,2}\} \in \mathbb{R}^{12\times12}$, $W_{i,2} \in \mathbb{R}^{12\times12}$ and $W_{i,3} \in \mathbb{R}^{6\times6}$, and then we have:

$$\begin{aligned}K_iT\bar{W} &= [K_{i,P}\ K_{i,D}\ K_{i,I}]\begin{bmatrix} T_1 & 0_{6\times12} & 0_{6\times6} \\ T_2 & 0_{6\times12} & 0_{6\times6} \\ 0_{6\times12} & 0_{6\times12} & I_6 \end{bmatrix} \times \begin{bmatrix} W_{i,1} & 0_{12\times12} & 0_{12\times6} \\ 0_{12\times12} & W_{i,2} & 0_{12\times6} \\ 0_{6\times12} & 0_{6\times12} & W_{i,3} \end{bmatrix} \\ &= [K_{i,P}T_1W_{i,1} + K_{i,D}T_2W_{i,1}\ 0_{4\times12}\ K_{i,I}W_{i,3}]\end{aligned} \qquad (4.21)$$

where:

$$\begin{aligned}&K_{i,P}T_1W_{i,1} = K_{i,P}[W_{i,1,1}\ 0_{6\times6}] = [K_{i,P}W_{i,1,1}\ 0_{6\times6}]K_{i,D}T_2W_{i,1} \\ &= K_{i,D}[0_{6\times6}\ W_{i,1,2}] = [0_{6\times6}\ K_{i,D}W_{i,1,2}]\end{aligned} \qquad (4.22)$$

By using the variable transformation technique

$$K_iT\bar{W}_i = [[Y_{i,1}\ Y_{i,2}]\ 0_{4\times12}\ \mathrm{Y}_{i,3}] = \bar{Y}_i \in \mathbb{R}^{4\times30} \qquad (4.23)$$

where:

$$Y_{i,1} = K_{i,P}W_{i,1,1}, Y_{i,2} = K_{i,D}W_{i,1,2}, Y_{i,3} = K_{i,I}W_{i,3}$$

Therefore, the RLIs in (4.20) can be formulated as follows:

$$\bar{W}_i\bar{Q}_i\bar{W}_i+\bar{W}_i\bar{A}_{i,k}^T+\bar{Y}_i^T\bar{B}_{i,k}^T+\bar{A}_{i,k}\bar{W}_i+\bar{B}_{i,k}\bar{Y}_i+\frac{1}{\rho_i}\bar{D}_{i,k}\bar{D}_{i,k}^T$$
$$+\bar{W}_i\bar{H}_{i,k}^T\bar{W}_i^{-1}\bar{H}_{i,k}\bar{W}_i+\lambda_i(\bar{W}\bar{M}_{i,k}^T\bar{W}_i^{-1}\bar{M}_{i,k}\bar{W}_i+\bar{W}_i\bar{M}_{i,k}^T+\bar{M}_{i,k}\bar{W}_i)\leq 0 \quad (4.24)$$
$$for\ i=1,...,N,\ k=1,...,L$$

By using Schur complement [28] to (4.24), we have the following LMIs:

$$\begin{bmatrix} \Phi_{i,k} & \bar{D}_{i,k} & \bar{W}_i\bar{Q}_i^{\frac{1}{2}} & \bar{W}_i\bar{H}_{i,k}^T & \bar{W}_i\bar{M}_{i,k}^T \\ * & -\rho_i I & 0 & 0 & 0 \\ * & * & -I & 0 & 0 \\ * & * & * & -\bar{W}_i & 0 \\ * & * & * & * & -\lambda_i^{-1}\bar{W}_i \end{bmatrix}\leq 0 \quad (4.25)$$

for $i=1,...,N$, and $k=1,...,L$, where $\Phi_{i,k}=\bar{W}_i\bar{A}_{i,k}^T+\bar{Y}_i^T\bar{B}_{i,k}^T+\bar{A}_{i,k}\bar{W}_i+\bar{B}_{i,k}\bar{Y}_i+\lambda_i(\bar{W}_i\bar{M}_{i,k}^T+\bar{M}_{i,k}\bar{W}_i)$.

Then, the above analyses can be concluded as the following main result.

Theorem 4.3 If we can solve $\bar{W}_i>0$ and $\bar{Y}_i>0$ from LMIs in (4.25) for a prescribed ρ_i, $i=1,...,N$, and $k=1,...,L$, then the decentralized H_∞ PID team formation tracking controllers of N quadrotor UAVs in (4.6) are designed as follows:

$$K_{i,P}=Y_{i,1}W_{i,1,1}^{-1}, K_{i,D}=Y_{i,2}W_{i,1,2}^{-1}, K_{i,I}=Y_{i,3}W_{i,3}^{-1},\ for\ i=1,...,N \quad (4.26)$$

By solving LMIs in (4.25) with the help of the LMI toolbox in MATLAB and based on Theorem 4.3, the PID controllers for the decentralized H_∞ PID team formation tracking control of N stochastic quadrotor UAVs are independently designed as (4.26). However, in practical applications, the control input is always restricted by physical mechanism and bounded by the saturation of actuator. To meet the real situation, the saturation control method in Reference [31] will be used as follows:

Suppose that $\bar{X}_i(t)$ is restricted to stay in an invariant ellipsoid $\varepsilon_r=\{\bar{X}_i(t)\in\mathbb{R}^{30}\mid \bar{X}_i^T(t)\bar{W}_i^{-1}\bar{X}_i(t)\leq 1\}$ for $t\geq 0$, and $\nu_i^2\bar{W}_i^{-1}\geq(K_iT)^TK_iT$ for $\nu_i\in\mathbb{R}^+$. We have [31]:

$$\begin{aligned}
&\max_{t\geq 0}\|u_i(t)\|_2\\
&=\max_{t\geq 0}\|K_iT\bar{X}_i(t)\|_2\\
&\leq\max_{\bar{X}_i(t)\in\varepsilon_r}\|K_iT\bar{X}_i(t)\|_2\\
&\leq\max_{\bar{X}_i(t)\in\varepsilon_r}\|\nu_i\bar{W}_i^{-\frac{1}{2}}\bar{X}_i(t)\|_2\\
&=\max_{\bar{X}_i(t)\in\varepsilon_r}\sqrt{\nu_i^2\bar{X}_i^T(t)\bar{W}_i^{-1}\bar{X}_i(t)}\\
&\leq\nu_i
\end{aligned}$$

that is, if $\nu_i^2\bar{W}_i^{-1} \geq (K_iT)^T K_iT$ holds or $\nu_i^2\bar{W}_i \geq (K_iT\bar{W}_i)^T K_iT\bar{W}_i = \bar{Y}_i^T\bar{Y}_i$ holds, then$\| u_i(t) \|_2 \leq \nu_i$ for all t.

Therefore, by using Schur complement [28], if the following LMIs hold,

$$\begin{bmatrix} \nu_i^2\bar{W}_i & \bar{Y}_i^T \\ * & I \end{bmatrix} \geq 0, \; for \; i=1,...,N \tag{4.27}$$

then we have the result,

$$\| u_i(t) \|_2 \leq \nu_i, \; for \; i=1,...,N \tag{4.28}$$

Thus, the design condition of the decentralized H_∞ PID team formation control for the team formation of stochastic quadrotor UAVs is also satisfied with the saturation property of actuator if the conditions in (4.25) and (4.27) are achieved simultaneously. Therefore, the optimal decentralized H_∞ saturation PID control design for the team formation of stochastic quadrotor UAVs can be designed by the following LMIs-constrained optimization problem for each quadrotor UAV

$$\begin{aligned} \rho_i^* &= \min_{\bar{W}_i > 0, \bar{Y}_i} \rho_i \\ & s.t. (4.25), (4.27) \end{aligned} \tag{4.29}$$

The LMIs-constrained optimization problem in (4.29) can be solved by decreasing ρ_i until there exists no feasible solution $\bar{W}_i > 0$ in LMIs.

Remark 4.4 The design complexity of proposed H$_\infty$ PID team formation tracking controller lies mainly in solving LMIs in (4.25). Therefore, based on Reference [28], the computational complexity is about $O(NLn(n+1)/2)$, where N is the number of quadrotor UAVs in the team formation, L is the number of fuzzy rules, and n is the dimension of $\bar{P}_i$ or $\bar{W}_i$, $n = 3 \times 12 = 36$ in the quadrotor UAV case.

Remark 4.5 Since quadrotor UAV is a highly nonlinear system as shown in (4.2), the H_∞ stabilization or tracking control design problem of quadrotor UAV needs to solve a corresponding HJI. In general, there still exists no analytic or numerical method to efficiently solve the HJI for PID team formation tracking control, except a very special system. To avoid this difficulty in solving HJI, the conventional PID control designs only consider a local linearized system of quadrotor UAV at some operation point, so that the HJI in (4.11) becomes a Riccati-like inequality (RLI) in (4.19), which can be transformed to an equivalent LMI in (4.25). However, the quadrotor UAV has to shift to many operation points during flying, for example, in taking-off, landing, and hovering. Therefore, the conventional PID control designs need to switch to an adequate PID controller as the operation point of quadrotor UAV shifts during flying. In the proposed design method, the T-S fuzzy interpolates L local linearized systems at L operation points by fuzzy interpolation method in (4.17) (in the simulation example $L = 72$). Therefore, the proposed PID controller not only can achieve the decentralized robust H_∞ team formation tracking control performance at L local linearized systems (i.e., L operation points) of quadrotor

UAVs but also has the ability to reject the worst-case effect of external disturbance, vortex coupling, and saturation of actuator. However, the proposed decentralized H_∞ PID team formation control design needs to solve L LMIs in (4.25) or the LMIs-constrained optimization problem in (4.29) for each UAV in the team formation tracking control design.

4.5 SIMULATION RESULTS OF TEAM FORMATION CONTROL OF 25 QUADROTOR UAVS

To illustrate the design procedure and validate the effectiveness of the proposed decentralized H_∞ PID team formation stochastic tracking control problem of N stochastic nonlinear quadrotor UAVs, a team formation control of 25 quadrotor UAVs is given as a design example in this section. To illustrate the implementation simplicity and validate the effectiveness of the decentralized H_∞ PID team formation tracking control, a comparison with conventional decentralized H_∞ T-S fuzzy control for team formation tracking of 25 quadrotor UAVs in Reference [15] will be showed too. Since the decentralized controller is equipped with the individual information of each quadrotor UAV and does not need the information of other quadrotor UAVs, the number of the quadrotor UAVs for the team formation can be easily increased to very large scale.

4.5.1 Parameters Setting of Large-Scale Quadrotor UAVs for Team Formation

The parameters of each stochastic dynamic of the quadrotor UAV system in (4.2) for the 25 quadrotor UAVs are identical and can be given as follows: $m_i = 1\text{kg}$ is the mass of each UAV. $J_{i,X} = J_{i,Y} = 1.25Ns^2\ /\ rad$ and $J_{i,Z} = 2.2Ns^2\ /\ rad$ are the moments of the inertia of ϕ_i, θ_i, and ψ_i, respectively. $d_{i,x} = d_{i,y} = d_{i,z} = 0.01Ns\ /\ m$ and $d_{i,\phi} = d_{i,\theta} = d_{i,\psi} = 0.013Ns\ /\ m$ are the aerodynamic coefficients of each quadrotor UAV. $g = 9.8m\ /\ s^2$ is the acceleration of gravity. $f_{i,j}(X_i(t))X_j(t-\tau_{i,j}(t)) = [0,0,0,0,0,0,x_{i,2}(t)\times 0.01x_{j,2}(t-0.1),0,z_{i,2}(t)\times 0.01z_{i,2}(t-0.1),0,0,0]^T$, $j \in N_i$ denotes that each quadrotor UAV suffers the trailing vortex coupling from neighboring quadrotor UAVs in N_i. External disturbance $v_i(t)$ for each quadrotor UAV is set as zero-mean Gaussian noise with unit variance. Twenty-five independent Wiener processes $\{\ w_i(t)\ \}_{i=1}^{25}$ in Figure 4.14 and Poisson counting process $\{\ p_i(t)\ \}_{i=1}^{25}$ in Figure 4.13 with Poisson counting intensity $\lambda_i = 1$ are set as sources of intrinsic fluctuations in 25 UAVs of the team formation.

To construct the T-S fuzzy model of each quadrotor UAV in (4.17), the premise variables are the state variables in (4.1). The operation point of $z_{i,2}$ is at 0.499 and 9.81. The operation point of $\phi_{i,1}$ is at -0.1 and 0.1. The operation point of $\phi_{i,2}$ is at -0.05 and 0.05. The operation point of $\theta_{i,1}$ is at -0.1 and 0.1. The operation point of $\theta_{i,2}$ is at -0.05 and 0.05. The operation point of $\psi_{i,1}$ is at 0. The operation point of $\phi_{i,2}$ is at 0. The membership function is selected as a trapezoidal function.

Based on the above setting, each UAV in the 25 quadrotor UAVs has 32 fuzzy rules, respectively. The kth T-S fuzzy model of the ith quadrotor UAV dynamic system in (4.16) is described as follows:

$$dX_i(t) = [(A_{i,k}X_i(t) + B_{i,k}u_i(t) + v_i(t) + \sum_{j \in N_i} F_{i,j,k}X_j(t - \tau_{i,j}(t)))dt \\ + H_{i,k}X_i(t)dw_i(t) + E_{i,k}X_i(t)dp_i(t)] \tag{4.30}$$

For the team formation tracking control problem of 25 quadrotor UAVs, both the position tracking performance and velocity tracking performance are important. Consequently, the weighting matrices are specified for each quadrotor UAV subsystem as follows:

$$Q_i = 0.01I_{12}, \text{ for i=1,...,25}$$

In order to make 25 UAVs maintain a diamond time-varying formation shape, the trajectory of the ith quadrotor UAV is constructed by a virtual leader $r(t)$ with a specific time-varying distance $d_i(t)$ in (4.3) for $i = 1,...,25$ and $A_{i,r} = -4I_{12}$, $C_{i,r} = 4I_{12}$ in (4.3). However, the diamond formation shape is difficult to directly design, so an indirect method is proposed. First, we design the square formation with time-varying distance $d_i^{'}(t)$ and then use the coordinate transformation to rotate $d_i^{'}(t)$ counterclockwise $\frac{\pi}{4}$ from the virtual $r(t)$ in the y-z plane to obtain the diamond formation shape with time-varying distance $d_i(t)$. Each single quadrotor UAV is first set with a specific time-varying distance $d_i^{'}(t)$ to represent a square formation shape and then varies the square shape with $0.5\sin(t)$ as follows:

$$\begin{aligned}
d_1^{'}(t) &= [0, -2 - 0.5\sin(t), -2 - 0.5\sin(t)] \\
d_2^{'}(t) &= [0, -2 - 0.5\sin(t), -1 - 0.5\sin(t)] \\
d_3^{'}(t) &= [0, -2 - 0.5\sin(t), 0] \\
d_4^{'}(t) &= [0, -2 - 0.5\sin(t), 1 + 0.5\sin(t)] \\
d_5^{'}(t) &= [0, -2 - 0.5\sin(t), 2 + 0.5\sin(t)] \\
d_6^{'}(t) &= [0, -1 - 0.5\sin(t), -2 - 0.5\sin(t)] \\
d_7^{'}(t) &= [0, -1 - 0.5\sin(t), -1 - 0.5\sin(t)] \\
d_8^{'}(t) &= [0, -1 - 0.5\sin(t), 0] \\
d_9^{'}(t) &= [0, -1 - 0.5\sin(t), 1 + 0.5\sin(t)] \\
d_{10}^{'}(t) &= [0, -1 - 0.5\sin(t), 2 + 0.5\sin(t)] \\
d_{11}^{'}(t) &= [0, 0, -2 - 0.5\sin(t)] \\
d_{12}^{'}(t) &= [0, 0, -1 - 0.5\sin(t)] \\
d_{13}^{'}(t) &= [0, 0, 0] \\
d_{14}^{'}(t) &= [0, 0, 1 + 0.5\sin(t)] \\
d_{15}^{'}(t) &= [0, 0, 2 + 0.5\sin(t)] \\
d_{16}^{'}(t) &= [0, 1 + 0.5\sin(t), -2 - 0.5\sin(t)]
\end{aligned}$$

$$d'_{17}(t) = [0, 1+0.5\sin(t), -1-0.5\sin(t)]$$
$$d'_{18}(t) = [0, 1+0.5\sin(t), 0]$$
$$d'_{19}(t) = [0, 1+0.5\sin(t), 1+0.5\sin(t)]$$
$$d'_{20}(t) = [0, 1+0.5\sin(t), 2+0.5\sin(t)]$$
$$d'_{21}(t) = [0, 2+0.5\sin(t), -2-0.5\sin(t)]$$
$$d'_{22}(t) = [0, 2+0.5\sin(t), -1-0.5\sin(t)]$$
$$d'_{23}(t) = [0, 2+0.5\sin(t), 0]$$
$$d'_{24}(t) = [0, 2+0.5\sin(t), 1+0.5\sin(t)]$$
$$d'_{25}(t) = [0, 2+0.5\sin(t), 2+0.5\sin(t)]$$

The $d_i(t)$ is obtained by rotating $d'_i(t)$ counterclockwise $\frac{\pi}{4}$ from the virtual leader quadrotor UAV $r(t)$ in the y-z plane and the diamond formation shape is shown in Figure 4.2 at $t = 0$. In the example, the trajectory of the virtual leader quadrotor UAV is specified with a radius 6 m in the x-y plane and is rising at a constant velocity at 2 m/s on the z-axis, as shown in Figure 4.3. The reference input $r(t)$ in (4.3)

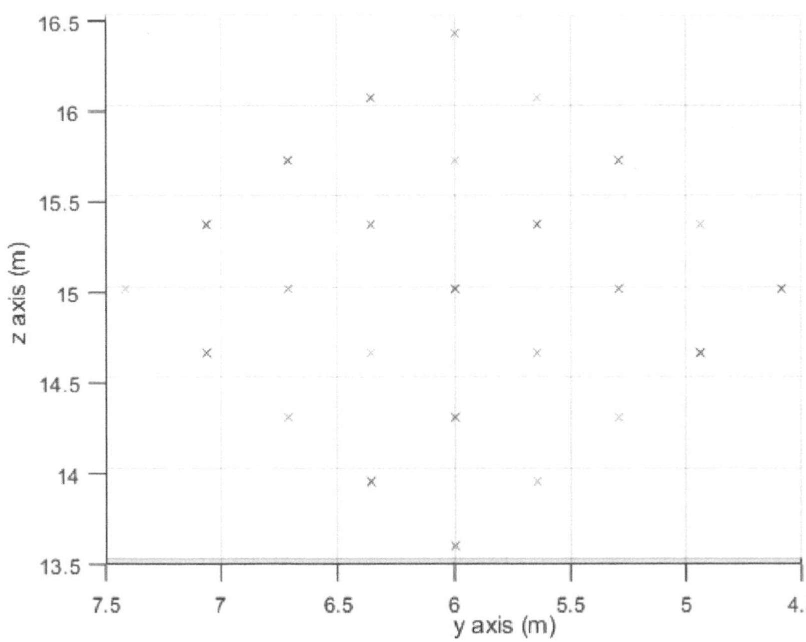

FIGURE 4.2 The diamond formation shape of 25 quadrotor UAVs for the team formation at the initial condition $t = 0$ from the y-z plane. In the figure, each point of 25 points represents one quadrotor UAV.

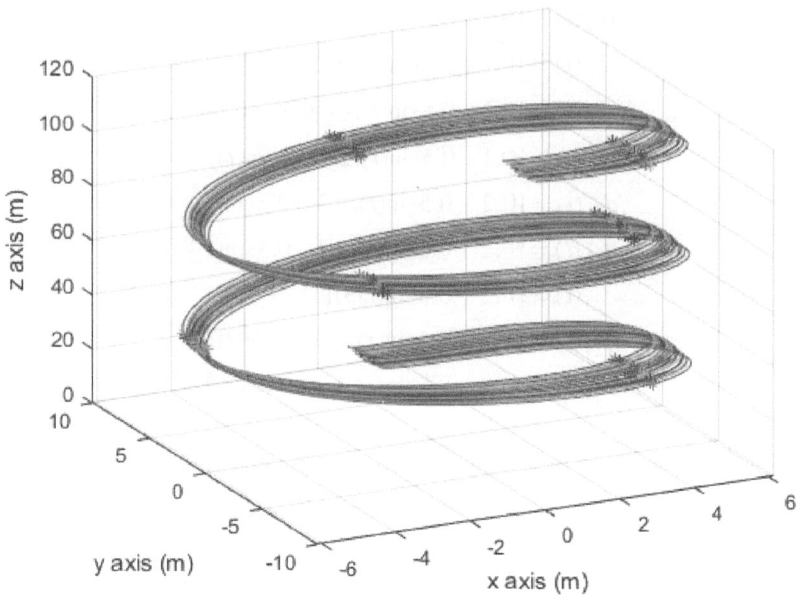

FIGURE 4.3 The flight trajectories of 25 quadrotor UAVs in 3D graph during 50 s. In this figure, each line represents the trajectory of one quadrotor UAV.

with three positions $[x_{r1}(t), y_{r1}(t), z_{r_1}(t)]^T$ and three attitudes $[\phi_{r1}(t), \theta_{r1}(t), \psi_{r1}(t)]^T$ is given as follows [3]:

$$r(t) = [x_{r1}(t), y_{r1}(t), z_{r_1}(t), \phi_{r1}(t), \theta_{r1}(t), \psi_{r1}(t),$$
$$x_{r2}(t), y_{r2}(t), z_{r2}(t), \phi_{r2}(t), \theta_{r2}(t), \psi_{r2}(t)]^T$$

where

$$x_{r1}(t) = 6\ \sin(0.1\pi t),\ y_{r1}(t) = 6\ \cos(0.1\pi t)$$
$$z_{r1}(t) = 15 + 2t, \psi_{r1}(t) = 0$$
$$\phi_{r1}(t) = \arcsin\Big(\frac{x_{r2}(t)\sin(\psi_{r1}(t)) - y_{r2}(t)\cos(\psi_{r1}(t))}{(x_{r2}^2(t) + y_{r2}^2(t) + (z_{r2}(t) + g)^2)^{\frac{1}{2}}}\Big)$$
$$\theta_{r1}(t) = \arcsin\Big(\frac{x_{r2}(t)\cos(\psi_{r1}(t)) + y_{r2}(t)\sin(\psi_{r1}(t))}{z_{r2}(t) + g}\Big)$$

On the other hand, the velocities of three positions and three attitudes are the differentiations of their three positions and three attitudes with time.

4.5.2 Simulation Results

Based on the above parameters setting, the simulation results for the proposed robust decentralized H_∞ PID team formation tracking of 25 quadrotor UAVs are shown in this section. Furthermore, the saturation constant ν_i in (4.28) for each quadrotor UAV is set as 100 (i.e., $\nu_i = 100, i = 1,...,25$) [32]. By solving 25 LMIs-constrained optimization problems in (4.29), the decentralized PID tracking controller for the 25 quadrotor UAVs in the team formation can be designed to achieve the optimal H_∞ decentralized PID team formation tracking performance under the external disturbance and interconnected coupling in the 25 quadrotor UAVs. The optimal attenuation level ρ_i^* of 25 UAVs is 0.92 when the ρ_i is iteratively decreased from 1.5 until there exists no solution $\bar{W}_i^* > 0$. The corresponding optimal PID control parameters $K_{i,P}^* = Y_{i,1}^* W_{i,1,1}^{*-1}, K_{i,I}^* = Y_{i,3}^* W_{i,3}^{*-1}, K_{i,D}^* = Y_{i,2}^* W_{i,1,2}^{*-1}$ are obtained by solving the LMIs-constrained optimization problem in (4.29).

The 3D flight trajectories of 25 quadrotor UAVs in the team formation are shown in Figure 4.3 with the time-varying formation shape within 50 s. It can be seen that each quadrotor UAV can track the desired formation trajectory $r(t) + d_i(t)$ where the virtual trajectory $r(t)$ is with radius 6m circle in the x-y plane and 2 m/s ascent rate on the z-axis. To highlight the time-varying formation shape, the snapshot of 3D flight trajectories of 25 quadrotor UAVs during 20–30 seconds is shown in Figure 4.4. In Figure 4.4, the formation shape for 25 quadrotor UAVs team formation is a diamond shape and becomes bigger during 20–25 s and smaller during 25–30 s.

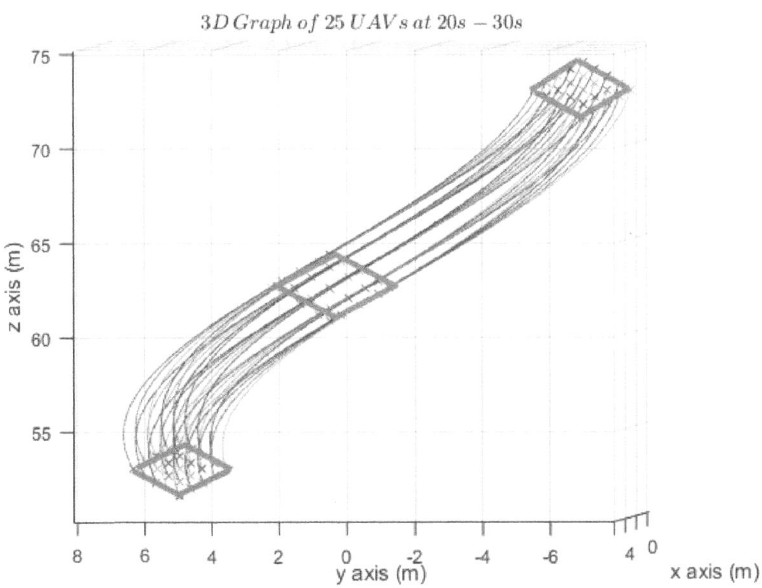

FIGURE 4.4 The flight trajectories of 25 quadrotor UAVs during 20–30 s. In the figure, left diamond denotes the formation shape of 25 quadrotor UAVs at 20 s, center diamond denotes the formation shape of 25 quadrotor UAVs at 25 s, and right diamond denotes the formation shape of 25 quadrotor UAVs at 30 s.

The three positions and the three velocities of 25 quadrotor UAVs are shown in Figures. 4.5–4.7. Due to the time-varying formation shape, it can be seen that the trajectories of y-axis and z-axis of any two quadrotor UAVs are derivated with a time-varying distance in Figures 4.6–4.7. For example, for 20–25 s in Figure 4.6, the

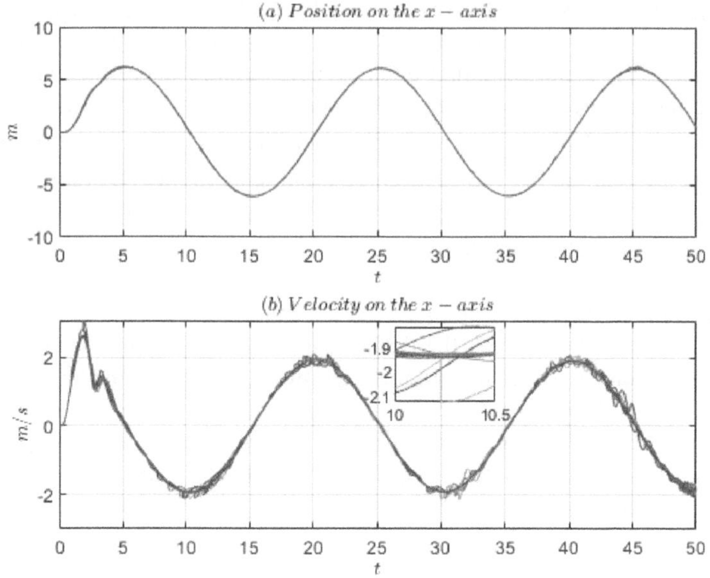

FIGURE 4.5 The position and velocity of the 25 quadrotor UAVs on the x-axis.

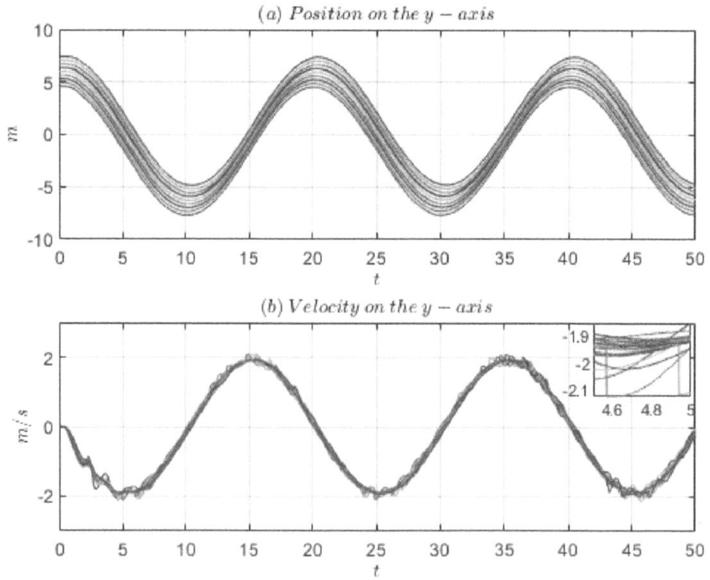

FIGURE 4.6 The position and velocity of the 25 quadrotor UAVs on the y-axis.

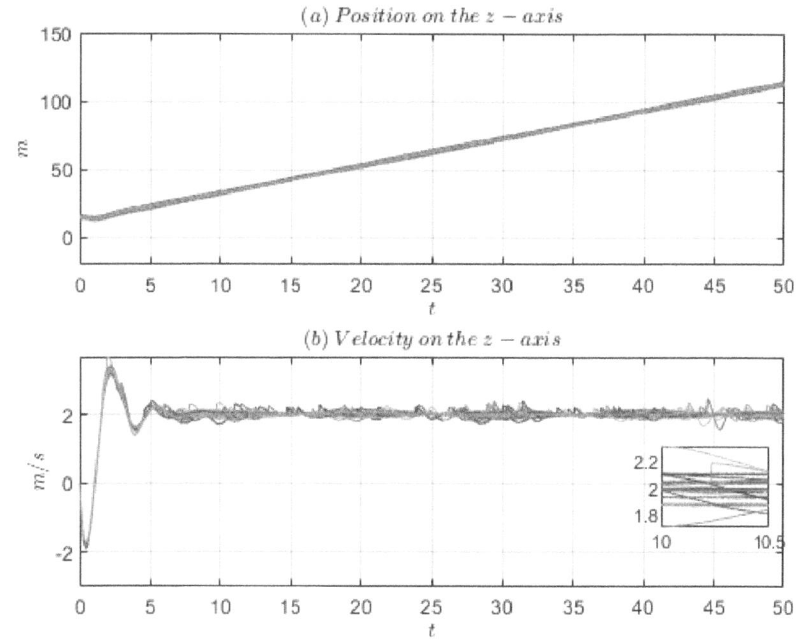

FIGURE 4.7 The position and velocity of the 25 quadrotor UAVs on the z-axis.

distance between any two quadrotor UAVs on the y-axis at 20 s is larger than the distance between any two quadrotor UAVs at 25 s. On the other hand, since the time-varying formation of 25 quadrotor UAVs only affects the trajectories of y-position and z-position, 25 quadrotor UAVs track the identical trajectory of x-position as shown in Figure 4.5. From Figures 4.5–4.7, by the proposed robust PID team formation tracking controller, the position tracking and the velocity tracking can be effectively achieved for 25 quadrotor UAVs only with some fluctuation caused by the stochastic effect. It is worth to point out that the stochastic effects on quadrotor UAV system are also efficiently reduced during the robust H_∞ PID team formation tracking control process. For example, the velocity on y-axis can be controlled to reference velocity when the discontinuous Poisson counting jump occurs in Figure 4.6.

The three attitude variables of 25 quadrotor UAVs with the corresponding angular velocities are shown in Figures 4.8–4.10, respectively. Caused by the different time-varying formation shapes for 25 quadrotor UAVs, the desired attitudes to be tracked for each quadrotor UAV are also different. Especially, due to the time-varying offset $d_i(t)$ of each quadrotor UAV, it makes the desired attitudes $\{\theta_{i,1}(t), \phi_{i,1}(t)\}$ have some ripples during the tracking process. From Figures. 4.8–4.10, there exist some large fluctuation caused by the transient response for three attitudes. Thereafter, the attitudes with the corresponding angular velocities for each quadrotor UAV can efficiently track the desired reference trajectories via the proposed decentralized H_∞ robust PID team formation tracking controller.

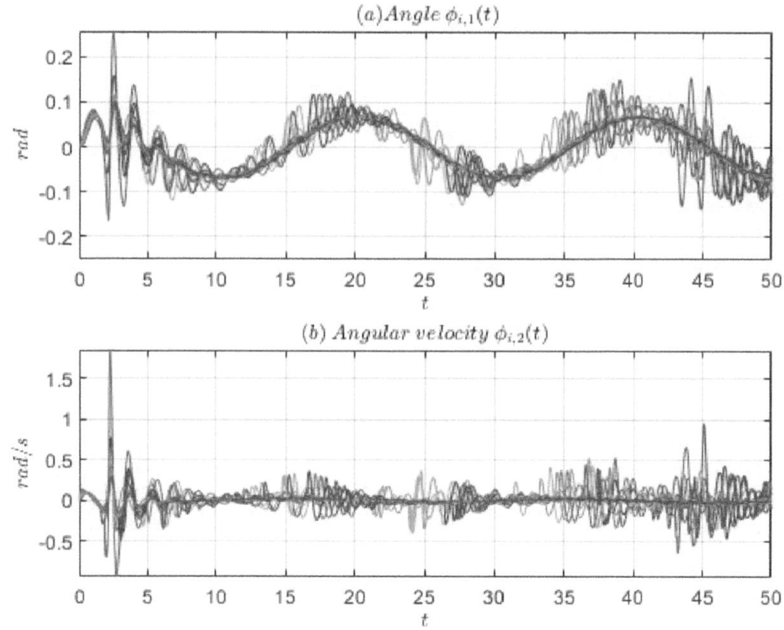

FIGURE 4.8 The angle $\phi_{i,1}(t)$ and angular velocity $\phi_{i,2}(t)$ of 25 quadrotor UAVs.

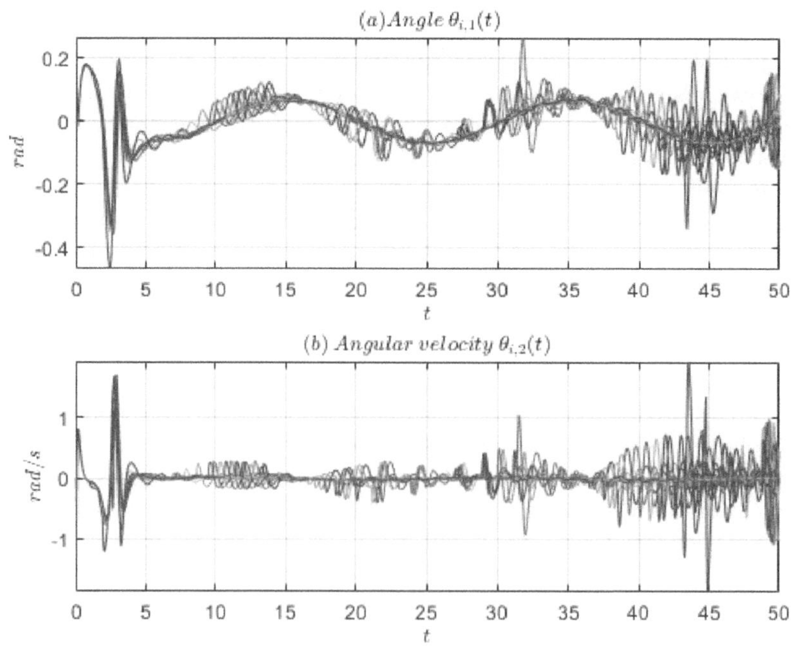

FIGURE 4.9 The angle $\theta_{i,1}(t)$ and angular velocity $\theta_{i,2}(t)$ of 25 quadrotor UAVs.

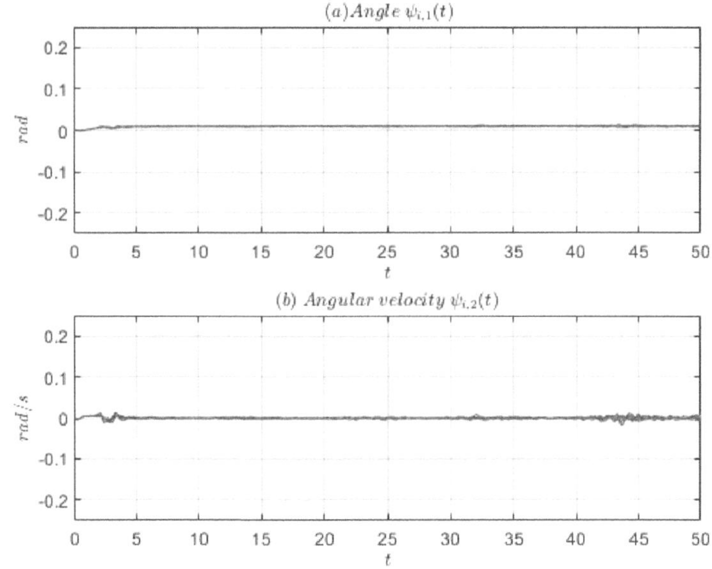

FIGURE 4.10 The angle $\psi_{i,1}(t)$ and angular velocity $\psi_{i,2}(t)$ of 25 quadrotor UAVs.

The control input $u_1(t)=[F_1(t),\tau_{1,\phi}(t),\tau_{1,\theta}(t),\tau_{1,\psi}(t)]^T$ of the first quadrotor UAV is shown in Figure 4.11. Under the effect of intrinsic continuous Wiener fluctuation, and discontinuous Poisson fluctuation on quadrotor UAV system, it can be seen that the control input has continuous fluctuation and some suddenly jumps. Indeed, once a Poisson jump occurs, the controller will use a relatively large control effort to reduce its effect. On the other hand, since the saturation effect is considered during the design, it is clear that the control input of first quadrotor UAV is below the saturation constant, that is, $\|u_1(t)\|\leq \nu_1=100,\ \forall t\in[0,50]$.

In Figure 4.12, in comparison with the 3D flight trajectory for 25 quadrotor UAVs based on the proposed decentralized H_∞ PID team formation tracking control strategy, the team formation tracking performance of the decentralized robust H_∞ T-S fuzzy tracking control design [15] is almost equal. Based on the above simulation result in Figures 4.5–4.11, the average decentralized H_∞ PID team formation tracking performance of 25 quadrotor UAVs in (4.5) is calculated as 0.118 and the average decentralized H_∞ fuzzy team formation tracking performance of 25 quadrotor UAVs is calculated as 0.107. However, the design and computation complexity of PID control and T-S fuzzy control [15] for 25 quadrotor UAVs are very different. The PID control for 25 quadrotor UAVs needs 72 fuzzy rules, and the T-S fuzzy control [15] for 25 quadrotor UAVs needs 1,944 fuzzy rules, that is, we need to solve 72 LMI $_s$ in (4.25) for PID control parameters of each UAV but 1,944 LMIs for fuzzy control parameters of each quadrotor UAV in Reference [15]. Furthermore, we only need to compute $u_i(t)=K_iT\bar{X}_i(t)$ in (4.6) at every instant time for PID control signal of each quadrotor UAV but $u_i(t)=\sum_{k=1}^{1944}h_k(p_i(t))K_{i,k}X_i(\hat{t})$ needs to be computed at every instant time for fuzzy control, where $h_k(p_i(t))$ is the compl$\hat{\text{e}}$x ith fuzzy interpolation

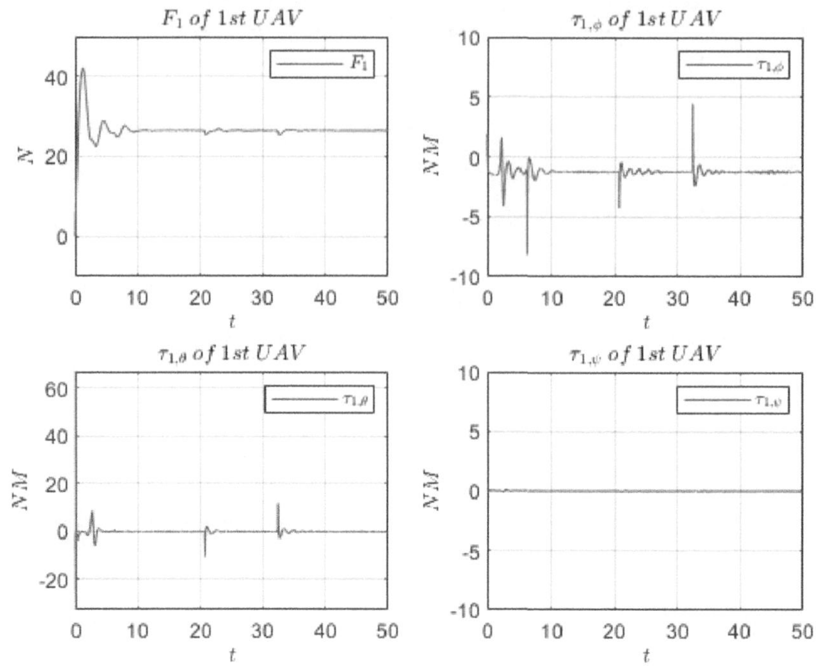

FIGURE 4.11 The control input for the first quadrotor UAV.

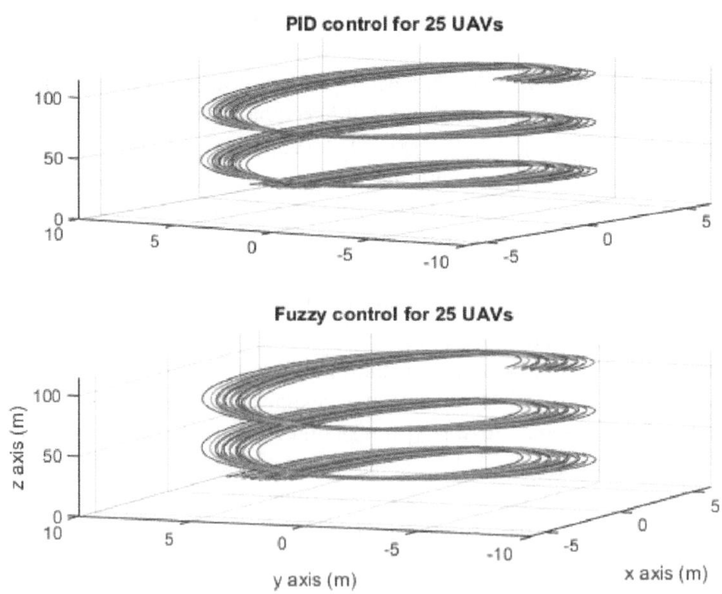

FIGURE 4.12 The 3D trajectories of 25 quadrotor UAVs of the proposed PID control in comparison with those of fuzzy control in Reference [15].

function. Based on the above discussion, the decentralized H_∞ PID team formation tracking control for 25 quadrotor UAVs can have less design and computation complexity than the decentralized H_∞ T-S fuzzy team formation tracking control for 25 quadrotor UAVs but with almost the same team formation tracking performance.

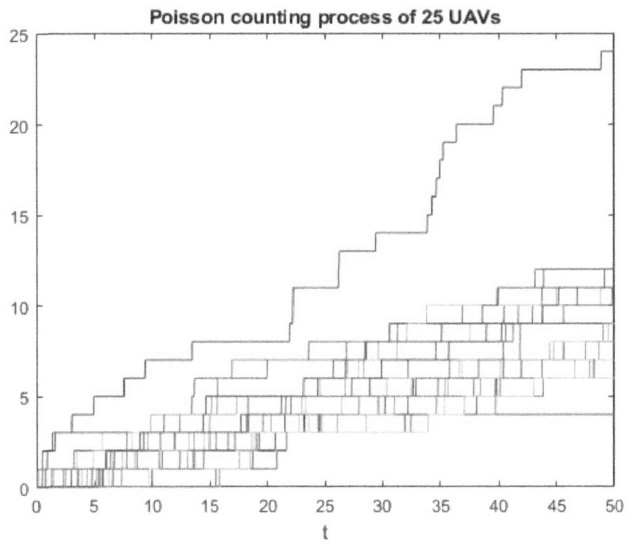

FIGURE 4.13 Poisson counting processes of 25 quadrotor UAVs with intensity $\lambda_i = 1$.

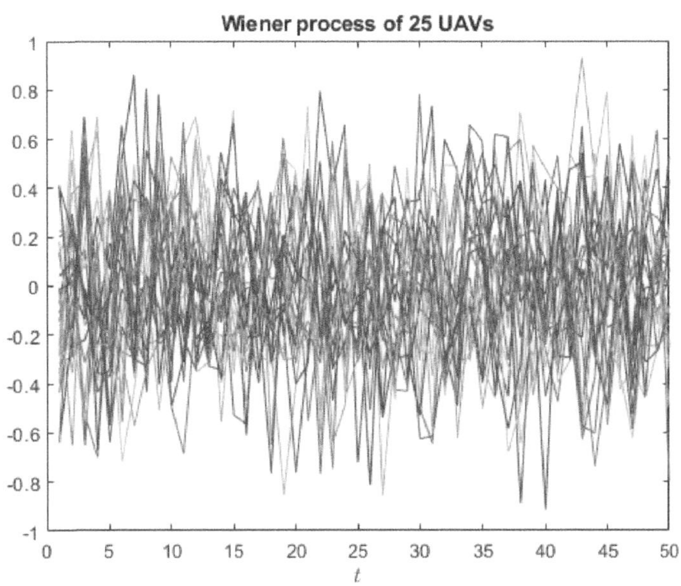

FIGURE 4.14 Wiener processes of 25 quadrotor UAVs.

Remark 4.6 The biggest barrier to apply the proposed algorithm to a real quadrotor UAV is the control input transformation. In the chapter, we considered four control inputs, that is, total thrust force of *i*th quadrotor UAV, and three rotation torques of quadrotor UAV. However, the quadrotor UAV is controlled by the four rotors of quadrotor UAV in the real situation. Therefore, the transformation between the theoretical control inputs and real control inputs should be tested by some practical experiments.

4.6 CONCLUSION

In this chapter, the decentralized H_∞ PID team formation tracking strategy is proposed for large-scale stochastic quadrotor UAVs under external disturbance, intrinsic stochastic fluctuation and trailing vortex coupling. By using the virtual leader concept, the reference trajectory of each quadrotor UAV is generated by the combination of the virtual leader trajectory with an adequate time-varying offset for each quadrotor UAV to form a desired flight formation shape. Then, by the proposed optimal H_∞ decentralized PID team formation controller for each quadrotor UAV, it can efficiently reject the effect of external disturbance and trailing vortex coupling from the neighboring quadrotor UAVs on the team formation reference tracking performance simultaneously. In order to avoid solving a complex nonlinear HJI for PID control of each quadrotor UAV, the T-S fuzzy interpolation method is employed to approximate the stochastic quadrotor UAV so that the HJI can be transformed to a set of BMIs. Furthermore, by using a novel variable transform method, the optimal decentralized H_∞ PID team formation tracking control design problem of large-scale quadrotor UAVs can be derived in terms of a set of independent LMIs-constrained optimization problems for each quadrotor UAV in the team formation, which can be efficiently solved in a single run via the current convex optimization technique of LMI toolbox in MATLAB without any tuning procedure of conventional PID designs to significantly simplify the design procedure. A simulation example of 25 team formation quadrotor UAVs is given to validate the reference tracking performance of the proposed method in comparison with the conventional decentralized H_∞ fuzzy tracking control scheme. From the simulation results, the decentralized H_∞ PID team formation control for 25 UAVs in this chapter has a less design and computation complexity than the fuzzy control for 25 quadrotor UAVs but with almost the same team formation tracking performance. Future works will focus on the decentralized H_∞ adaptive or attack-tolerant PID team formation tracking control of large-scale quadrotor UAVs under event-triggered control mechanism and Dos attacks [33,34].

Appendix 4A

Proof of Theorem 4.1

Proof. (i) Considering the numerator of the decentralized robust H_∞ PID team formation tracking strategy of the ith quadrotor UAV system in (4.8) with the Lyapunov function $\bar{V}_i(\bar{X}_i(t)) \in C^2$, $\bar{V}_i(\bar{X}_i(t)) > 0$, and $\bar{V}_i(0) = 0$, we have the following inequality:

$$\begin{aligned}
&E\{\int_0^{t_f}[\bar{X}_i^T(t)\bar{Q}_i\bar{X}_i(t)]dt\} \\
&= E\{\bar{V}_i(\bar{X}_i(0))\} - E\{\bar{V}_i(\bar{X}_i(t_f))\} \\
&\quad + E\{\int_0^{t_f}[\bar{X}_i^T(t)\bar{Q}_i\bar{X}_i(t)]dt + \int_0^{t_f} d\bar{V}_i(\bar{X}_i(t))\} \\
&\leq E\{\bar{V}_i(\bar{X}_i(t))\} + E\{\int_0^{t_f}[\bar{X}_i^T(t)\bar{Q}_i\bar{X}_i(t)]dt + \int_0^{t_f} \bar{V}_i(\bar{X}_i(t))\}
\end{aligned} \tag{4.31}$$

By using the Itô-Lévy formula in (4.9) with the fact $E\{dw_i(t)\} = 0$, and $E\{dp_i(t)\} = \lambda_i dt$, the (4.31) can be rewritten as follows:

$$\begin{aligned}
&E\{\int_0^{t_f}[\bar{X}_i^T(t)\bar{Q}_i\bar{X}_i(t)]dt\} \\
&\leq E\{\bar{V}_i(\bar{X}_i(0))\} + E\{\int_0^{t_f}[\bar{X}_i^T(t)\bar{Q}_i\bar{X}_i(t) \\
&+\bar{V}_{i,\bar{X}}^T(\bar{X}_i(t))(\bar{F}_i(X_i(t)) + \bar{G}(\bar{X}_i(t), K_i)) \\
&+\bar{V}_{i,\bar{X}}^T(\bar{X}_i(t))\bar{D}_i(\bar{X}_i(t))\bar{v}_i(t) + \frac{1}{2}\bar{\sigma}_i^T(\bar{X}_i(t))\bar{V}_{i,\bar{X}\bar{X}}\bar{\sigma}_i(\bar{X}_i(t)) \\
&+\lambda_i(\bar{V}_i(\bar{X}_i(t) + \bar{E}_i(\bar{X}_i(t)) - \bar{V}_i(\bar{X}_i(t)))]dt\}
\end{aligned} \tag{4.32}$$

By applying Lemma 4.2 with $G = \frac{1}{2}\bar{D}_i^T(\bar{X}_i(t))\bar{V}_{i,\bar{X}}(\bar{X}_i(t))$, $H = \bar{v}_i(t)$, and $Z = \rho_i I$ for $\rho_i \in \mathbb{R}^+$, the term $\bar{V}_{i,\bar{X}}^T(\bar{X}_i(t))\bar{D}_i(\bar{X}_i(t))\bar{v}_i(t)$ from (4.32) could be bounded above as follows:

$$\begin{aligned}
&\bar{V}_{i,\bar{X}}^T(\bar{X}_i(t))\bar{D}_i(\bar{X}_i(t))\bar{v}_i(t) \\
&\leq \frac{1}{4\rho_i}\bar{V}_{i,\bar{X}}^T(\bar{X}_i(t))\bar{D}_i(\bar{X}_i(t)) \times \bar{D}_i^T(\bar{X}_i(t))\bar{V}_{i,\bar{X}}(\bar{X}_i(t)) + \rho_i\bar{v}_i^T(t)\bar{v}_i(t)
\end{aligned} \tag{4.33}$$

With the inequality in (4.33), we can rewrite (4.32) as follows:

$$
\begin{aligned}
&E\{\textstyle\int_0^{t_f}[\bar{X}_i^T(t)\bar{Q}_i\bar{X}_i(t)]dt\} \\
&\leq E\{\bar{V}(\bar{X}_i(0))\} + E\{\textstyle\int_0^{t_f}[(\bar{X}_i^T(t)\bar{Q}_i\bar{X}_i(t) + \bar{V}_{i,\bar{X}}^T(\bar{F}_i(X_i(t)) + \bar{G}(\bar{X}_i(t), K_i)) \\
&+\frac{1}{4\rho_i}\bar{V}_{i,\bar{X}}^T(\bar{X}_i(t))\bar{D}_i(\bar{X}_i(t)) \times \bar{D}_i^T(\bar{X}_i(t))\bar{V}_{i,\bar{X}}(\bar{X}_i(t)) + \rho_i\bar{v}_i^T(t)\bar{v}_i(t) \\
&+\frac{1}{2}\bar{\sigma}_i^T(\bar{X}(t))\bar{V}_{i,\bar{X}\bar{X}}(\bar{X}_i(t))\bar{\sigma}_i(\bar{X}_i(t)) + \lambda_i(\bar{V}_i(\bar{X}_i(t) + \bar{E}_i(\bar{X}_i(t)) \\
&-\bar{V}_i(\bar{X}_i(t)))]dt\}
\end{aligned} \tag{4.34}
$$

Then, if the following HJI is satisfied:

$$
\begin{aligned}
&E\{\bar{X}_i^T(t)\bar{Q}_i\bar{X}_i(t) + \bar{V}_{i,\bar{X}}^T(\bar{F}_i(\bar{X}_i(t)) + \bar{G}(\bar{X}_i(t), K_i)) \\
&+\frac{1}{4\rho_i}\bar{V}_{i,\bar{X}}^T(\bar{X}_i(t))\bar{D}_i(\bar{X}_i(t))\bar{D}_i^T(\bar{X}_i(t))\bar{V}_{i,\bar{X}}(\bar{X}_i(t)) \\
&+\frac{1}{2}\bar{\sigma}_i^T(\bar{X}_i(t))\bar{V}_{i,\bar{X}\bar{X}}(\bar{X}_i(t))\bar{\sigma}_i(\bar{X}_i(t)) \\
&+\lambda_i(\bar{V}_i(\bar{X}_i(t) + \bar{E}_i(\bar{X}_i(t))) - \bar{V}_i(\bar{X}_i(t)))\} \leq 0
\end{aligned} \tag{4.35}
$$

we obtain the following result:

$$
E\{\textstyle\int_0^{t_f}[\bar{X}_i^T(t)\bar{Q}_i\bar{X}_i(t)]dt\} \leq E\{\bar{V}_i(\bar{X}_i(0))\} + E\{\textstyle\int_0^{t_f}\{\bar{v}_i^T(t)\bar{v}_i(t)dt\} \tag{4.36}
$$

that is, $J_{i,\infty}(K_{i,P}, K_{i,D}, K_{i,I}) \leq \rho_i$ in (4.8) is satisfied. (ii) Since $\bar{X}_i(0)$ and $E\{\bar{V}_i(X_i(0))\}$ are finite and we assume $E\{\rho_i\int_0^{\infty}\bar{v}_i^T(t)\bar{v}_i(t)dt\} < \infty$, the right-hand side of (4.36) is finite. Therefore, $\bar{X}_i(t) \to 0$ as $t \to \infty$ in the mean-square sense. Accordingly, $X_i(t) \to X_{i,r}(t)$, that is, the asymptotical team formation tracking ability is achieved in the mean square sense.

Appendix 4B
Proof of Theorem 4.2

Proof. (i) By using the Itô-Lévy formula in (4.9), the stochastic fuzzy quadrotor UAV system in (4.17), and the Lyapunov function in (4.18), we have:

$$\begin{aligned}
&E\{\textstyle\int_0^{t_f}[\bar{X}_i^T(t)\bar{Q}_i\bar{X}_i(t)]dt\} \\
&= E\{\bar{V}(\bar{X}_i(0))\} - E\{\bar{V}(\bar{X}_i(t_f))\} \\
&\quad + E\{\textstyle\int_0^{t_f}[\bar{X}_i^T(t)\bar{Q}_i\bar{X}_i(t)]dt + \int_0^{t_f} d\bar{V}_i(\bar{X}_i(t))\} \\
&\le E\{\bar{V}_i(\bar{X}_i(0))\} + [\sum_{k=1}^{L} h_k(p_i(t))(\bar{X}_i^T(t)\times(\bar{Q}_i + (\bar{A}_{i,k}+\bar{B}_{i,k}K_iT)^T\bar{P}_i \qquad (4.37)\\
&+\bar{P}_i(\bar{A}_{i,k}+\bar{B}_{i,k}K_iT) + \bar{H}_{i,k}^T\bar{P}_i\bar{H}_{i,k} \\
&+\lambda_i(\bar{M}_{i,k}^T\bar{P}_i\bar{M}_{i,k} + \bar{M}_{i,k}^T\bar{P}_i + \bar{P}_i\bar{M}_{i,k}))\bar{X}_i(t) \\
&+\bar{X}_i^T(t)\bar{P}_i\bar{D}_{i,k}\bar{v}_i(t) + \bar{v}_i^T\bar{D}_{i,k}^T\bar{P}_i\bar{X}_i(t))]dt\}] \\
&\text{for } i=1,\dots,N,\ k=1,\dots,L
\end{aligned}$$

By applying Lemma 4.2 with $G=\bar{D}_{i,k}^T\bar{P}_i\bar{X}_i(t)$, $H=\bar{v}_i(t)$ and $Z=\rho_i I$ for $\rho_i\in\mathbb{R}^+$, the term $\bar{X}_i^T(t)\bar{P}_i\bar{D}_{i,k}\bar{v}_i(t)+\bar{v}_i^T\bar{D}_{i,k}^T\bar{P}_i\bar{X}_i(t)$ from (4.32) could be bounded above as follows:

$$\begin{aligned}
&E\{\textstyle\int_0^{t_f}[\bar{X}_i^T(t)\bar{Q}_i\bar{X}_i(t)]dt\} \\
&= E\{\bar{V}(\bar{X}_i(0))\} - E\{\bar{V}(\bar{X}_i(t_f))\} \\
&\quad + E\{\textstyle\int_0^{t_f}[\bar{X}_i^T(t)\bar{Q}_i\bar{X}_i(t)]dt + \int_0^{t_f} d\bar{V}_i(\bar{X}_i(t))\} \\
&\le E\{\bar{V}_i(\bar{X}_i(0))\} + [\sum_{k=1}^{L} h_k(p_i(t))\bar{X}_i^T(t)\times(\bar{Q}_i + (\bar{A}_{i,k}+\bar{B}_{i,k}K_iT)^T\bar{P}_i \qquad (4.38)\\
&+\bar{P}_i(\bar{A}_{i,k}+\bar{B}_{i,k}K_iT) + \frac{1}{\rho_i}\bar{P}_i\bar{D}_{i,k}\bar{D}_{i,k}^T\bar{P}_i + \bar{H}_{i,k}^T\bar{P}_i\bar{H}_{i,k} \\
&+\lambda_i(\bar{M}_{i,k}^T\bar{P}_i\bar{M}_{i,k} + \bar{M}_{i,k}^T\bar{P}_i + \bar{P}_i\bar{M}_{i,k})\bar{X}_i(t) + \rho_i\bar{v}_i^T(t)\bar{v}_i(t))]dt\}] \\
&\text{for } i=1,\dots,N,\ k=1,\dots,L
\end{aligned}$$

If the following algebraic RLIs are satisfied:

$$\begin{aligned} &E\{\bar{Q}_i + (\bar{A}_{i,k} + \bar{B}_{i,k}K_i T)^T \bar{P} + \bar{P}(\bar{A}_{i,k} + \bar{B}_{i,k}K_i T) \\ &+ \frac{1}{\rho_i}\bar{P}_i\bar{D}_{i,k}\bar{D}_{i,k}^T\bar{P}_i + \bar{H}_{i,k}^T\bar{P}_i\bar{H}_{i,k} \\ &+ \lambda_i(\bar{M}_{i,k}^T\bar{P}_i\bar{M}_{i,k} + \bar{M}_{i,k}^T\bar{P}_i + \bar{P}_i\bar{M}_{i,k})\} \leq 0 \\ &\text{for } i = 1,\ldots,N,\ k = 1,\ldots,L \end{aligned} \tag{4.39}$$

then we immediately have the result

$$E\{\int_0^{t_f} \bar{X}_i^T(t)\bar{Q}_i\bar{X}_i(t)\} \leq E\{\bar{V}_i(\bar{X}_i(0))\} + E\{\int_0^{t_f} \rho_i\bar{v}_i^T(t)\bar{v}_i(t)dt \tag{4.40}$$

, that is, $J_{i,\infty}(k_{i,P}, K_{i,D}, K_{i,I}) \leq \rho_i$ is satisfied.

(ii) Since $\bar{X}_i(0)$ is finite in (4.8), $E\{\bar{V}_i(\bar{X}_i(0))\}$ is finite. Since we assume $v_i(t) \in \mathcal{L}_{\mathcal{F}}^2[0;\infty)$, $E\{\rho_i \int_0^\infty \bar{v}_i^T(t)\bar{v}_i(t)dt\} < \infty$. Therefore, the right-hand side of (4.7) is finite because ρ_i is finite. According to the definition of $\bar{X}_i(t)$ in (4.7), it implies $X_i(t) \to X_{i,r}(t)$ as $t \to \infty$ in the mean square sense.

Q.E.D.

REFERENCES

[1] N. H. Motlagh, M. Bagaa, and T. Taleb, "UAV-based IoT platform: A crowd surveillance use case," *IEEE Communications Magazine*, vol. 55, no. 2, pp. 128–134, Feb. 2017.

[2] K. Dorling, J. Heinrichs, G. G. Messier, and S. Magierowski, "Vehicle routing problems for drone delivery," *IEEE Transactions on Systems, Man, and Cybernetics: Systems*, vol. 47, no. 1, pp. 70–85, Jan. 2017.

[3] B. Zhao, B. Xian, Y. Zhang, and X. Zhang, "Nonlinear robust adaptive tracking control of a quadrotor UAV via immersion and invariance methodology," *IEEE Transactions on Industrial Electronics*, vol. 62, no. 5, pp. 2891–2902, May 2015.

[4] M. Bisheban and T. Lee, "Geometric adaptive control with neural networks for a quadrotor in wind fields," *IEEE Transactions on Control Systems Technology*, vol. 29, no. 4, pp. 1533–1548, July 2021.

[5] D. J. Almakhles, "Robust backstepping sliding mode control for a quadrotor trajectory tracking application," *IEEE Access*, vol. 8, pp. 5515–5525, Dec. 2020.

[6] J. Sun, Y. Wang, Y. Yu, and C. Sun, "Nonlinear robust compensation method for trajectory tracking control of quadrotors," *IEEE Access*, vol. 7, pp. 26766–26776, Mar. 2019.

[7] A. C. Satici, H. Poonawala, and M. W. Spong, "Robust optimal control of quadrotor UAVs," *IEEE Access*, vol. 1, pp. 79–93, May 2013.

[8] D. Shi, Z. Wu, and W. Chou, "Generalized extended state observer based high precision attitude control of quadrotor vehicles subject to wind disturbance," *IEEE Access*, vol. 6, pp. 32349–32359, May 2018.

[9] H. Du, W. Zhu, G. Wen, Z. Duan, and J. Lu, "Distributed formation control of multiple quadrotor aircraft based on nonsmooth consensus algorithms," *IEEE Transactions on Cybernetics*, vol. 49, no. 1, pp. 342–353, Jan. 2019.

[10] Z. Hou and I. Fantoni, "Interactive leader–follower consensus of multiple quadrotors based on composite nonlinear feedback control," *IEEE Transactions on Control Systems Technology*, vol. 26, no. 5, pp. 1732–1743, Sep. 2018.
[11] B. S. Chen, C. P. Wang, and M. Y. Lee, "Stochastic robust team tracking control of multi-UAV networked system under Wiener and Poisson random fluctuations," *IEEE Transactions on Cybernetics*, vol. 51, no. 12, pp. 5486–5799, Dec. 2020.
[12] M. Y. Lee, B. S. Chen, Y. Chang, and C. L. Hwang, "Stochastic robust team formation tracking design of multi-VTOL-UAV networked control system in smart city under time-varying delay and random fluctuation," *IEEE Access*, vol. 8, pp. 131310–131326, July 2020.
[13] H. Luan, Y. Xu, D. Liu, Z. Du, H. Qian, X. Liu, and X. Tong, "Energy efficient task cooperation for multi-UAV networks: A coalition formation game approach," *IEEE Access*, vol. 8, pp. 149372–149384, Apr. 2020.
[14] J. Choi, Y. Song, S. Lim, C. Kwon, and H. Oh, "Decentralized multi-subgroup formation control with connectivity preservation and collision avoidance," *IEEE Access*, vol. 8, pp. 71525–71534, Apr. 2020.
[15] M. Y. Lee, B. S. Chen, C. Y. Tsai, and C. L. Hwang, "Stochastic H_∞ robust decentralized tracking control of large-scale team formation UAV network system with time-varying delay and packet dropout under interconnected couplings and wiener fluctuations," *IEEE Access*, vol. 9, pp. 41976–41997, Mar. 2021.
[16] M. H. Amoozgar, A. Chamseddine, and Y. Zhang, "Fault-tolerant fuzzy gain-scheduled PID for a Quadrotor helicopter tested in the presence of actuator faults," *IFAC Proceedings*, vol. 45, no. 3, pp. 282–287, 2012.
[17] F. Goodarzi, D. Lee, and T. Lee, "Geometric nonlinear PID control of a quadrotor UAV on SE(3)," in *Proceeding of the European Control Conference*, Zurich, Switzerland, June 2013, pp. 3845–3850.
[18] A. Alaimo, V. Milazzo, C. L. R. Milazzo, and A. Ricciardello, "PID controller applied to hexacopter flight," *Journal of Intelligent and Robotic Systems*, vol. 73, no. 1–4, pp. 261–270, Oct. 2013.
[19] B. Kada and Y. Ghazzawi, "Robust PID controller design for an UAV flight control system," in *Proceeding of the WCECS2011*, San Francisco, Vol. II 1–6. ISSN: 2078-0966 (Online), 2011.
[20] J. Moreno-Valenzuela, R. Perez-Alcocer, M. Guerrero-Medina, and A. Dzul, "Nonlinear PID-type controller for quadrotor trajectory tracking," *IEEE/ASME Transactions on Mechatronics*, vol. 23, no. 5, pp. 2436–2447, Oct. 2018.
[21] B. S. Chen, Y. M. Cheng, and C. H. Lee, "A genetic approach to mixed H2/H_∞ optimal PID control," *IEEE Control Systems Magazine*, vol. 15, no. 5, pp. 51–60, Oct. 1995.
[22] C. S. Tseng and B. S. Chen, "Multiobjective PID control design in uncertain robotic systems using neural network elimination scheme," *IEEE Transactions on Systems, Man, and Cybernetics – Part A: Systems and Humans*, vol. 31, no. 6, pp. 632–644, Nov. 2001.
[23] M. A. S. Teixeira, F. Neves-Jr, A. Koubaa, L. V. R. De Arruda, and A. S. De Oliveira, "A quadral-fuzzy control approach to flight formation by a fleet of unmanned aerial vehicles," *IEEE Access*, vol. 8, pp. 64366–64381, Apr. 2020.
[24] W. Zhang and B. S. Chen, "State feedback H_∞ control for a class of nonlinear stochastic systems," *SIAM Journal on Control and Optimization*, vol. 44, no. 6, pp. 1973–1991, 2006.
[25] W. Zhang, L. Xie, and B. S. Chen, *Stochastic H2/H_∞ Control: A Nash Game Approach*, Boca Raton, FL: CRC Press, 2017.
[26] C. S. Tseng, B. S. Chen, and H. J. Uang, "Fuzzy tracking control design for nonlinear dynamic systems via T-S fuzzy model," *IEEE Transactions on Fuzzy Systems*, vol. 9, no. 3, pp. 381–392, June 2001.

[27] T. Takagi and M. Sugeno, "Fuzzy identification of systems and its applications to modeling and control," *IEEE Transactions on Systems, Man, and Cybernetics*, vol. SMC-15, no. 1, pp. 116–132, Feb. 1985.
[28] S. Boyd, L. E. Ghaoui, E. Feron, and V. Balakrishnan, *Linear Matrix Inequalities in System and Control Theory*, Philadelphia, PA: SIAM, 1994.
[29] B. K. Osendal and A. Sulem, *Applied Stochastic Control of Jump Diffusions*, 2nd ed., Berlin: Springer, 2007.
[30] C.-S. Tseng, B.-S. Chen, and H.-J. Uang, "Fuzzy tracking control design for nonlinear dynamic systems via T–S fuzzy model," *IEEE Transactions on Fuzzy Systems*, vol. 9, no. 3, pp. 381–392, June 2001.
[31] B. S. Chen, Y. Y. Chen, and C. L. Lin, "Nonlinear fuzzy H/sub /spl infin// guidance law with saturation of actuators against maneuvering targets," *IEEE Transactions on Control Systems*, vol. 10, no. 6, pp. 769–779, Nov. 2002.
[32] Z. Yu, Y. Qu, and Y. Zhang, "Distributed fault-tolerant cooperative control for multi-UAVs under actuator fault and input saturation," *IEEE Transactions on Control Systems Technology*, vol. 27, no. 6, pp. 2417–2429, Nov. 2019.
[33] L. Zhang, W. W. Che, B. Chen, and C. Lin, "Adaptive fuzzy output-feedback consensus tracking control of nonlinear multiagent systems in prescribed performance," *IEEE Transactions on Cybernetics*, vol. 53, no. 3, pp. 1932–1943, May 2022.
[34] Y. S. Ma, W. W. Che, C. Deng, and Z. G. Wu, "Observer-based event-triggered containment control for MASs under DoS attacks," *IEEE Transactions on Cybernetics*, vol. 52, no. 12, pp. 13156–13167 Aug. 2021.

5 Stochastic Decentralized H_∞ Tracking Control of Large-Scale Team Formation UAV Networked Control System with Time-Varying Delay and Packet Dropout under Interconnected Coupling and Wiener Fluctuation

5.1 INTRODUCTION

For this generation, it is a landmark that we are going to the next industry revolution toward automation and data exchange in manufacturing technologies and smart cities. The involved processes include cyber-physical systems (CPS) [1], the Internet of Things (IoT) [2], the Industrial Internet of Things (IIoT) [3], cloud computing [4], and the references within them. To meet the wide application of industrial needs in the next era, the quadrotor unmanned aerial vehicle (UAV) technology has become a popular research issue in industry and academia area due to its fast-growing market and vast applications such as military, searching, rescuing, delivery, and wireless communication [5–7]. Due to the wide prospect of new industry requirements and relative technologies mentioned above, a large-scale quadrotor UAV system connected by the wireless communication network is expected for various fields [8], and a large-scale quadrotor UAV formation flight is an important prospective technology to construct a smart city in the future. Quadrotor UAV is one of the popular rotor UAVs with four equally spaced rotors, usually arranged at the corners of a square body. With four independent rotors, a quadrotor UAV has six degrees of freedom (DoF) and four independent control inputs.

DOI: 10.1201/9781003545965-5

Some linear and nonlinear control schemes have been proposed to deal with the attitude control and position control of the quadrotor UAV in recent years. For example, proportional-derivative (PD) attitude control has been proposed in Reference [9]. Proportional-integral-derivative (PID) attitude control was studied in References [10] and [11], which includes two parts—the inner loop control (attitude control) and the outer loop control (position control). An adaptive control was proposed by the immersion and invariance methodology in Reference [12] and by the reconfiguration control scheme in Reference [13]. In Reference [14], a quadrotor UAV control system was augmented with the multilayer neural networks, and the weights of the neural networks was adjusted online according to an adaptive law. A backstepping approach was proposed for quadrotor UAV with a nonlinear control strategy in Reference [15], and the corresponding back stepping controller was divided into three subcontrollers, that is, attitude controller, altitude controller, and position controller. A combining integral sliding mode and backstepping sliding mode controller in a double-loop control structure was proposed for quadrotor UAV in Reference [16]. Some other robust tracking control schemes were also proposed for quadrotor UAV; a robust signal compensator was introduced for tracking control under the effect of nonlinearities and uncertainties in Reference [17]; an L_1 optimal tracking control is proposed in Reference [18]; an H_∞ adaptive fuzzy tracking control was given in Reference [19].

With the widespread application of multi-quadrotor UAVs in the future, the team formation flight control design problem of multi-quadrotor UAV has recently become a hot topic. The common team formation control method is to establish a control strategy for a team formation of quadrotor UAVs to achieve their desired target formation tracking, such as switching interaction topologies method in Reference [20], leader-following method with a consensus formation control algorithm in Reference [21], interactive leader-follower configuration in Reference [22] with leader to react with the followers via a virtual structure [23]. However, the number of quadrotor UAVs in the simulation of multi-quadrotor UAV team formation NCS is almost less than five quadrotor UAVs in most literature. The centralized control architecture has been usually used in the conventional multi-quadrotor UAV formation control. In Reference [25], the robust follower tracking control design of stochastic multi-quadrotor UAV team formation NCS is transformed into an equivalent linear matrix inequality (LMI)-constrained problem to contain the team information of all quadrotor UAVs. By adopting the centralized control method for a large-scale quadrotor UAV NCS, the matrix size of the LMI will become extremely large and complex, which will lead to a heavy calculation loading and even an unsolvable problem in the T-S fuzzy controller solving process. For a large-scale UAV team formation system, other centralized control methods also have the unsolvable or unscalable difficulty due to high computational complexity. For the physical configuration and high size of the interconnection in the large-scale team formation quadrotor UAV networked control system, centralized control is neither economically feasible nor necessary if the scale of team formation is more than five quadrotor UAVs.

During the actual quadrotor UAV flight process, the state of the quadrotor UAV will inevitably be disturbed by the wind external disturbance [26] and [27], and intrinsic random fluctuation due to nonideal component characteristics. Moreover, in a nonideal large-scale team formation flight environment, quadrotor UAVs have the opportunity to suffer couplings from other quadrotor UAVs which induce coupling effect among subsystems in the large-scale quadrotor UAV team system such as communication co-channel interference [28–32]. In addition, the forward quadrotor UAV will induce a trailing vortex effect on the rear quadrotor UAVs in the close formation flight, that is, the aerodynamic force and torque on the relative rear quadrotor UAVs [33]. Consequently, the dynamic of team formation quadrotor UAVs will be affected by the couplings from other quadrotor UAVs in the large-scale quadrotor UAV system, which leads the decentralized controller design to a more difficult problem. If the design of decentralized controller does not consider the above coupling effects, the coupling terms in the quadrotor UAV NCS system may cause instability or even collision, especially for a large-scale quadrotor UAV team formation system. How to decouple the above couplings among large-scale quadrotor UAV NCS is a challenge of decentralized team formation tracking control design.

Different from the traditional feedback control system of point-to-point connection, in this chapter, the control signal is transmitted through a wireless network from the ground control system (GCS) in future smart cities. The team members of large-scale quadrotor UAVs are connected to the networked control system (NCS). By using the network control configuration, the control signal can be calculated at the server on the ground instead of the quadrotor UAV itself. In addition, the demands of wireless network will become more and more inflated with the rapid expansion of network applications, and data traffic will also increase dramatically in the future smart cities. Obviously, the large-scale quadrotor UAV team formation control through wireless transmission in NCS will cause network congestion or even paralysis. In order to make the proposed team formation tracking control design of quadrotor UAVs more comprehensive and more in line with actual communication requirements, a discrete event-triggered mechanism [34–37] is also adopted in this chapter to effectively use transmission resources within a limited bandwidth. Consequently, the network-induced delay and packet dropout are inevitable for the NCS and wireless transmission [46], which will make the control performance worse or even invalid. A team formation control design for multi-quadrotor UAV NCS with event-triggered mechanism in NCS has been proposed in Reference [37]. However, the effect of packet dropout may not be considered. Moreover, the control design scheme is not only non-scalable when the number of quadrotor UAVs increases but also infeasible for a large-scale quadrotor UAV team formation networked control system due to the high computational complexity.

In this chapter, the decentralized robust team formation tracking control of large-scale quadrotor UAV NCS is proposed under wind gust external disturbance and internal random fluctuation modeled by Wiener process. In addition, the coupling effect caused from other quadrotor UAVs in large-scale quadrotor UAV NCS is considered in the team formation configuration as well, especially for the practical implementation of robust team formation tracking control of large-scale UAV NCS. Due to the network requirement, the event-triggered mechanism is adopted to save transmission bandwidth and power consumption in smart city, and the network-induced delay and packet dropout both are also considered in the large-scale quadrotor UAV NCS, too. In this chapter, the most important thing is to prove a robust decentralized H_∞ tracking control to efficiently decouple the coupling effect and significantly reduce the computational complexity of team formation control of large-scale quadrotor UAV NCS. As a result, the proposed robust H_∞ decentralized team formation tracking control scheme is not only comprehensive for actual flight and network environment but also scalable and feasible for controller design of the large-scale quadrotor UAV team formation NCS while the number of quadrotor UAV team formation NCS increases.

By using Itô formula, the decentralized robust stochastic H_∞ team formation tracking control design problem for large-scale quadrotor UAV NCS is transformed into an independent Hamilton–Jacobi inequality (HJI)-constrained optimization problems for each quadrotor UAV. Since the HJI is a nonlinear partial differential inequality, there is no numerical or analytical method to effectively solve the HJI. To overcome the difficulty in solving HJI, Takagi-Sugeno (T-S) fuzzy interpolation method is proposed to approximate nonlinear stochastic quadrotor UAV system by interpolating multiple local linear stochastic systems. Therefore, the HJI-constrained optimization problem of the H_∞ robust decentralized team formation tracking design of the large-scale quadrotor UAV NCS can be converted into an independent linear matrix inequality (LMI)-constrained optimization problem for each UAV in the team formation NCS. The MATLAB LMI toolbox can be easily employed to solve the LMI-constrained optimization for the decentralized H_∞ team formation tracking control design of large-scale quadrotor UAV NCS.

The main purposes of this chapter are described as follows:

(i) In the previous multi-quadrotor UAV team formation control design problem in References [25,38,39], the interconnected coupling effect is often ignored, which will lead to quadrotor UAVs collision during flight. In this chapter, the interconnected coupling effect from other UAVs in the large-scale team formation NCS is modelled as a state-dependent fluctuation. Based on the proposed novel decentralized H_∞ team formation tracking control strategy, the effect of interconnected coupling on team formation tracking performance can be efficiently attenuated to simplify the design procedure of decentralized control of large-scale team formation of quadrotor UAVs.

(ii) A fuzzy model-based decentralized H_∞ stochastic team formation reference tracking control strategy for the nonlinear large-scale quadrotor UAV network control system is constructed by solving a low-order LMIs-constrained optimization problem instead of a high-order nonlinear partial differential HJI-constrained optimization problem of centralized H_∞ stochastic team formation tracking control. This can overcome the difficulty of high computational complexity or even infeasible team formation design problem of large-scale UAVs by using a centralized control scheme. Moreover, compared with conventional UAV formation tracking control methods [12,14,16], the proposed decentralized H_∞ T-S fuzzy controller can be easily implemented without any auxiliary dynamic equation for controller synthesis, for example, the dynamic adaptive learning law in References [12,14] and dynamic stabilization function in Reference [16]. Therefore, the proposed decentralized H_∞ T-S fuzzy team formation tracking control is applicable for very large-scale team formation network control of quadrotor UAVs.

(iii) Compared with the previous networked-based UAV formation control issues [40,41], the other uncertain effects of external wind gust disturbance, intrinsic Wiener fluctuation, time-varying network-induced delay, and packet dropout on team formation tracking performance are all simultaneously attenuated by the proposed decentralized H_∞ fuzzy team formation controller for each quadrotor UAV to efficiently achieve team formation of large-scale UAV NCS. Moreover, the event-triggered mechanism is adopted for network bandwidth saving. Therefore, it is a very practical team formation tracking control design of large-scale quadrotor UAV NCS in future smart cities.

The remainder of this chapter is organized as follows. The preliminary of a quadrotor UAV system, event-triggered mechanism, and network configuration is introduced in Section 5.2. Section 5.3 presents the problem formulation of the virtual leader-follower team formation construction to form a robust decentralized H_∞ team formation tracking control problem for each UAV NCS in the large-scale quadrotor UAV NCS. Further, the design condition is transformed to an equivalent HJI-based problem. In Section 5.4, the T-S fuzzy model is introduced to deal with the stochastic nonlinear virtual leader-follower H_∞ robust decentralized team formation tracking control problem of large-scale quadrotor UAV NCS. A design example of 25 quadrotor UAV team formation NCS is given to illustrate the design procedure and to demonstrate the robust H_∞ team formation tracking performance of the proposed virtual leader-follower robust decentralized team formation control strategy in Section 5.5. At the end, the conclusion of this chapter is made in Section 5.6.

Notation: I_n: the n-dimension identity matrix; A^T: the transpose of matrix A; $A \geq 0 (A > 0)$: symmetric positive semidefinite(symmetric positive-definite) matrix A; Z^+: a set of positive integers; E: the expectation operator; and $\mathfrak{L}^2_{\mathcal{F}}(\mathbb{R}^+;\mathbb{R}^n)$: the space of nonanticipative stochastic process $y(t) \in \mathbb{R}^n$ with respect to an increasing σ-algebras $\{\mathcal{F}_t\}_{t\geq 0}$ satisfies $\|y(t)\|_{\mathfrak{L}^2(\mathbb{R}^+;\mathbb{R}^n)} \triangleq E\{\int_0^\infty y^T(t)y(t)dt\}^{\frac{1}{2}} < \infty$; C^n: the space contains the functions which have n th continuous derivatives.

5.2 PRELIMINARY OF A LARGE-SCALE QUADROTOR UAV NCS AND NETWORK CONFIGURATION

5.2.1 Stochastic Quadrotor UAV Dynamic Model

The quadrotor UAV dynamics is derived by introducing two frames as shown in Figure 5.1. The inertial frame is defined by the ground which is spanned by an orthonormal basis $(b_x, b_y, b_z,)$, with gravity pointing in the negative z-direction. The body frame is defined by the orientation of the quadrotor which is spanned by an orthonormal basis (b_1, b_2, b_3), with the rotor axes pointing in the positive b_3 and the arms pointing in the b_1 and b_2 directions.

We define the position and velocity of the quadrotor in the inertial frame as $I = (x, y, z)^T$ and $\dot{I} = (\dot{x}, \dot{y}, \dot{z})^T$, respectively. Similarly, we define the roll, pitch, and yaw angles of the quadrotor in the body frame as $B = (\phi, \theta, \psi)^T$, with corresponding

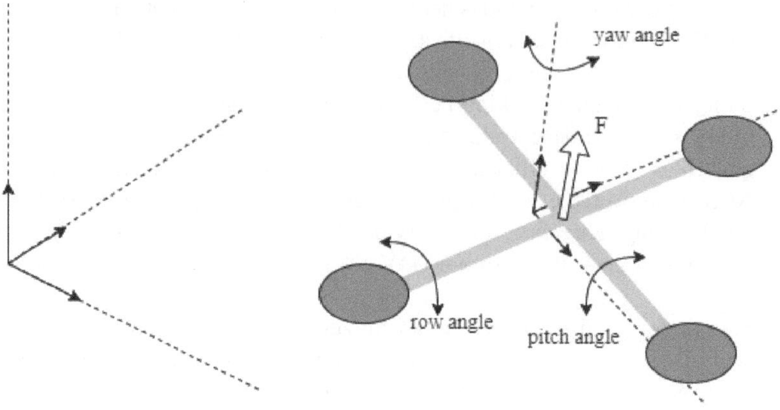

FIGURE 5.1 Inertial frame and body frame of quadrotor UAV.

angular velocities ... The body and inertial frame can be related by a rotation matrix that the position and velocity of the quadrotor in the inertial frame can be obtained from the roll, pitch, and yaw angles in the body frame. As a result, the ith quadrotor UAV dynamic of the large-scale quadrotor UAV networked control system can be described as follows [24]:

$$
\begin{aligned}
&\dot{x}_1^i(t) = x_2^i(t) \\
&\dot{x}_2^i(t) = -\frac{K_x^i}{m^i} x_2^i(t) + (\cos\phi_1{}^i(t)\sin\theta_1^i(t)\cos\psi_1^i(t)) \\
&\qquad + \sin\phi_1^i(t)\sin\psi_1^i(t))\frac{F^i(t)}{m^i} + \upsilon_x^i(t) \\
&\dot{y}_1^i(t) = y_2^i(t) \\
&\dot{y}_2^i(t) = -\frac{K_y^i}{m^i} y_2^i(t) + (\cos\phi_1{}^i(t)\sin\theta_1^i(t)\sin\psi_1^i(t)) \\
&\qquad - \sin\phi_1^i(t)\cos\psi_1^i(t))\frac{F^i(t)}{m^i} + \upsilon_y^i(t) \\
&\dot{z}_1^i(t) = z_2^i(t) \\
&\dot{z}_2^i(t) = -\frac{K_z^i}{m^i} z_2^i(t) - g + \cos\phi_1{}^i(t)\cos\theta_1^i(t)\frac{F^i(t)}{m^i} + \upsilon_z^i(t) \\
&\dot{\phi}_1^i(t) = \phi_2^i(t) \\
&\dot{\phi}_2^i(t) = \frac{J_y^i - J_z^i}{J_x^i}\theta_1{}^i(t)\psi_1^i(t) - \frac{K_\phi^i}{J_x^i}\phi_2^i(t) + \frac{1}{J_x^i}\tau_\phi^i(t) + \upsilon_\phi^i(t) \\
&\dot{\theta}_1^i(t) = \theta_2^i(t)
\end{aligned}
\tag{5.1}
$$

$$\dot{\theta}_2^i(t) = \frac{J_z^i - J_x^i}{J_y^i}\phi_1{}^i(t)\psi_1^i(t) - \frac{K_\theta^i}{J_y^i}\theta_2^i(t) + \frac{1}{J_y^i}\tau_\theta^i(t) + v_\theta^i(t)$$

$$\dot{\psi}_1^i(t) = \psi_2^i(t)$$

$$\dot{\psi}_2^i(t) = \frac{J_x^i - J_y^i}{J_z^i}\phi_1{}^i(t)\theta_1^i(t) - \frac{K_\psi^i}{J_z^i}\psi_2^i(t) + \frac{1}{J_z^i}\tau_\psi^i(t) + v_\psi^i(t)$$

where $x_1^i(t), y_1^i(t), z_1^i(t) \in \mathbb{R}^1$ are the position states of the mass center of the ith quadrotor UAV on the Cartesian coordinate with respect to the inertial frame. $\phi_1{}^i(t)$, $\theta_1^i(t)$, $\psi_1^i(t)$ are the Euler angles of rotation which are fixed to the body frame of the ith quadrotor UAV. These three angles respectively represent roll angle $(-\pi/2 < \phi_1^i(t) < \pi/2)$, pitch angle $(-\pi/2 < \theta_1^i(t) < \pi/2)$, and yaw angle $(-\pi < \psi_1^i(t) < \pi)$ related to the orientation of the quadrotor UAV to be controlled by changing the rotational speed of four rotors. The total thrust $F^i(t)$ on the ith quadrotor UAV in the body frame is the sum of the thrust magnitude of four motors. $\tau_\phi^i(t), \tau_\theta^i(t)$, $\tau_\psi^i(t)$ are the rotation forces of the ith quadrtor UAV. $\{v_x^i(t), v_y^i(t), v_z^i(t), v_\phi^i(t), v_\theta^i(t), v_\psi^i(t)\}$ denotes the disturbance on three velocities and three angular velocities of the ith UAV, g is the gravity constant, J_x^i, J_y^i, J_z^i are the moments of inertia of the ith UAV, $K_x^i, K_y^i, K_z^i, K_\phi^i, K_\theta^i, K_\varphi^i$ represent the aerodynamic damping coefficients of the ith UAV.

A large-scale UAV team formation NCS is denoted as a set $S = \{S_1, S_2, \ldots, S_N\}$, which is composed of N single UAV subsystem $\{S_i\}_1^N$. From the practical point of view, every single UAV is not only under external disturbance due to wind gust and internal random fluctuation induced from UAV components, but it also suffered interconnected couplings from aerodynamics or nonideal communication from other quadrotor UAVs in their neighborhood. Taking these considerations into account, the ith quadrotor UAV subsystem S_i can be represented as follows:

$$\begin{aligned} S_i : dX_i(t) = {} & [f_i(X_i(t)) + g_i(X_i(t))U_i(t) \\ & + \sum_{j \in N_i} f_{ij}(X_i(t))X_j(t - \tau_{ij}(t)) + v_i(t)]dt \\ & + \vartheta_i(X_i(t))dW_i(t), \; for\; i = 1, \cdots, N \end{aligned} \tag{5.2}$$

where $X_i(t) = [x_1^i(t), x_2^i(t), y_1^i(t), y_2^i(t), z_1{}^i(t), z_2^i(t), \phi_1^i(t), \quad \phi_2^i(t), \theta_1{}^i(t), \theta_2^i(t), \psi_1^i(t), \psi_2^i(t)]^T$, $U_i(t) = [F^i(t), \tau_\phi^i(t), \tau_\theta^i(t), \tau_\varphi^i(t)]^T$ and $v_i(t) = [0, v_x^i(t), 0, v_y{}^i(t), 0, v_z^i(t), 0, v_\phi^i(t), 0, v_\theta^i(t), 0, v_\psi^i(t)]^T$ with

$$\begin{bmatrix} x_2^i(t) \\ -\frac{K_x^i}{m^i}x_2^i(t) \\ y_2^i(t) \end{bmatrix}$$

$$f_i(X_i(t)) = \begin{bmatrix} -\dfrac{K_y^i}{m^i} y_2^i(t) \\ z_2^i(t) \\ -\dfrac{K_z^i}{m^i} z_2^i(t) - g \\ \phi_2^i(t) \\ \dfrac{J_y^i - J_z^i}{J_x^i} \theta_1^i(t)\psi_1^i(t) - \dfrac{K_\phi^i}{J_x^i} \phi_2^i(t) \\ \theta_2^i(t) \\ \dfrac{J_z^i - J_x^i}{J_y^i} \phi_1^i(t)\psi_1^i(t) - \dfrac{K_\theta^i}{J_y^i} \theta_2^i(t) \\ \Psi_2^i(t) \\ \dfrac{J_x^i - J_y^i}{J_z^i} \phi_1^i(t)\theta_1^i(t) - \dfrac{K_\psi^i}{J_z^i} \psi_2^i(t) \end{bmatrix}$$

$$g_i(X_i(t)) = \begin{bmatrix} 0 & 0 & 0 & 0 \\ \begin{matrix}(\cos\phi_1^i(t)\sin\theta_1^i(t)\cos\psi_1^i(t) \\ +\sin\phi_1^i(t)\sin\psi_1^i(t))\dfrac{1}{m^i}\end{matrix} & 0 & 0 & 0 \\ 0 & 0 & 0 & 0 \\ \begin{matrix}(\cos\phi_1^i(t)\sin\theta_1^i(t)\sin\psi_1^i(t) \\ -\sin\phi_1^i(t)\sin\psi_1^i(t))\dfrac{1}{m^i}\end{matrix} & 0 & 0 & 0 \\ 0 & 0 & 0 & 0 \\ \dfrac{1}{m^i}\cos\phi_1^i(t)\cos\theta_1^i(t) & 0 & 0 & 0 \\ 0 & 0 & 0 & 0 \\ 0 & \dfrac{1}{J_x^i} & 0 & 0 \\ 0 & 0 & 0 & 0 \\ 0 & 0 & \dfrac{1}{J_y^i} & 0 \\ 0 & 0 & 0 & 0 \\ 0 & 0 & 0 & \dfrac{1}{J_z^i} \end{bmatrix}$$

where $X_i(t) \in \mathbb{R}^{12}$ and $U_i(t) \in \mathbb{R}^4$ represent the state vector and control input, respectively. $f_i(X_i(t))$ denotes the nonlinear internal dynamic interactive vector, and $g_i(X_i(t)) \in \mathbb{R}^{12\times 4}$ denotes the input matrix satisfied with Lipschitz continuity. $f_{ij}(X_i(t)) \in \mathbb{R}^{12}$ denotes the coupling matrix between the ith quadrotor UAV and the jth UAV. $X_j(t-\tau_{ij}(t)) \in \mathbb{R}^{12}$ represents the coupling terms from the jth UAV to the ith quadrotor UAV with a time-varying delay $\tau_{ij}(t) \in \mathbb{R}^+$. $\sum_{j\in N_i} f_{ij}(X_i(t))X_j(t-\tau_{ij}(t)) \in \mathbb{R}^{12}$ represents the interconnected couplings such as trailing vortex effect caused by neighborhood quadrotor UAVs of the ith quadrotor UAV. $v_i(t) \in \mathfrak{L}^2_{\mathcal{F}}(\mathbb{R}^+;\mathbb{R}^{12})$ is the finite energy external disturbance on the ith single quadrotor UAV S_i. $W_i(t)$ represents the Wiener process to model the intrinsic uncertain effect of the ith quadrotor UAV. $\vartheta_i(X_i(t))dW_i(t)$ is the continuous intrinsic random fluctuation due to nonideal component characteristics in the ith quadrotor UAV subsystems. N_i represents a neighborhood set of the quadrotor UAVs nearing the ith quadrotor UAV.

5.2.2 A Large-Scale Quadrotor UAV Networked Control System with Discrete Event-Triggered Mechanism

In order to make signal transmission in NCS more efficient, a discrete event-triggered mechanism is adopted as a part of tracking control for the large-scale quadrotor UAV team formation NCS in this chapter. Based on the event-triggered mechanism, the control signal is only updated when a state-dependent triggering condition is violated. The conspicuous advantage of the event-triggered mechanism is that the number of transmitted signals and the calculation time of controller can be reduced significantly while retaining a satisfactory closed-loop team formation tracking performance. Therefore, the event-triggered mechanism is applied to saving the limited communication resources, especially for the large-scale team formation NCS which may cause considerable signal transmissions among the wireless network in this chapter.

The event generator can be briefly composed of a sampler, a register, and a comparator, as shown in Figure 5.2. At first, the state signals are sampled every h second by a sampler. The time-triggered sampling instant of the ith UAV with period h is denoted as a sampling sequence set $S_{i1} = \{0, h, 2h, ..., sh\}, s \in Z_+$. We assume that $X_i(sh)$ with its time stamp s can be encapsulated into a sampled data packet $(s, X_i(sh))$ which meets the related network protocol. The comparator is used to determine whether the sampled signal needs to be transmitted according to the following event-triggered mechanism, that is, the release time instant t^i_{k+1} h can be determined by:

$$
\begin{aligned}
&t^i_{k+1}h \\
&= t^i_k h + \min_{j\in Z^+}\{jh | E\{[X_i((t_k^{\ i}+j)h) - X_i(t^i_k h)]^T \Phi_i \\
&\times [X_i((t^i_k + j)h) - X_i(t^i_k h)] \geq \sigma_i X_i^T(t^i_k h)\Phi_i X_i(t^i_k h)\}
\end{aligned}
\tag{5.3}
$$

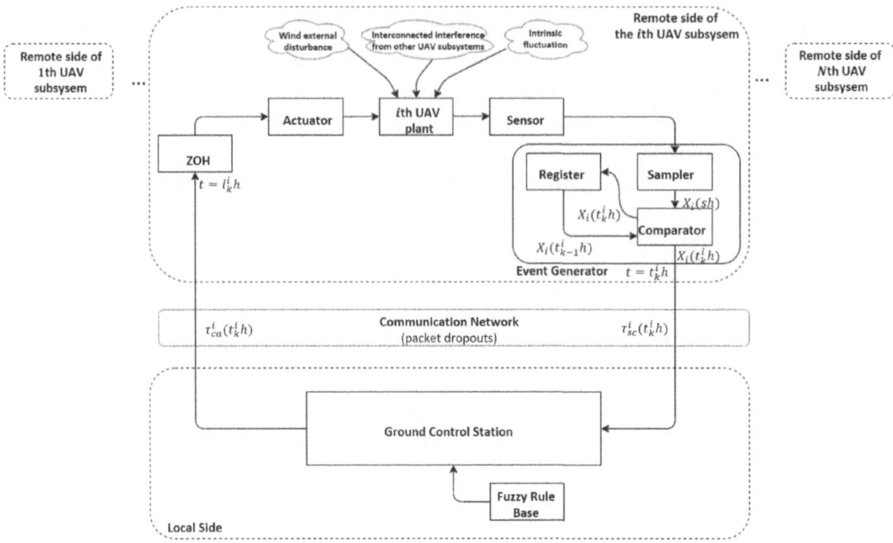

FIGURE 5.2 The configuration and signal flows of a large-scale quadrotor UAV network control system with event-triggered mechanism. under time-varying network-induced delays and packet dropouts.

where $t_k^i \in Z_+$ denotes the time stamp of the sampled data packet $(t_k^i,\ X_i(t_k^i h))$, which violates the triggered condition, and the sampled data packet will be stored at the register of quadrotor UAV. The transmission sequence set of the ith quadrotor UAV can be represented as $S_{i2} = \{0, t_1^i h, t_2^i h, ..., t_k^i h\} \subset S_{i1}$. $\sigma_i \in (0,1]$ is a threshold, and $\Phi_i > 0$ is a weighting matrix. Once the triggering condition is violated, the current sampled data packet is immediately released and transmitted to the ground control system (GCS); otherwise, it is discarded off right away. According to the discrete event-triggered mechanism, a discrete event-driven zero-order hold (ZOH) is set on the quadrotor UAV, as shown in Figure 5.2. Once the packet with discrete control information arrives at the UAV, the discrete control signal is transformed as a continuous signal immediately by ZOH as an input signal to the actuator of the quadrotor UAV.

Until now, the ith quadrotor UAV subsystem of large-scale quadrotor UAVs with discrete event-triggered mechanism and ZOH can be described as:

$$\begin{aligned} S_i : dX_i(t) &= [f_i(X_i(t)) + g_i(X_i(t))U_i(t) \\ &\quad + \sum_{j \in N_i} f_{ij}(X_i(t))X_j(t - \tau_{ij}(t)) + v_i(t)]dt \\ &\quad + \vartheta_i(X_i(t))dW_i(t), \\ U_i(t) &= K_i(X_i(t_k^i h)) \\ &\quad for\ i = 1, \cdots, N,\ \ t \in [t_k^i h, t_{k+1}^i h) \end{aligned} \tag{5.4}$$

with the nonlinear team formation controller gain to be designed $K_i(\cdot)$, for $i = 1,\cdots,N$.

Remark 5.1 σ_i in (5.3) is the threshold parameter of the ith quadrotor UAV, which represents the level of the event trigger. While $\sigma_i = 0$, the event-triggered embedded quadrotor UAV NCS will be reduced to a periodically time-triggered UAV NCS. Subsequently, if a small event-triggered level σ_i is chosen, the event-triggered mechanism in (5.3) is more easily triggered and it will lead to a better tracking control performance. However, due to the large number of signal transmission, the communication load will increase. On the other hand, if a large event-triggered level σ_i is chosen, the event-triggered mechanism in (5.3) is hardly triggered during the tracking control process. In this case, the communication load is relaxed. However, the control performance will be degraded. As a result, for the choice of the value of event-triggered level, there is a trade-off between control performance and communication load.

The main purpose of this chapter is to design a comprehensive decentralized event-triggered control scheme for a robust team formation of large-scale quadrotor UAV NCS with consideration of system uncertain effects such as external disturbance, intrinsic random fluctuation, coupling from neighboring quadrotor UAVs, network effect such as time-varying channel delay, and packet dropout. Under the above discrete event triggering mechanism, several sampled data packets will be discarded, which means that the wireless communication network loads will be greatly reduced. Hence, a good network service can be still maintained under large-scale quadrotor UAV team formation NCS in this chapter. Also, since quadrotor UAV is a power limited system, the power used to transmit signals will also be reduced to achieve a longer flight time of quadrotor UAV. However, the time-varying network-induced delays and packet dropouts are unavoidable in wireless transmission and network configuration. In the next section, the ith quadrotor UAV subsystem in (5.4) will be reformulated to include time-varying network-induced delay and packet dropout.

5.2.3 Discrete Event-Triggered Mechanism in the Large-Scale Quadrotor UAV NCS with Time-Varying Delay

In order to illustrate the large-scale quadrotor UAV NCS under network-induced time-varying delay and packet dropout with a discrete event-triggered mechanism, the overall system configuration with the signal flow is shown in Figure 5.2.

Consider the network-induced round-trip delay $\tau^i_{t_k}$ of the ith quadrotor UAV from sensor to controller and controller to actuator at time instant t^i_k. The round-trip delay $\tau^i_{t_k}$ in the network is defined as $\tau^i_{t_k} = \tau^i_{sc}(t^i_k h) + \tau^i_{ca}(t^i_k h)$, where $\tau^i_{sc}(t^i_k h)$ and $\tau^i_{ca}(t_k{}^i h)$ represent the network-induced delay of the kth transmitted signal of the ith quadrotor UAV from sensor to controller and controller to actuator, respectively. The delay upper bound of the ith quadrotor UAV is defined as $\tau^i_M := \max\{\tau^i_{t_k} \mid k = 1,2,\cdots\}$. The time instant of the data packet $(t^i_k,\ X_i(t^i_k h))$ arriving at the ZOH on the ith UAV is denoted as $l^i_k,\ \ k \in Z^+$.

Due to the uncertain wireless channel, the problem of packet disorder is inevitable during the signal transmission process. To deal with this problem [41–43], once the disorder signal is received by the actuator, the signal is immediately discarded and the information of actuator will not be updated, that is, the actuator only updates the signals in order. In this case, the updating sequence $\{l_k^i\}_{k\in\mathbb{N}}$ becomes an increasing sequence in the actuator, that is, $l_k^i < l_n^i$, $\forall k < n$, $k,n \in \mathbb{N}$, and the packet $(t_k^i$, $X_i(t_k^i h))$ is delayed by $\tau_k^i = l_k^i - t_k^i h$ in NCS.

Furthermore, the packet dropout has the chance to occur in transmission at any time. In the event-triggered mechanism, the signal is only transmitted when the state has changed over a certain level compared to the last transmitted signal. If the packet dropouts occur, the control input can not arrive at the desperately control-needed quadrotor UAV subsystem, which may cause the performance to be extremely terrible or even make the system unstable. To implement a more general decentralized control design in quadrotor UAV NCS, the effect of packet dropouts is also considered in this chapter.

At the updating instant l_k^i, the number of accumulated data packet dropouts since the last updating instant l_{k-1}^i is denoted as δ_k^i. We assume that the maximum number of data packet dropouts is $\bar{\delta}_i$, that is, $\delta_k^i \le \bar{\delta}_i$, where $\delta_k^i \in \{0,1,\cdots,\bar{\delta}_i\}$. Now, we define γ_k^i to construct the time interval between l_k^i and l_{k+1}^i with packet dropouts as follows:

$$\gamma_k^i := \min\{j \mid l_k^i + jh \ge l_{k+1}^i,\ j = 0,1,2,\cdots\} \tag{5.5}$$

then it is clear that $\gamma_k^i \ge 1$ because $l_k^i < l_{k+1}^i$. The interval $[l_k^i, l_{k+1}^i)$ can thus be written as:

$$[l_k^i, l_{k+1}^i) = \bigcup_{j=1}^{\gamma_k^i} I_j^i \tag{5.6}$$

where $I_j^i = [l_k^i + (j-1)h,\ l_k^i + jh)$, $I_{\gamma_k^i}^i = [l_k^i + (\gamma_k^i - 1)h,\ l_{k+1}^i)$, $j = 1,2,\cdots,\gamma_k^i - 1$.

Now, we define the two functions $\eta_i(t)$ and $e_i(t)$ on $[l_k{}^i, l_{k+1}^i)$ as:

$$\eta_i(t) := \begin{cases} t - t_k^i h, & t \in I_1^i \\ t - t_k^i h - h, & t \in I_2^i \\ \vdots & \vdots \\ t - t_k^i h - (\gamma_k^i - 1)h, & t \in I_{\gamma_k^i}^i \end{cases} \tag{5.7}$$

and

$$e_i(t) := \begin{cases} X_i(t_k^i h) - X_i(t_k^i h), & t \in I_1^i \\ X_i(t_k^i h + h) - X_i(t_k^i h), & t \in I_2^i \\ \vdots & \vdots \\ X_i(t_k^i h + (\gamma_k^i - 1)h) - X_i(t_k^i h), & t \in I_{\gamma_k^i}^i \end{cases} \tag{5.8}$$

By taking the effect of time-varying network-induced delay and the packet dropout in the network into consideration over the time interval $[l_k{}^i, l_{k+1}^i)$, from (5.5) –(5.8), the transmitted control signal with time-varying network-induced delay under packet dropout can be represented as:

$$U_i(t) = K_i(X_i(t_k^i h)) = K_i(X_i(t-\eta_i(t)) - e_i(t)),\ t \in [l_k^i, l_{k+1}^i) \tag{5.9}$$

Thus, we have a new expression of control signal $U_i(t)$. The control signal is considered as a sampled signal with a discrete event-triggered mechanism and time-varying network-induced delay and packet dropout in NCS, which is related to the error function $e_i(t)$ and the delayed packet $X_i(t-\eta_i(t))$.

In addition, the function $\eta_i(t)$ is satisfied with

$$0 \le \eta_i(t) \le h + \tau_k^i + \overline{\delta}_i h \le h + \tau_M^i + \overline{\delta}_i h \le \overline{\eta}_i \tag{5.10}$$

where τ_M^i is the upper bound of time-varying network-induced delay τ_k^i, and $\overline{\eta}_i$ is the upper bound of $\eta_i(t)$.

The ith quadrotor UAV subsystem with discrete event-triggered control mechanism in NCS with time-varying delay and packet dropout can be described as:

$$\begin{aligned} S_i : dX_i(t) = {} & [f_i(X_i(t)) + g_i(X_i(t))U_i(t) \\ & + \sum_{j\in N_i} f_{ij}(X_i(t))X_j(t-\tau_{ij}(t)) + v_i(t)]dt \\ & + \vartheta_i(X_i(t))dW_i(t),\quad for\ i = 1,\ldots,N \end{aligned} \tag{5.11}$$

where $t \in [l_k^i, l_{k+1}^i)$. For the system in (5.9) to (5.11) we supplement the initial condition of the state $X_i(t)$ in $[-\tau_M^i, 0)$ as $X_i(t) = \chi_i(t)$, $t \in [-\tau_M^i, 0)$ where $\chi_i(t)$ is a continuous function in $[-\tau_M^i, 0)$.

5.3 DESIGN PROBLEM DESCRIPTION

The purpose of this chapter is to provide a robust decentralized H_∞ team formation tracking control with a discrete event-triggered mechanism for a large-scale quadrotor UAV in the team formation networked control system involving time-varying network-induced delay and packet dropout.

5.3.1 Team Formation Tracking of Large-Scale Quadrotor UAV NCS by Virtual Leader-Follower Structure

The team formation shape of the large-scale UAVs is constructed by one virtual leader and N UAVs. The desired position, velocity, and attitude of the virtual leader could be generated by a reference model as follows:

$$dX_v(t) = (A_v X_v(t) + r(t))dt \tag{5.12}$$

where $A_v \in \mathbb{R}^{12\times 12}$, $r(t) \in \mathbb{R}^{12}$, and $X_v(t) \in \mathbb{R}^{12}$ are the desired trajectories of the virtual leader.

Remark 5.2 The reference model in (5.12) is to generate the desired flight trajectory of the virtual leader. The matrix A_v denotes a specific asymptotically stable matrix. $r(t)$ denotes a bounded reference input which meets a physically reasonable flight trajectory. It is clear that $X_v(t) = -A_v^{-1}r(t)$ at the steady state. If we choose $A_v = -I$, the desired trajectory $X_v(t)$ is equivalent to the reference input $r(t)$ at the steady state. If this is the case, $X_v(t)$ will converge to $r(t)$ at the steady state, and $r(t)$ can be considered as the given target to be tracked in this case.

Remark 5.3 To avoid fixed and known obstacles, the trajectory of virtual leader ($X_v(t)$ in (5.12)) can be carefully designed beforehand to achieve obstacle avoidance. Moreover, to achieve the real-time obstacle avoidance, the quadrotor UAV can take the geometric information of flight environment by the equipped sensors and send the geometric information like line-of-sight vectors and relative velocity vectors between virtual leader and obstacles to the ground control station via wireless network. Based on this obstacle information, the designer can plan the reference input $r(t)$ in (5.12) for a guidance of quadrotor UAV team by some developed collision avoidance algorithms (e.g., [44]) to modify the trajectory of virtual leader on time to avoid obstacles. Since there are many collision avoidance strategies for quadrotor UAV in team formation flight, how to combine these collision avoidance algorithms with virtual leader in (5.12) as a guidance law for quadrotor UAV team formation will be a further research topic [45].

In order to construct a team formation of large-scale quadrotor UAV NCS, each quadrotor UAV dynamic should be shifted a specified distance d_i from the virtual leader, respectively (i.e., the shifted trajectory of the ith quadrotor UAV is $X_i(t) - d_i$), then the ith UAV will track the trajectory of the virtual leader with a shift d_i to form a specific formation shape. Therefore, the team formation problem of large-scale N UAVs can be transformed as N quadrotor decentralized reference tracking control problems of every single quadrotor UAV.

By the trajectory information of the virtual leader and specified shifted distance of formation, every quadrotor UAV in a large-scale quadrotor UAV NCS will track the trajectory of virtual leader with a relative shifted distance to form a team formation during the team flight process. A novel decentralized H_∞ tracking control strategy with the formation tracking error $X_{i,e}(t) = X_v(t) - (X_i(t) - d_i)$ is considered in the following to efficiently attenuate the effect of an arbitrary reference input signal, external disturbance, and couplings induced from other quadrotor UAVs on the team formation tracking performance of the stochastic large-scale quadrotor UAV NCS with the discrete event-triggered mechanism under time-varying network-induced delay and packet dropout:

$$J_\infty^i(U_i(t)) = \sup_{r_i(t), v_i(t), \sum_{j \in N_i \%} X_j(t-\tau_{ij}(t))} \frac{E\{\int_0^{t_f} [X_{i,e}^T(t) Q_i X_{i,e}(t) - V_i(X_{i,e}(0))] dt\}}{E\{\int_0^{t_f} r_i^T(t) r_i(t) + v_i^T(t) v_i(t) + \sum_{j \in N_i} X_j^T(t - \tau_{ij}(t)) X_j\}} \tag{5.13}$$

$$for\ i = 1, \cdots, N$$

$V_i(X_{i,e}(0))$ denotes the effect of initial condition and it should be excluded in (5.13) and $Q_i \in \mathbb{R}^{12\times 12}$ is a positive definition matrix. The terms $\sum_{j\in N} X_j^T(t-\tau_{ij}(t))X_j(t-\tau_{ij}(t))$ of the denominator in (5.13) are proposed in the above decentralized H_∞ team formation tracking strategy to attenuate the coupling effect from other neighboring quadrotor UAVs on the team formation tracking of the ith quadrotor UAV.

Remark 5.4 For the robust decentralized H_∞ team formation tracking control performance indices in (5.13), the designer will specify the corresponding parameters of weighting matrix according to their design purpose. Especially, if the jth state tracking is more significant for the designer in the ith quadrotor UAV, then the designer will set a large parameter in the (j,j) term of Q_i. In contrast, the designer will set a small parameter in the (k,k) term of Q_i if he considers the kth state tracking is insignificant.

The ith quadrotor UAV shifted subsystem and reference model of virtual leader can be formulated as the new ith augmented system with the augmented state $\bar{X}_i(t) = [X_v^T(t), X_{i,e}^T(t)]^T$ for each quadrotor UAV. If every $\bar{X}_i(t)$ can be robust H_∞ stabilized at the origin, then the ith quadrotor UAV can robustly track on their own desired trajectory with the specific formation d_i of large-scale quadrotor UAV NCS by the decentralized control.

As a result, the stochastic dynamic of the ith augmented system $\bar{X}_i(t)$ is composed of virtual leader trajectory $X_v(t)$ in (5.12) and the shifted tracking error dynamic $X_{i,e}(t)$ of the ith quadrotor UAV as follows:

$$\begin{aligned} d\bar{X}_i(t) &= [\bar{f}_i(\bar{X}_i(t)) + \bar{g}_i(\bar{X}_i(t))\bar{U}_i(t) \\ &\quad + \sum_{j\in N_i} \bar{f}_{ij}(\bar{X}_i(t))X_j(t-\tau_{ij}(t)) \\ &\quad + H\bar{v}_i(t)]dt + \bar{\vartheta}_i(\bar{X}_i(t))d\bar{W}_i(t) \\ \bar{X}_i(t) &= \bar{\chi}_i(t),\ t\in[-\tau_M^i, 0), \text{for i=1,}\cdots\text{,N} \end{aligned} \tag{5.14}$$

where $\bar{v}_i(t) = [r^T(t), v_i^T(t)]^T$.

$$\bar{f}_i(\bar{X}_i(t)) = \begin{bmatrix} A_r X_v(t) \\ A_r X_v(t) - f_i(X_v(t) - X_{i,e}(t) + d_i) \end{bmatrix}$$

$$\bar{g}_i(\bar{X}_i(t)) = \begin{bmatrix} 0_{12\times 4} \\ -g_i(X_v(t) - X_{i,e}(t) + d_i) \end{bmatrix},$$

$$\bar{f}_{ij}(\bar{X}_i(t)) = \begin{bmatrix} 0_{12\times 12} \\ -f_{ij}(X_v(t) - X_{i,e}(t) + d_i) \end{bmatrix}$$

$$H = \begin{bmatrix} I_{12} & 0_{12\times 12} \\ I_{12} & -I_{12} \end{bmatrix},\ d\bar{W}_i(t) = \begin{bmatrix} 0_{12\times 1} \\ dW_i(t) \end{bmatrix}$$

$$\bar{\vartheta}_i(\bar{X}_i(t)) = \begin{bmatrix} 0_{12\times 12} & 0_{12\times 12} \\ 0_{12\times 12} & -\vartheta_i(X_v(t) - X_{i,e}(t) + d_i) \end{bmatrix}$$

We define the new variable $\bar{e}_i(t)=[\tilde{e}_v^T(t),\tilde{e}_i^T(t)]^T$ for the ith UAV augmented system in the following:

$$\tilde{e}_v(t):=\begin{cases} X_v(t_k^ih)-X_v(t_k^ih), & t\in I_1^i \\ X_v(t_k^ih+h)-X_v(t_k^ih), & t\in I_2^i \\ \vdots & \vdots \\ X_v(t_k^ih+(\gamma_k^i-1)h)-X_v(t_k^ih), & t\in I_{\gamma_k^i}^i \end{cases} \tag{5.15}$$

$$\tilde{e}_i(t):=\begin{cases} X_{i,e}(t_k^ih)-X_{i,e}(t_k^ih), & t\in I_1^i \\ X_{i,e}(t_k^ih+h)-X_{i,e}(t_k^ih), & t\in I_2^i \\ \vdots & \vdots \\ X_{i,e}(t_k^ih+(\gamma_k^i-1)h)-X_{i,e}(t_k^ih), & t\in I_{\gamma_k^i}^i \end{cases} \tag{5.16}$$

By the similar mathematical expression in (5.9) under time-varying delay and packet dropout with new definition for the ith quadrotor UAV networked tracking error subsystem in (5.14), the decentralized control signal of the ith quadrotor UAV networked tracking subsystem (5.14) is given as follows:

$$\bar{U}_i(t)=\bar{K}_i(\bar{X}_i(t_k^ih))=\bar{K}_i(\bar{X}_i(t-\eta_i(t))-\bar{e}_i(t)),\ t\in[l_k^i,l_{k+1}^i) \tag{5.17}$$

and it is clear that (5.14) can be rewritten as:

$$\begin{aligned} d\bar{X}_i(t)=&[\bar{f}_i(\bar{X}_i(t))+\bar{g}_i(\bar{X}_i(t))\bar{K}_i(\bar{X}_i(t-\eta_i(t)) \\ &-\bar{e}_i(t))+\bar{f}_{-i}(\bar{X}_i(t))\bar{X}_{-i}(t-\tau_{-i}(t)) \\ &+H\bar{v}_i(t)]dt+\bar{\vartheta}_i(\bar{X}_i(t))d\bar{W}_i(t) \\ &for\ i=1,\cdots,\mathrm{N} \end{aligned} \tag{5.18}$$

where $\bar{f}_{-i}(\bar{X}_i(t))\in\mathbb{R}^{24\times12N_i}$ and $\bar{X}_{-i}(t-\tau_{-i}(t))\in\mathbb{R}^{12N_i}$ are given as follows:

$$\bar{f}_{-i}(\bar{X}_i(t))=\begin{bmatrix}\cdots & 0_{12\times12} & \cdots \\ \cdots & -f_{ij}(X_v(t)-X_{i,e}(t)) & \cdots\end{bmatrix},\ j\in N_i$$

$$\bar{X}_{-i}(t-\tau_{-i}(t))=\begin{bmatrix}\cdots & X_j^T(t-\tau_{ij}(t)) & \cdots\end{bmatrix}^T,\ j\in N_i$$

Since packet contains information about the virtual leader and tracking error in (5.18), the released time of event-triggered mechanism is reformulated as follows:

$$\begin{aligned} t_{k+1}^ih=&t_k^ih+\min_{\mathrm{j}\in Z^+}\{\mathrm{jh}|\mathrm{E}\{[\bar{\mathrm{X}}_\mathrm{i}((\mathrm{t}_\mathrm{k}^\mathrm{i}+\mathrm{j})\mathrm{h})-\bar{\mathrm{X}}_\mathrm{i}(\mathrm{t}_\mathrm{k}^\mathrm{i}\mathrm{h})]^\mathrm{T}\bar{\Phi}_\mathrm{i} \\ &\times[\bar{X}_i((t_k^i+j)h)-\bar{X}_i(t_k^ih)]\geq\sigma_i\bar{X}_i^T(t_k^ih)\bar{\Phi}_i\bar{X}_i(t_k^ih)\} \end{aligned} \tag{5.19}$$

According to the network configuration and event-triggered mechanism mentioned in the previous section and considering the reference information to be a part for the controller design, the decentralized controller can be designed to achieve the H_∞ team formation tracking objective in (5.13) of large-scale quadrotor UAVs.

To simplify the design procedure, the decentralized H_∞ team formation tracking strategy in (5.13) for each UAV in the large-scale quadrotor UAVs, UAV team formation NCS can be rewritten as the following decentralized H_∞ stabilization problem of the augmented tracking error system in (5.13):

$$J_\infty^i(\bar{U}_i(t)) = \sup_{\bar{v}_i(t),\, \bar{X}_{-i}(t-\tau_{-i}(t))} \frac{E\{\int_0^{t_f}[\bar{X}_i^T(t)\bar{Q}_i\bar{X}_i(t)]dt - \bar{V}(\bar{X}_i(0))\}}{E\{\int_0^{t_f}\bar{X}_{-i}^T(t-\tau_{-i}(t))\bar{X}_{-i}(t-\tau_{-i}(t)) + \bar{v}_i^T(t)\bar{v}_i(t)dt\}} \leq \rho_i^2 \; for\; i = 1,\cdots,N \tag{5.20}$$

where $\bar{Q}_i = diag\{0_{12\times12}, Q_i\}$, $\bar{V}_i(\bar{X}_i(0))$ is the effect of initial condition that should be excluded in (5.20) and ρ_i denotes the upper bound of $J_\infty{}^i$.

Lemma 5.1 For any matrices (or vectors) M_1 and M_2 with appropriate dimensions, we have:

$$M_1^T M_2 + M_2^T M_1 \leq M_1^T R^{-1} M_2 + M_2^T R M_1$$

where R is any positive definite matrix.

Lemma 5.2 Given a Lyapunov function $V(\bar{X}_i(t))$ which satisfies (i) $V(\bar{X}_i(t)) \geq 0$, (ii) $V(0) = 0$, and (iii) $V(\bar{X}_i(t)) \in C^2$. Then, by Itô formula [46,47], the increment of Lyapunov function $V(\bar{X}_i(t))$ with respect to the ith nonlinear stochastic augmented quadrotor UAV network control subsystem in (5.13) can be derived as:

$$\begin{aligned} dV(\bar{X}_i(t)) &= [(\frac{\partial V(\bar{X}_i(t))}{\partial \bar{X}_i(t)})^T(\bar{f}_i(\bar{X}_i(t)) + \bar{g}_i(\bar{X}_i(t))\bar{U}_i(t) \\ &+ \bar{f}_{-i}(\bar{X}_i(t))\bar{X}_{-i}(t-\tau_{-i}(t)) + H\bar{v}_i(t)) \\ &+ \frac{1}{2}\bar{\vartheta}_i^T(\bar{X}_i(t))(\frac{\partial V^2(\bar{X}_i(t))}{\partial \bar{X}_i^2(t)})\bar{\vartheta}_i(\bar{X}_i(t))]dt + \bar{\vartheta}_i(\bar{X}_i(t))d\bar{W}_i(t) \end{aligned} \tag{5.21}$$

Theorem 5.1 For the decentralized H_∞ team formation tracking control of the ith stochastic nonlinear UAV network control subsystem in (5.13) with event-triggered mechanism, if one could specify a nonlinear tracking controller $\bar{U}_i(t) = \bar{K}_i(\bar{X}_i(t-\eta_i(t)) - \bar{e}_i(t))$ and the event-triggered weighting matrix $\bar{\Phi}_i > 0$ such that the following decoupled HJI has a positive solution $V_i(\bar{X}_i(t)) > 0$ with $V_i(0) = 0$ for each quadrotor UAV:

$$
\begin{aligned}
&\bar{X}_i^T(t)\bar{Q}_i\bar{X}_i(t)+\Big(\frac{\partial V_i(\bar{X}_i(t))}{\partial \bar{X}_i(t)}\Big)^T\bar{f}_i(\bar{X}_i(t))\\
&+\Big(\frac{\partial V_i(\bar{X}_i(t))}{\partial \bar{X}_i(t)}\Big)^T\bar{g}_i(\bar{X}_i(t))\bar{U}_i(t)+\frac{1}{4\rho_i^2}\Big(\Big(\frac{\partial V_i(\bar{X}_i(t))}{\partial \bar{X}_i(t)}\Big)^T\\
&\times\bar{f}_{-i}(\bar{X}_i(t))\bar{f}_{-i}^T(\bar{X}_i(t))\frac{\partial V_i(\bar{X}_i(t))}{\partial \bar{X}_i(t)}+\Big(\frac{\partial V_i(\bar{X}_i(t))}{\partial \bar{X}_i(t)}\Big)^T HH^T\\
&\times\frac{\partial V_i(\bar{X}_i(t))}{\partial \bar{X}_i(t)}\Big)+\frac{1}{2}\bar{\vartheta}_i^T(\bar{X}_i(t))\Big(\frac{\partial V_i^2(\bar{X}_i(t))}{\partial \bar{X}_i^2(t)}\Big)\bar{\vartheta}_i(\bar{X}_i(t))\\
&+\sigma_i\bar{X}_i^T(t-\eta_i(t))\bar{\Phi}_i\bar{X}_i(t-\eta_i(t))\\
&-2\sigma_i\bar{X}_i^T(t-\eta_i(t))\bar{\Phi}_i\bar{e}_i(t)+(\sigma_i-1)\bar{e}_i^T(t)\bar{\Phi}_i\bar{e}_i(t)\le 0,\\
&for\ i=1,\cdots,N
\end{aligned}
\tag{5.22}
$$

then the ith quadrotor UAV networked decentralized control subsystem with a discrete event-triggered mechanism in (5.19) of large-scale quadrotor UAVs team formation NCS with time-varying delay and packet dropout can achieve the decentralized H$_\infty$ robust team formation tracking performance with a prescribed attenuation level ρ_i in (5.13) or (5.20).

Proof. See Appendix 5A.

For the decoupled HJI in (5.22) for the H_∞ decentralized team formation tracking control of large-scale quadrotor UAV NCS, the term $\frac{1}{4\rho_i^2}\Big(\Big(\frac{\partial V_i(\bar{X}_i(t))}{\partial \bar{X}_i(t)}\Big)^T\bar{f}_{-i}(\bar{X}_i(t))\bar{f}_{-i}^T(\bar{X}_i(t))\frac{\partial V_i(\bar{X}_i(t))}{\partial \bar{X}_i(t)}$ is due to decoupling the neighborhood quadrotor UAVs' coupling and the terms with σ_i are due to the effect of event-triggered mechanism of team formation NCS.

5.4 DECENTRALIZED H$_\infty$ FUZZY REFERENCE TRACKING CONTROL DESIGN IN LARGE-SCALE TEAM FORMATION QUADROTOR UAV NCS

In general, it is very difficult to solve the nonlinear partial differential HJI in (5.22) for the robust H$_\infty$ decentralized team formation tracking control problem of each quadrotor UAV NCS. Thus, T-S fuzzy model is employed to interpolate several local stochastic linear systems to approximate the ith augmented system in (5.14) to simplify the H_∞ decentralized team formation control design problem of large-scale quadrotor UAVs NCS in the following section.

5.4.1 T-S Fuzzy Model for Stochastic Large-Scale Quadrotor UAV Team Formation NCS

In this section, the sth Plant Rule of the nonlinear large-scale UAVs quadrotor NCS in (5.14) is constructed as follows [48,49]:

Plant Rule s:

$$\text{If } z_{i,1}(t) \text{ is } F_{s,1}, z_{i,2}(t) \text{ is } F_{s,2}, \cdots, \text{and } z_{i,p}(t) \text{ is } F_{s,p},$$

then,

$$\begin{aligned} d\bar{X}_i(t) &= [\bar{A}_{is}\bar{X}_i(t)) + \bar{B}_{is}\bar{U}_i(t) + \bar{F}_{-is}\bar{X}_{-i}(t-\tau_{-i}(t)) + H\bar{v}_i(t)]dt \\ &+ \bar{C}_{is}\bar{X}_i(t)dW_i(t), s = 1,\ldots,S \end{aligned} \tag{5.23}$$

where,

$$\bar{A}_{is} = \begin{bmatrix} A_r & 0 \\ A_r - A_{is} & A_{is} \end{bmatrix}, \bar{B}_{is} = \begin{bmatrix} 0 \\ -B_{is} \end{bmatrix}$$

$$\bar{C}_{is} = \begin{bmatrix} 0 & 0 \\ C_{is} & -C_{is} \end{bmatrix}, \bar{F}_{-is} = \begin{bmatrix} 0_{12\times 12N_i} \\ F_{-is} \end{bmatrix}$$

where S is the number of the fuzzy rules, $F_{-is} = [\cdots F_{ijs} \cdots] \in \mathbb{R}^{12\times 12N_i}$, $j \in N_i$. $\bar{A}_{is}$, $\bar{B}_{is}$, $\bar{C}_{is}$, and $\bar{F}_{-is}$ are the local linearization matrices of $\bar{f}_i(\bar{X}_i(t))$, $\bar{g}_i(\bar{X}_i(t))$, $\bar{\vartheta}_i(\bar{X}_i(t))$, and $\bar{f}_{ij}(\bar{X}_i(t))$, respectively. $z_{i,1}(t),\cdots,$ $z_{i,p}(t)$ are the premise variables related to the state of the ith quadrotor UAV, $F_{s,1},\cdots,F_{s,p}$ are the fuzzy sets, and p is the number of the premise variable of each quadrotor UAV.

The membership function $M_s(z_i(t))$ is defined as [48]:

$$M_s(z_i(t)) = \prod_{l=1}^{p} F_{s,l}(z_{i,l}(t)) \geq 0, \text{for s = 1,}\cdots\text{S} \tag{5.24}$$

where $F_{s,l}(z_{i,l}(t))$ is the membership grade of $z_{i,l}(t)$ in $F_{s,l.}$ $z_i(t) = [z_{i,1}(t),\cdots z_{i,g}(t)]^T$, and $\sum_{s=1}^{S} M_s(z_i(t)) \geq 0$ for $s = 1,2,\ldots,S$.

The interpolation functions of the ith quadrotor UAV can be inferred as follows [48]:

$$h_s(z_i(t)) = \frac{M_s(z_i(t))}{\sum_{s=1}^{S} M_s(z_i(t))} \geq 0 \tag{5.25}$$

, for $s = 1,2,\ldots,S$, and it is clear that:

$$\sum_{s=1}^{S} h_s(z_i(t)) = 1 \tag{5.26}$$

Based on T-S fuzzy models (5.23)–(5.26), the ith augmented quadrotor UAV team formation NCS with network-induced delays and packet dropouts in (5.14) or (5.13) can be represented as:

$$\begin{aligned} d\bar{X}_i(t) &= \sum_{s=1}^{S} h_s(z_i(t))[\bar{A}_{is}\bar{X}_i(t)) + \bar{B}_{is}\bar{U}_i(t) \\ &+ \bar{F}_{-is}\bar{X}_{-i}(t-\tau_{-i}(t)) + H\bar{v}_i(t)]dt + \bar{C}_{is}\bar{X}_i(t)dW_i(t) \end{aligned} \tag{5.27}$$

Also, the wth rule control rule for the discrete event-triggered mechanism-based controller for the ith team formation tracking quadrotor UAV network control system can be described as follows:

Control Rule w:

$$\text{If } z_{i,1}(t) \text{ is } F_{w,1},\ z_{i,2}(t) \text{ is } F_{w,2},\cdots, \text{ and } z_{i,p}(t) \text{ is } F_{w,p},$$

then,

$$\begin{aligned}\bar{U}_i(t) &= K_{v,iw}X_v(t_kh) + K_{iw}X_{i,e}(t_kh)\\ &= K_{v,iw}(X_v(t-\eta_i(t)) - \tilde{e}_v(t))\\ &\quad + K_{iw}(X_{i,e}(t-\eta_i(t)) - \tilde{e}_i(t))t \in [l_k^i, l_{k+1}^i)\\ &\textit{for } w = 1,\cdots,S\end{aligned} \tag{5.28}$$

where $K_{v,iw}$, $K_{iw} \in \mathbb{R}^{4\times 12}$, $\bar{U}_i(t) \in \mathbb{R}^4$, and the decentralized fuzzy tracking controller for the i th augmented quadrotor UAV in (5.27) is given as follows:

$$\begin{aligned}\bar{U}_i(t) = \sum_{w=1}^{S} h_w(z_i(t_kh))K_{v,iw}(X_v(t-\eta_i(t) - \tilde{e}_v(t))\\ + K_{iw}(X_{i,e}(t-\eta_i(t) - \tilde{e}_i(t)),\ t \in [l_k^{\,i}, l_{k+1}^i)\end{aligned} \tag{5.29}$$

By combining T-S fuzzy system of the ith augmented quadrotor UAV team formation NCS in (5.27) with the decentralized fuzzy controller in (5.29), the i th augmented quadrotor UAV team formation tracking error dynamic system with event-triggered mechanism in (5.14) or (5.13) can be rewritten as follows:

$$\begin{aligned}d\bar{X}_i(t) = \{&\sum_{s=1}^{S} h_s(z_i(t))\sum_{w=1}^{S} h_w(z_i(t_kh))[\bar{A}_{is}\bar{X}_i(t)\\ &+ \bar{B}_{is}\bar{K}_{iw}(\bar{X}_i(t-\eta_i(t)) - \bar{e}_i(t))\\ &+ \bar{F}_{-is}\bar{X}_{-i}(t-\tau_{-i}(t))] + H\bar{v}_i(t)\}dt\\ &+ \bar{C}_{is}\bar{X}_i(t)dW_i(t),\ t \in [l_k^i, l_{k+1}^{\;\,i})\end{aligned} \tag{5.30}$$

where $\bar{K}_{iw} = \left[K_{v,iw}, K_{iw}\right]$.

As a result, the H_∞ decentralized large-scale UAV team formation tracking control design problem can be solved by the decentralized H_∞ stabilizing strategy in (5.20) of the augmented T-S fuzzy tracking error system of the ith quadrotor UAV network control system in (5.30) with the decentralized fuzzy reference tracking control in (5.29).

5.4.2 Decentralized H_∞ Fuzzy Reference Tracking Control Design for Team Formation Control Design of Large-Scale UAVs

By using the T-S fuzzy interpolation method, the decentralized H_∞ tracking controller design for large-scale team formation UAV NCS will be transformed to an equivalent LMIs-constrained problem, which can be easily solved by using MATLAB LMI toolbox. To begin with, the Lyapunov function for the ith augmented UAV tracking error subsystem in (5.14) is chosen as $V_i(\bar{X}_i(t)) = \bar{X}_i^T(t)P_i\bar{X}_i(t)$, where P_i is a positive definite symmetric matrix $P_i = P_i^T > 0$, and we immediately have the following results.

Lemma 5.3 By the fact that $E\{\bar{X}_i(t) - \bar{X}_i(t-\eta_i(t))\} = E\{\int_{t-\eta_i(t)}^{t} d\bar{X}_i(s)ds\}$, the ith augmented stochastic T-S fuzzy quadrotor UAV tracking error subsystem in (5.30) can be rewritten as:

$$
\begin{aligned}
E\{d\bar{X}_i(t)\} = E\{&\sum_{w,s=1}^{S} h_s(z_i(t))h_w(z_i(t_k h)) \\
&\times[(\bar{A}_{is} + \bar{B}_{is}\bar{K}_{iw})\bar{X}_i(t) - \bar{B}_{is}\bar{K}_{iw}\int_{t-\eta_i(t)}^{t} d\bar{X}_i(s)ds - \bar{B}_{is}\bar{K}_{iw}\bar{e}_i(t) \\
&+ \bar{F}_{-is}\bar{X}_{-i}(t-\tau_{-i}(t)) + H\bar{v}_i(t)]dt + \bar{C}_{is}\bar{X}_i(t)dW_i(t)\},
\end{aligned}
$$

Lemma 5.4 By Itô formula in (5.21) and Lemma 5.3, if the Lyapunov function of the ith augmented stochastic T-S fuzzy quadrotor UAV network tracking error subsystem in (5.30) of the large-scale quadrotor UAVs team formation NCS is chosen as $V_i(\bar{X}_i(t)) = \bar{X}_i^T(t)P_i\bar{X}_i(t)$ with the positive matrix P_i, then the corresponding increment can be derived as [51]:

$$
\begin{aligned}
E\{dV_i(\bar{X}_i(t))\} = E\{&\sum_{w,s=1}^{S} h_s(z_i(t))h_w(z_i(t_k h))\{[(\bar{A}_{is} + \bar{B}_{is}\bar{K}_{iw})\bar{X}_i(t) \\
&- \bar{B}_{is}\bar{K}_{iw}\int_{t-\eta_i(t)}^{t} d\bar{X}_i(s)ds - \bar{B}_{is}\bar{K}_{iw}\bar{e}_i(t) \\
&+ \bar{F}_{-is}\bar{X}_{-i}(t-\tau_{-i}(t)) + H\bar{v}_i(t)]^T P_i\bar{X}_i(t) \\
&+ \bar{X}_i^T(t)P_i[(\bar{A}_{is} + \bar{B}_{is}\bar{K}_{iw})\bar{X}_i(t) \\
&- \bar{B}_{is}\bar{K}_{iw}\int_{t-\eta_i(t)}^{t} d\bar{X}_i(s)ds - \bar{B}_{is}\bar{K}_{iw}\bar{e}_i(t) \\
&+ \bar{F}_{-is}\bar{X}_{-i}(t-\tau_{-i}(t)) + H\bar{v}_i(t)] \\
&+ (\bar{C}_{is}\bar{X}_i(t))^T P_i\bar{C}_{is}\bar{X}_i(t)\}dt\}
\end{aligned}
$$

Theorem 5.2 For the stochastic augmented tracking error dynamic of the ith quadrotor UAV in (5.14) with the given scalars $0 < \sigma_i < 1$, $\rho_i > 0$, delay upper bound $\tau_M^i > 0$, the maximum number of packet dropout $\delta_i > 0$ and weighting matrix $Q_i > 0$, if there exist matrices $W_{i1} > 0$, $W_{i2} > 0$, $\breve{\Phi}_{i1} > 0$, $\breve{\Phi}_{i2} > 0$, $Y_{v,iw}$ and Y_{iw} with appropriate dimension as the solution of the following linear matrix inequalities (LMIs):

$$\begin{bmatrix}
\Omega_{i,11}^{s,w} & \Omega_{i,12}^{s,w} & -\tau_M^i W_{i1} & 0 & 0 & 0 & \bar{\delta}_i W_{i1} & 0 & 0 & 0 & 0_{12} & \frac{1}{\rho_i} I & 0 & 0 & W_{i1} C_{is}^T \\
* & \Omega_{i,22}^{s,w} & \Omega_{i,23}^{s,w} & \Omega_{i,24}^{s,w} & \Omega_{i,25}^{s,w} & \Omega_{i,26}^{s,w} & 0 & \bar{\delta}_i W_{i2} & 0 & W_{i2} Q_i^{\frac{1}{2}} & \frac{1}{\rho_i} F_{-is} & \frac{1}{\rho_i} I & -\frac{1}{\rho_i} I & 0 & -W_{i2} C_{is}^T \\
* & * & -2\tau_M^i W_{i1} & 0 & 0 & 0 & \Omega_{i,37}^{s,w} & 0 & 0 & 0 & 0 & 0 & 0 & 0 & 0 \\
* & * & * & -2\tau_M^i W_{i2} & 0 & 0 & 0 & \Omega_{i,48}^{s,w} & 0 & 0 & 0 & 0 & 0 & 0 & 0 \\
* & * & * & * & \Omega_{i,55}^{s,w} & 0 & -\sigma_i \breve{\Phi}_{i1} & 0 & 0 & 0 & 0 & 0 & 0 & 0 & 0 \\
* & * & * & * & * & \Omega_{i,66}^{s,w} & 0 & -\sigma_i \breve{\Phi}_{i2} & 0 & 0 & 0 & 0 & 0 & 0 & 0 \\
* & * & * & * & * & * & \Omega_{i,77}^{s,w} & 0 & 0 & 0 & 0 & 0 & 0 & 0 & 0 \\
* & * & * & * & * & * & * & \Omega_{i,88}^{s,w} & 0 & 0 & 0 & 0 & 0 & 0 & 0 \\
* & * & * & * & * & * & * & * & -I & 0 & 0 & 0 & 0 & 0 & 0 \\
* & * & * & * & * & * & * & * & * & -I & 0 & 0 & 0 & 0 & 0 \\
* & * & * & * & * & * & * & * & * & * & -I & 0 & 0 & 0 & 0 \\
* & * & * & * & * & * & * & * & * & * & * & -I & 0 & 0 & 0 \\
* & * & * & * & * & * & * & * & * & * & * & * & -I & 0 & 0 \\
* & * & * & * & * & * & * & * & * & * & * & * & * & -W_{i1} & 0 \\
* & * & * & * & * & * & * & * & * & * & * & * & * & * & -W_{i2}
\end{bmatrix} \leq 0 \quad (5.31)$$

$$for\ s = 1, 2, \cdots, S.\ w = 1, 2, \cdots, S$$

where,

$$
\begin{aligned}
&\Omega_{i,11}^{s,w} = A_r W_{i1} + W_{i1} A_r^T \\
&\Omega_{i,12}^{s,w} = W_{i1} A_r^T - W_{i1} A_{is}^T - Y_{v,iw}^T B_{is}^T \\
&\Omega_{i,22}^{s,w} = A_{is} W_{i2} + W_{i2} A_{is}^T - B_{is} Y_{iw} - Y_{iw}^{\ T} B_{is}^T \\
&\Omega_{i,23}^{s,w} = B_{is} Y_{v,iw},\ \Omega_{i,24}^{s,w} = B_{is} Y_{iw} - \tau_M^i W_{i2} \\
&\Omega_{i,25}^{s,w} = B_{is} Y_{v,iw},\ \Omega_{i,26}^{s,w} = B_{is} Y_{iw} \\
&\Omega_{i,37}^{s,w} - (\tau_M^i + \bar{\delta}_i) W_{i1}, \Omega_{i,48}^{\ s,w} = -(\tau_M^i + \bar{\delta}_i) W_{i2} \\
&\Omega_{i,55}^{s,w} = (\sigma_i - 1)\breve{\Phi}_{i1},\ \Omega_{i,66}^{s,w} = (\sigma_i - 1)\breve{\Phi}_{i2} \\
&\Omega_{i,77}^{s,w} = \sigma_i \breve{\Phi}_{i1} - 2\bar{\delta}_i W_{i1},\ \Omega_{i,88}^{s,w} = \sigma_i \breve{\Phi}_{i2} - 2\bar{\delta}_i W_{i2}
\end{aligned}
$$

for $s = 1,2,\cdots$ S, $w = 1,2,\cdots$ S, $i = 1,2,\cdots$, N, then the decentralized H$_\infty$ team formation tracking performance in (5.13) of large-scale quadrotor team formation NCS with event-triggered mechanism under time-varying network-induced delay and packet dropout can be guaranteed under an attenuation level ρ_i for any arbitrary reference input, external disturbance and interconnected coupling from other UAVs. Moreover, the weighting matrix of event-triggered mechanism $\bar{\Phi}_i = diag\{\bar{\Phi}_{i1}, \bar{\Phi}_{i2}\}$ can be obtained from $\bar{\Phi}_{i1} = W_{i1}^{-1} \breve{\Phi}_{i1} W_{i1}^{-1}$, $\bar{\Phi}_{i2} = W_{i2}^{-1} \breve{\Phi}_{i2} W_{i2}^{-1}$. The decentralized fuzzy controller gains of each quadrotor UAV in team formation NCS can be obtained from $K_{v,iw} = Y_{v,iw} W_{i1}^{-1}$, $K_{iw} = Y_{iw} W_{i2}^{-1}$, and $\bar{K}_{iw} = \left[K_{v,iw}\ K_{iw}\right]$.

Proof. See Appendix 5B.

Remark 5.5 In general, due to the characteristics of system structure of NCS, the designer may care about the effect of maximum delay τ_M^i and maximum dropout $\bar{\delta}_i$ in the designed control strategy of NCS. Hence, these two factors are considered in the LMI constraints in (5.31). Intuitively, the controller is hard to be designed if these two factors are both large. According to the properties of LMI, the LMI constraints in (5.31) will become infeasible if .. and $\bar{\delta}_i$ are both large, that is, it may not find a set of design variables to make (5.31) hold for large τ_M^i and $\bar{\delta}_i$. As a result, the feasibility of LMIs in (5.31) can be regarded as a tradeoff between the maximum delay τ_M^i and maximum dropout $\bar{\delta}_i$ in the design procedure of the decentralized H_∞ robust team formation tracking control of large-scale quadrotor NCS in Theorem 5.2. Especially, to increase the feasibility and reduce the computational complexity in solving LMIs, these two factors will be given beforehand by the designer while solving the LMIs in (5.31).

To obtain a better robust tracking performance for the large-scale quadrotor UAV team formation NCS, the optimal H_∞ decentralized team formation tracking control design problem can be formulated as the following N independent minimization problems for each quadrotor UAV:

$$
\begin{aligned}
&\min_{\{W_{i1}, W_{i2}, \breve{\Phi}_{i1}, \breve{\Phi}_{i2}, Y_{v,iw}, Y_{iw}\}} \rho_i^2 \\
&s.t.\ W_{i1} > 0, W_{i2} > 0, \breve{\Phi}_{i1} > 0, \breve{\Phi}_{i2} > 0 \text{ and } (5.31) \\
&for\ i = 1,\cdots,N
\end{aligned}
\tag{5.32}
$$

Based on the above design analysis, a design procedure is proposed for the optimal H_∞ decentralized robust virtual leader-follower team formation tracking control of large-scale quadrotor UAV formation team NCS in the following:

Step 1. Construct the virtual leader dynamic system in (5.12) to specify the trajectory of virtual-leader with the desired reference path $x_r(t)$.

Step 2. Set up the shifted distance d_i of each quadrotor UAV subsystem of team formation NCS with respect to the virtual leader to form a specified formation of large-scale UAVs team formation NCS in (5.11). According to the actual network environment to choose a maximum number of successive packet dropouts $\bar{\delta}_i$ and event-triggered level σ_i.

Step 3. Augment the virtual leader dynamics (5.12) and the ith UAV network tracking error subsystem (5.14).

Step 4. Construct the T-S fuzzy model of each quadrotor UAV subsystem of team formation NCS in (5.30) to approximate the nonlinear decentralized network team formation tracking subsystem of team formation NCS in (5.18) by interpolating several local linearized systems.

Step 5. Based on the local linearized matrices by T-S fuzzy method, the LMI-constrained optimization problem in (5.32) is constructed for the optimal robust decentralized controller design embedded with event-triggered mechanism for large-scale quadrotor UAVs team formation NCS with time-varying network-induced delay and packet dropout.

Step 6. Solve N LMI-constrained optimization problems in (5.32) to obtain W_{i1}, W_{i2}, $Y_{v,iw}$, Y_{iw}, $\breve{\Phi}_{i1}$, $\breve{\Phi}_{i2}$ for decentralized fuzzy control gains and weighting matrix of event-triggered mechanism by $K_{v,iw} = Y_{v,iw} W_{i1}^{-1}$, $K_{iw} = Y_{iw} W_{i2}^{-1}$ in (5.29) and $\bar{\Phi}_i = diag\{\bar{\Phi}_{i1}, \bar{\Phi}_{i2}\}$ from $\bar{\Phi}_{i1} = W_{i1}^{-1} \breve{\Phi}_{i1} W_{i1}^{-1}$, $\bar{\Phi}_{i2} = W_{i2}^{-1} \breve{\Phi}_{i2} W_{i2}^{-1}$ for $w = 1,2,\cdots,S$ and $i = 1,2,\cdots,N$.

Step 7. Construct the decentralized control law $U_i(t)$ in (5.29) for each quadrotor UAV subsystem in the large-scale UAV team formation NCS.

5.5 DESIGN EXAMPLE

To illustrate the design procedure and confirm the effectiveness of the proposed H_∞ decentralized robust controller for large-scale quadrotor UAVs team formation NCS, a 25-quadrotor UAV team formation task is provided in this section under network configuration which is almost impossible by the conventional centralized controller. For the practical network environment, time-varying network-induced delay and packet dropout in network configuration are considered in this simulation, and the advantage of the discrete event-triggered mechanism is also discussed.

5.5.1 Large-scale Quadrotor UAV Team Formation NCS Model Setting

In this example, we consider a 25-quadrotor UAV team formation NCS. Since the decentralized controller is applied, it could be easily extended to a team with very large-scale quadrotor UAVs.

The parameters of each stochastic tracking error dynamic of the quadrotor UAV subsystem in (5.2) for the 25-quadrotor UAV formation NCS are specified as follows:

$$K_x^i = K_y^i = K_z^i = 0.01,\ K_\phi^i = K_\theta^i = K_\psi^i = 0.13$$
$$J_x^i = J_y^i = 1.25,\ J_z^i = 2.2,\ m^i = 3, g = 9.81$$
$$i = 1,\cdots,25$$

The Wiener process is used to model the system parameter variation of the quadrotor UAV during flight process caused by disturbed component characteristics in the quadrotor UAV team formation NCS. We set 25 independent Wiener processes $\{W_i(t)\}_{i=1}^{25}$ for each UAV in the stochastic 25-quadrotor UAV team formation NCS. The wind gust is considered as an external disturbance of quadrotor UAV, which is set to be zero-mean Gaussian noise with unit variance in each quadrotor UAV.

To construct the T-S fuzzy model of each quadrotor UAV team formation NCS in (5.11), the premise variables and their operation points of T-S fuzzy model are given in Table 5.1. The grade of the membership function is selected as a trapezoidal function.

TABLE 5.1
The Premise Variables and Their Operation Points of T-S Fuzzy Method

Premise Variables	Operation Points
$z_{i,1} = -g / \psi_1^i$	$-6.4452, -6.2452, -6.0452$
$z_{i,2} = \dfrac{J_y^i - J_z^i}{J_x^i}\theta_1^i$	$-0.07, 0.001, 0.07$
$z_{i,3} = \dfrac{J_z^i - J_x^i}{J_y^i}\phi_1^i$	$-0.07, 0.001, 0.07$
$z_{i,4} = \begin{matrix}(\cos\phi_1^i \sin\theta_1^i \cos\psi \\ +\sin\phi_1^i \sin\psi_1^i)\dfrac{1}{m^i}\end{matrix}$	$-0.03, 0.001, 0.03$
$z_{i,5} = \begin{matrix}(\cos\phi_1^i \sin\theta_1^i \sin\psi_1^i \\ -\sin\phi_1^i \sin\psi_1^i)\dfrac{1}{m^i}\end{matrix}$	$-0.03, 0.001, 0.03$
$z_{i,6} = \begin{matrix}(\cos\phi_1^i \sin\theta_1^i \sin\psi_1^i \\ -\sin\phi_1^i \sin\psi_1^i)\dfrac{1}{m^i}\end{matrix}$	$0.3331, 0.3334$
$z_{i,7} = 0.01x_2^i$	$-30, 30$
$z_{i,8} = 0.01z_2^i$	$10, 100$

Based on the above setting, each quadrotor UAV in the 25-quadrotor UAV team formation NCS has 1,944 fuzzy rules, respectively. The qth fuzzy model of the ith UAV tracking error dynamic system in (5.30) is described as follows:

$$\begin{aligned} d\bar{X}_i(t) &= [\bar{A}_{iq}\bar{X}_i(t)+\bar{B}_{iq}\bar{K}_{iq}(\bar{X}_i(t-\eta_i(t))-\bar{e}_i(t)) \\ &+\bar{F}_{-iq}\bar{X}_{-i}(t-\tau_{-i}(t))+H\bar{v}_i(t)]dt \\ &+\bar{C}_{iq}\bar{X}_i(t)dW_i(t),\ t\in[l_k^i, l_{k+1}{}^i) \end{aligned}$$

For the 25-quadrotor UAV team formation tracking control problem, the position tracking performance is more important than the velocity tracking control performance. Consequently, the weighting on the position tracking error should be greater than the weighting on the velocity tracking error. The weighting matrices are given for every single networked tracking error of each quadrotor UAV subsystem as follows:

$$Q_i = diag\{1,0.01,1,0.01,1,0.01,1,0.01,1,0.01,1,0.01\}\ for\ i=1,\cdots,25$$

and the event-triggered level σ_i is given as 10^{-5} for each quadrotor UAV in this design example.

Due to the great amount of quadrotor UAVs among 25-quadrotor UAV team formation NCS, several interconnected coupling effects will emerge such as co-channel interference in communication and trailing vortex by aerodynamics in close formation. In this example, the coupling terms are constructed to model the state variation caused by aerodynamic forces induced by a trailing vortex. In the actual flight process, the trailing dynamics of UAV is affected by the leading quadrotor UAVs' trailing vortex. According to the characteristics of the trailing vortex in which the strength of the trailing vortex will be proportional to the movement speed in the related direction of the ahead quadrotor UAV [32], the trailing vortex effect is modeled in the following to approximate the trailing vortex with respect to the specified formation shape and flight path. In this simulation case

$$\begin{aligned} &f_{ij}(X_i(t))X_j(t-\tau_{ij}(t)) = [0, x_2^i(t)\times 0.01x_2^j(t-0.1),0,0,0, \\ &z_2^i(t)\times 0.01z_2^j(t-0.1),0,0,0,0,0,0]^T,\ j\in N_i \end{aligned} \tag{5.33}$$

where the leading UAVs that make the i th quadrotor UAV suffer the trailing vortex effect from neighboring quadrotor UAVs are in N_i.

The flight path of 25-quadrotor UAV team formation NCS is based on the trajectory of virtual leader which is constructed by $r(t)$ with $A_v=-I_{12\times 12}$ in (5.12). Every single quadrotor UAV will hold a certain distance in Table 5A.1 (in Appendix 5C) from the trajectory of the virtual leader $X_v(t)$ during the flying process, as shown in Figure 5.3. In this design example, we construct the reference path of virtual leader

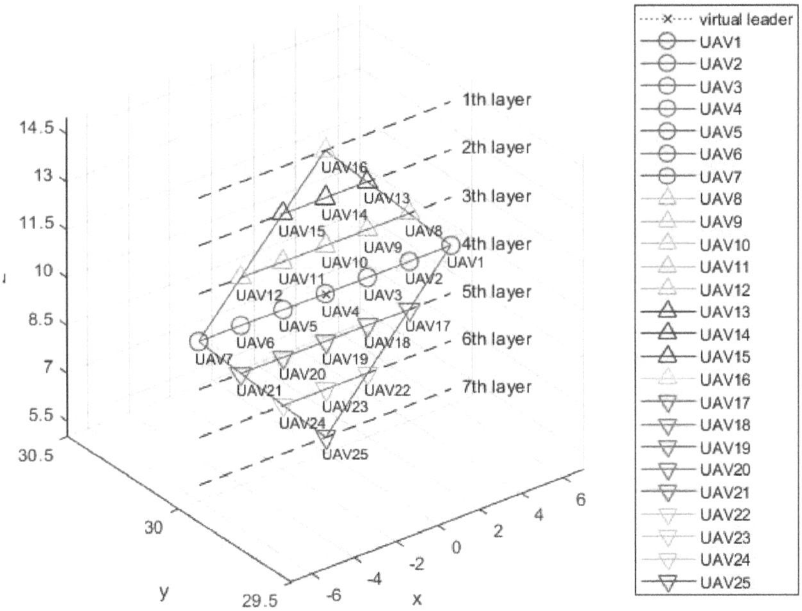

FIGURE 5.3 The formation of 25-UAV team.

as a circle with a radius of 30 m in the *x-y* plane, rising at a constant velocity on the z-axis as (5.34). The reference input vector in (5.12) is given as follows:

$$r(t) = [x_r(t), \dot{x}_r(t), y_r(t), \dot{y}_r(t), z_r(t), \dot{z}_r(t), \phi_r(t), \dot{\phi}_r(t),\\ \theta_r(t), \dot{\theta}_r(t), \psi_r(t), \dot{\psi}_r(t)]^T$$

where reference position and attitude are generated as the following according to the planned mission:

$$\begin{aligned} & x_r(t) = 30\sin(0.1t),\ y_r(t) = 30\cos(0.1t) \\ & z_r(t) = 10 + 0.5t,\ \psi_r(t) = \pi/2 \\ & \phi_r = \arcsin\left(\frac{\dot{x}_r \sin(\psi_r) - \dot{y}_r \cos(\psi_r)}{(\dot{x}_r^2 + \dot{y}_r^2 + (\dot{z}_r + g)^2)^{\frac{1}{2}}}\right) \\ & \theta_r = \arctan\left(\frac{\dot{x}_r \cos(\psi_r) + \dot{y}_r \sin(\psi_r)}{\dot{z}_r + g}\right) \end{aligned} \tag{5.34}$$

To make the flying process easier to be viewed in the 3D flying process, the architecture of 25-quadrotor UAV formation is constructed as a diamond shape in the *y-z* plane from top to bottom as shown in Figure 5.3.

According to the actual network environment, the upper bound of time-varying network-induced delays $\tau^i_{M\_}$ is set to be 8 milliseconds (5ms) and the maximum number of packet dropouts δ_i is set to be within 2 successive packet dropouts with the successfully transmitted probability 0.99 for each network tracking control subsystem in this simulation. Solving 25 LMIs-constrained optimization problems in (5.32), the decentralized tracking controller for each UAV in the 25-UAV team formation NCS can be obtained to achieve the optimal H_∞ team formation tracking performance to suppress the effect of external disturbance and interconnected coupling on the 25-quadrotor UAV team formation NCS. The average optimal disturbance and coupling attenuation level of the 25-quadrotor UAV team formation NCS is obtained as $\rho^* = 2.3112$, and the corresponding fuzzy controller is given in (5.29). At the same time, the event-triggered weighting matrix of each quadrotor UAV team formation NCS is also obtained under the given event-triggered level. The 3-D trajectories of 25-quadrotor UAV team formation flight process are shown in Figure 5.4 with a discrete event-triggered mechanism under time-varying network-induced delay and packet dropout. The trajectory of each quadrotor UAV converges to the desired reference $X_v(t)$ of the virtual leader with a shifted distance quickly in the 25-quadrotor UAV team formation NCS with the proposed decentralized robust H_∞ team formation tracking controller. During the flight process, the entire 25-quadrotor UAV NCS maintains the formation we have specified in Figure 5.3 quite well. To make the flight process easier to be observed in a 3D graph, the markers on the Figure 5.4 represent 25 quadrotor UAVs for every other 10 s. Furthermore, a zoom-in 3D graph is displayed in Figure 5.5 when 25 quadrotor

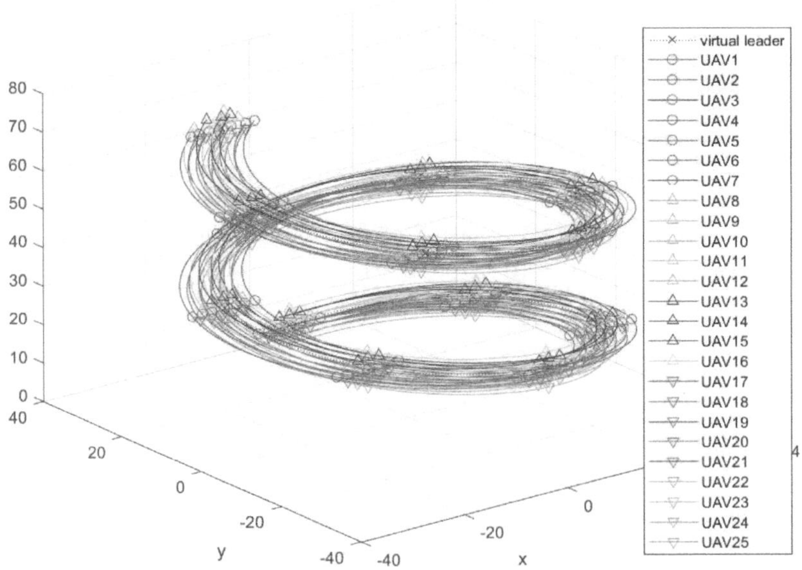

FIGURE 5.4 The flight process of 25-quadrotor UAV team formation NCS

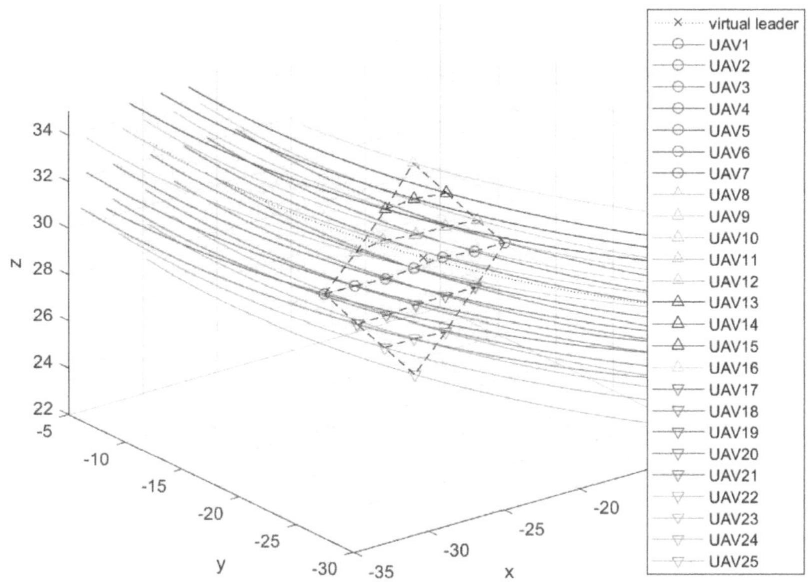

FIGURE 5.5 The zoom in 3D graph at $t = 50$ s.

UAVs are flying at the 50 s. The position and velocity trajectories of every quadrotor UAV subsystems and virtual leader of 25-quadrotor UAV formation team NCS are shown in Figures 5.A1–5.A6 in Appendix 5C, and attitude trajectories are also shown in Figures 5.A7–5.A12 in Appendix 5C as well.

5.5.2 Event-Triggered Performance of NCS

To reduce the network burden and energy consumption, the discrete event-triggered mechanism is adopted. As a result, the number of transmitted signals will be reduced to achieve a more effective bandwidth saving. In the discrete event-triggered mechanism, the sampled period h is set as 0.01 s in this simulation, and the step-size of Runge-Kutta is set to be 0.002 for each continuous quadrotor UAV subsystem. Saving rate is defined as the performance of the event-triggered mechanism, which represents how much signal is transmitted by comparing the number of original sampled signals with triggered signals as follows:

$$\text{saving rate} = \frac{\text{triggered signals}}{\text{sampled signals}}$$

To illustrate the performance of the event-triggered mechanism, the saving rate of the event-triggered mechanism of each UAV in the 25-quadrotor UAV team formation NCS is provided in Table 5.2. The average saving rate for the large-scale quadrotor UAV NCS is 0.1685, which means it can reduce 83.15% power to the transmitted signal

TABLE 5.2
Saving Rate of Each Quadrotor UAV

The saving rate	UAV1	UAV2	UAV3	UAV4	UAV5
	0.1638	0.1657	0.1678	0.1688	0.1705
The saving rate	UAV6	UAV7	UAV8	UAV9	UAV10
	0.1697	0.1705	0.1620	0.1644	0.1645
The saving rate	UAV11	UAV12	UAV13	UAV14	UAV15
	0.1658	0.1659	0.1603	0.1614	0.1621
The saving rate	UAV16	UAV17	UAV18	UAV19	UAV20
	0.1580	0.1690	0.1717	0.1737	0.1730
The saving rate	UAV21	UAV22	UAV23	UAV24	UAV25
	0.1728	0.1766	0.1765	0.1762	0.1813

as well as transmission bandwidth in the NCS. In addition, the release instants and time intervals of UAV1 based on event-triggered mechanism are shown in Figure 5.A13, and the decision process with event-triggered condition in (5.3) is shown in Figure 5.A14. To view the precise event-triggered process, the zoom-in figures are depicted in Figures 5.A15–5.A16.

5.5.3 Time-Varying Network-Induced Delays in NCS

Considering the related applications of the fifth generation (5G) mobile network, the requirement of latency (end-to-end round-trip delay) is less than 1 millisecond (ms). In this simulation, we suppose the upper bound of delay is 8 ms, which can withstand an additional 7 ms delay compared with the original requirement under 5G communication. Figure 5.A17 shows the phenomenon of delay of quadrotor UAV1 in this simulation environment.

5.5.4 Packet Dropouts

Considering the requirement of packet dropout rate in the 5G network, the packet dropout rate is lower than 10^{-4} without the guaranteed quality, and even lower than 10^{-7} with the guaranteed quality. We suppose the probability of packets successfully transmitted through the network is 0.99, and the maximum number of successive lost packet is 2. The occurrence of packet dropout of each quadrotor UAV by the proposed decentralized H_∞ team formation tracking control strategy is shown in Figure 5.A18, and the actual numbers of packet dropout and transmitted signals strategy are in Table 5A.2.

5.6 CONCLUSION

In this chapter, a novel decentralized reference tracking control scheme for large-scale quadrotor UAVs team formation NCS is proposed to overcome the interconnected coupling and the non-scalable or infeasible difficulty due to high computation complexity by the conventional centralized control. Under the network configuration, a discrete event-triggered mechanism is embedded in the decentralized H_∞ team formation tracking controller design scheme to save the network traffic bandwidth especially for the large-scale quadrotor UAV team formation NCS case. For the actual flight process, the wind disturbance and intrinsic fluctuation caused by nonideal component characteristics in the quadrotor UAV team formation NCS are also considered. Furthermore, the coupling effect induced by other quadrotor UAVs in large-scale UAV team formation NCS is taken into account too. To simplify the design procedure, an augmented tracking error quadrotor UAV subsystem is constructed so that the H_∞ decentralized team formation UAV tracking design problem of large-scale UAV NCS could be transformed to an H_∞ stabilization problem of each UAV-augmented tracking error subsystem. Next, the T-S fuzzy method is applied such that the difficult HJI-constrained optimization problem for each decentralized optimal H_∞ tracking control problem of UAV team formation NCS can be transformed into an LMIs-constrained optimization problem, which can be solved efficiently by MATLAB LMI toolbox. In the simulation example, a 25-quadrotor UAV team formation is constructed with a diamond formation shape in team formation NCS under time-varying network-induced delays and packet dropouts. The proposed robust H_∞ decentralized tracking controller embedded with event-triggered mechanism can not only have good team formation tracking performance for the large-scale quadrotor UAV NCS but also lead to a scalable and feasible team formation control scheme to overcome an infeasible dilemma of centralized control and maintain a good network service at the same time when the number of UAVs is increased or decreased. In the future, to strengthen the flexibility of the team formation, the desired fixed formation distance of each quadrotor UAV will be changed to time-varying formation distance $d_i(t)$. In this case, the team formation of multi-quadrotor UAVs NCS can be employed to achieve more complicated tasks. Since $d_i(t)$ and $\dot{d}_i(t)$ will appear in the T-S fuzzy system in (5.13), more efforts are needed for the robust decentralized H_∞ team formation tracking control design of large-scale quadrotor UAVs NCS. As a result, the formation tracking control of multi-quadrotor UAV NCS with time-vary formation distance will be addressed in the following chapters.

Appendix 5A
Proof of Theorem 5.1

From the numerator of decentralized H_∞ team formation tracking performance of the i th UAV NCS subsystem in (5.20), we get:

$$\begin{aligned}E\{\textstyle\int_0^{t^f}[\bar{X}_i^T(t)\bar{Q}_i\bar{X}_i(t)]dt &= E\{V_i(\bar{X}_i(0))-V_i(\bar{X}_i(t_f)\} \\ &\quad + E\{\textstyle\int_0^{t^f}[\bar{X}_i^T(t)\bar{Q}_i\bar{X}_i(t)]dt + dV_i(\bar{X}_i(t))\}\end{aligned} \tag{5.35}$$

By Lemma 5.2, and the fact $E\{d\bar{W}_i(t)\}=0$, we get

$$\begin{aligned}&E\{\textstyle\int_0^{t^f}[\bar{X}_i^T(t)\bar{Q}_i\bar{X}_i(t)]dt\} \\ &\leq E\{V_i(\bar{X}_i(0))\} + E\{\textstyle\int_0^{t^f}[\bar{X}_i^T(t)\bar{Q}_i\bar{X}_i(t) \\ &\quad + (\frac{\partial V_i(\bar{X}_i(t))}{\partial \bar{X}_i(t)})^T(\bar{f}_i(\bar{X}_i(t)) + \bar{g}_i(\bar{X}_i(t))\bar{U}_i(t) \\ &\quad + \bar{f}_{-i}(\bar{X}_i(t))\bar{X}_{-i}(t-\tau_{-i}(t)) \\ &\quad + H\bar{v}_i(t) + \frac{1}{2}\bar{\vartheta}_i^T(\bar{X}_i(t))(\frac{\partial V_i^2(\bar{X}_i(t))}{\partial \bar{X}_i^2(t)})\bar{\vartheta}_i(\bar{X}_i(t)))]dt\}\end{aligned} \tag{5.36}$$

By using Lemma 5.1, the following two inequalities hold:

$$\begin{aligned}&(\frac{\partial V_i(\bar{X}_i(t))}{\partial \bar{X}_i(t)})^T H\bar{v}_i(t) \\ &\leq \frac{1}{4\rho_i^2}(\frac{\partial V_i(\bar{X}_i(t))}{\partial \bar{X}_i(t)})^T HH^T \frac{\partial V_i(\bar{X}_i(t))}{\partial \bar{X}_i(t)} + \rho_i^2\bar{v}_i^T(t)\bar{v}_i(t)\end{aligned} \tag{5.37}$$

and

$$\begin{aligned}&(\frac{\partial V_i(\bar{X}_i(t))}{\partial \bar{X}_i(t)})^T \bar{f}_{-i}(\bar{X}_i(t))\bar{X}_{-i}(t-\tau_{-i}(t)) \\ &\leq \frac{1}{4\rho_i^2}(\frac{\partial V_i(\bar{X}_i(t))}{\partial \bar{X}_i(t)})^T \bar{f}_{-i}(\bar{X}_i(t))\bar{f}_{-i}^T(\bar{X}_i(t))\frac{\partial V_i(\bar{X}_i(t))}{\partial \bar{X}_i(t)} \\ &\quad + \rho_i^2\bar{X}_{-i}^T(t-\tau_{-i}(t))\bar{X}_{-i}(t-\tau_{-i}(t))\end{aligned} \tag{5.38}$$

with $\rho_i > 0$. According to the event-triggered mechanism, the control signal will only be triggered when the condition in (5.19) is violated. The states of the ith quadrotor

UAV decentralized team formation tracking control system between the two transmitted sequential instants must satisfy the following inequality:

$$\begin{aligned}&E\{\bar{X}_i^T(t-\eta_i(t))\sigma_i\bar{\Phi}_i\bar{X}_i(t-\eta_i(t))\\&-\bar{X}_i^T(t-\eta_i(t))\sigma_i\bar{\Phi}_i\bar{e}_i(t)-\bar{e}_i^T(t)\sigma_i\bar{\Phi}_i\bar{X}_i(t-\eta_i(t))\\&+\sigma_i\bar{e}_i^T(t)\bar{\Phi}_i\bar{e}_i(t)-\bar{e}_i^T(t)\bar{\Phi}_i\bar{e}_i(t)\}\geq 0,\ t\in[l_k^{\,i},l_{k+1}^i)\end{aligned} \tag{5.39}$$

By (5.37), (5.38) and the requirement of event-triggered mechanism in (5.39), we obtain the following inequality:

$$\begin{aligned}&E\{\textstyle\int_0^{t_f}[\bar{X}_i^T(t)\bar{Q}_i\bar{X}_i(t)]dt\}\\&\leq E\{V_i(\bar{X}_i(0))\}E\{\textstyle\int_0^{t_f}[\bar{X}_i^{\,T}(t)\bar{Q}_i\bar{X}_i(t)\\&+(\frac{\partial V_i(\bar{X}_i(t))}{\partial\bar{X}_i(t)})^T\bar{f}_i(\bar{X}_i(t))+(\frac{\partial V_i(\bar{X}_i(t))}{\partial\bar{X}_i(t)})^T\bar{g}_i(\bar{X}_i(t))\bar{U}_i(t)\\&+\frac{1}{4\rho_i^2}((\frac{\partial V_i(\bar{X}_i(t))}{\partial\bar{X}_i(t)})^T\bar{f}_{-i}(\bar{X}_i(t))\bar{f}_{-i}^T(\bar{X}_i(t))\frac{\partial V_i(\bar{X}_i(t))}{\partial\bar{X}_i(t)}\\&+(\frac{\partial V_i(\bar{X}_i(t))}{\partial\bar{X}_i(t)})^T HH^T\frac{\partial V_i(\bar{X}_i(t))}{\partial\bar{X}_i(t)})+\sigma_i\bar{X}_i^T(t-\eta_i(t))\bar{\Phi}_i\bar{X}_i(t-\eta_i(t))\\&-2\sigma_i\bar{X}_i^T(t-\eta_i(t))\bar{\Phi}_i\bar{e}_i(t)+(\sigma_i-1)\bar{e}_i^T(t)\bar{\Phi}_i\bar{e}_i(t)\\&+\rho_i^2(\bar{X}_{-i}^T(t-\tau_{-i}(t))\bar{X}_{-i}(t-\tau_{-i}(t))+\bar{\upsilon}_i^T(t)\bar{\upsilon}_i(t))\\&+\frac{1}{2}\bar{\vartheta}_i^T(\bar{X}_i(t))(\frac{\partial V^2(\bar{X}_i(t))}{\partial\bar{X}_i^2(t)})\bar{\vartheta}_i(\bar{X}_i(t))]dt\}\end{aligned} \tag{5.40}$$

As a result, if the HJI in (5.22) is satisfied, the above inequality can be reduced to:

$$\begin{aligned}&E\{\textstyle\int_0^{t_f}[\bar{X}_i^T(t)\bar{Q}_i\bar{X}_i(t)]dt\}\leq E\{\textstyle\int_0^{t_f}\rho_i^2[\,\bar{\upsilon}_i^T(t)\bar{\upsilon}_i(t))+\bar{X}_{-i}^T(t-\tau_{-i}(t))\\&+\bar{X}_{-i}(t-\tau_{-i}(t))]dt\}+V_i(\bar{X}_i(0))\end{aligned} \tag{5.41}$$

then the H_∞ decentralized team formation tracking performance of the large-scale UAV NCS in (5.20) is guaranteed. The proof is completed.

Appendix 5B
Proof of Theorem 5.2

With the Lyapunov function and Lemma 5.4, we have:

$$
\begin{aligned}
&E\{\int_0^{t^f} \bar{X}_i^T(t)\bar{Q}_i\bar{X}_i(t)dt\} \\
&= E\{\bar{V}_i(\bar{X}_i(0)) - \bar{V}_i(\bar{X}_i(t_f))\} + E\{\int_0^{t^f}[\bar{X}_i^T(t)\bar{Q}_i\bar{X}_i(t)dt + dV(\bar{X}_i(t))]\} \\
&\le E\{\bar{V}_i(\bar{X}_i(0)) + E\{\int_0^{t^f}[\sum_{k=0}^{\infty}\sum_{w,s=1}^{S} h_w(z_i(t_k h)) \\
&\times h_s(z_i(t))\Theta_k(t)\{\bar{X}_i^T(t)[\bar{Q}_i + P_i(\bar{A}_{is} + \bar{B}_{is}\bar{K}_{iw}) \\
&+ (\bar{A}_{is} + \bar{B}_{is}\bar{K}_{iw})^T P_i]\bar{X}_i(t) - [\bar{B}_{is}\bar{K}_{iw}\int_{t-\eta_i(t)}^{t} d\bar{X}_i(s)ds]^T P_i\bar{X}_i(t) \\
&- \bar{X}_i^T(t)P_i[B_{is}\bar{K}_{iw}\int_{t-\eta_i(t)}^{t} d\bar{X}_i(s)ds] - [\bar{B}_{is}\bar{K}_{iw}\bar{e}_i(t)]^T P_i\bar{X}_i(t) \\
&- \bar{X}_i^T(t)P_i[\bar{B}_{is}\bar{K}_{iw}\bar{e}_i(t)] + [\bar{F}_{-is}\bar{X}_{-i}(t-\tau_{-i}(t))]^T P_i\bar{X}_i(t) \\
&+ \bar{X}_i^T(t)P_i[\bar{F}_{-is}\bar{X}_{-i}(t-\tau_{-i}(t))] + [H\bar{\upsilon}_i(t)]^T P_i\bar{X}_i(t) \\
&+ \bar{X}_i^T(t)P_i[H\bar{\upsilon}_i(t)] + (\bar{C}_{is}\bar{X}_i(t))^T P_i\bar{C}_{is}\bar{X}_i(t)\}dt\}
\end{aligned}
\tag{5.42}
$$

where:

$$
\Theta_k(t) = \begin{Bmatrix} 1, \ t \in [l_k^i, l_{k+1}^i) \\ 0, \ \text{o.w.} \end{Bmatrix}
$$

By Lemma 5.1, the terms in (5.42) can be bounded as follows:

$$
\begin{aligned}
&[\bar{F}_{-is}\bar{X}_{-i}(t-\tau_{-i}(t))]^T P_i\bar{X}_i(t) + \bar{X}_i^T(t)P_i[\bar{F}_{-is}\bar{X}_{-i}(t-\tau_{-i}(t))] \\
&\le \rho_i^2\bar{X}_{-i}^T(t-\tau_{-i}(t))\bar{X}_{-i}(t-\tau_{-i}(t)) + \frac{1}{\rho_i^2}\bar{X}_i^T(t)P_i\bar{F}_{-is}\bar{F}_{-is}{}^T P_i\bar{X}_i(t)
\end{aligned}
\tag{5.43}
$$

and

$$
\begin{aligned}
&[H\bar{\upsilon}_i(t)]^T P_i\bar{X}_i(t) + \bar{X}_i^T(t)P_i[H\bar{\upsilon}_i(t)] \le \rho_i^2\bar{\upsilon}_i^T(t)\bar{\upsilon}_i(t) \\
&+ \frac{1}{\rho_i^2}\bar{X}_i^T(t)P_iHH^T P_i\bar{X}_i(t)
\end{aligned}
\tag{5.44}
$$

with scalar $\rho_i > 0$. Then we obtain the following inequality from (5.42) to (5.44):

$$
\begin{aligned}
& E\{\int_0^{t_f} \bar{X}_i^T(t)\bar{Q}_i\bar{X}_i(t)dt\} \\
& \leq E\{V(\bar{X}_i(0)) + E\{\int_0^{t_f} \sum_{k=0}^{\infty}\sum_{s,w=1}^{S} h_s(z_i(t))h_w(z_i(t_k h)) \\
& \times \Theta_k(t)\{\bar{X}_i^T(t)[\bar{Q}_i + P_i(\bar{A}_{is} + \bar{B}_{is}\bar{K}_{iw}) + (\bar{A}_{is} + \bar{B}_{is}\bar{K}_{iw})^T P_i]\bar{X}_i(t) \\
& -[\bar{B}_{is}\bar{K}_{iw}\int_{t-\eta_i(t)}^{t} d\bar{X}_i(s)ds]^T P_i\bar{X}_i(t) - \bar{X}_i^T(t)P_i[\bar{B}_{is}\bar{K}_{iw}\int_{t-\eta_i(t)}^{t} d\bar{X}_i(s)ds] \\
& -[\bar{B}_{is}\bar{K}_{iw}\bar{e}_i(t)]^T P_i\bar{X}_i(t) - \bar{X}_i^T(t)P_i[\bar{B}_{is}\bar{K}_{iw}\bar{e}_i(t)] \\
& +\rho_i^2\bar{X}_{-i}^T(t-\tau_{-i}(t))\bar{X}_{-i}(t-\tau_{-i}(t)) + \frac{1}{\rho_i^2}\bar{X}_i^T(t)P_i\bar{F}_{-is}\bar{F}_{-is}{}^T P_i\bar{X}_i(t) \\
& +\rho_i^2\bar{v}_i^T(t)\bar{v}_i(t) + \frac{1}{\rho_i^2}\bar{X}_i^T(t)P_iHH^T P_i\bar{X}_i(t) + (\bar{C}_{is}\bar{X}_i(t))^T P_i\bar{C}_{is}\bar{X}_i(t)\}dt\}
\end{aligned}
\tag{5.45}
$$

By adopting the event-triggered mechanism, the energy of the ith quadrotor UAV state before the next transmitted instant must additionally meet the following requirement:

$$
\begin{aligned}
& E\{\sigma_i\bar{X}_i^T(t-\eta_i(t))\bar{\Phi}_i\bar{X}_i(t-\eta_i(t)) - \sigma_i\bar{X}_i^T(t-\eta_i(t))\bar{\Phi}_i\bar{e}_i(t) \\
& -\sigma_i\bar{e}_i^T(t)\bar{\Phi}_i^T\bar{X}_i(t-\eta_i(t)) + (\sigma_i - 1)\bar{e}_i^T(t)\bar{\Phi}_i\bar{e}_i(t)\} \geq 0,\ t \in [l_k^i, l_{k+1}^i)
\end{aligned}
\tag{5.46}
$$

By the Newton-Leibniz formula in Reference [50] for the network tracking error of the ith quadrotor UAV subsystem in (5.13), the following equation is satisfied:

$$
\begin{aligned}
& 2[\bar{X}_i(t) - \bar{X}_i(t-\eta_i(t)) - \int_{t-\eta_i(t)}^{t} d\bar{X}_i(s)ds]^T \\
& \times P_i[\bar{\delta}_i\bar{X}_i(t-\eta_i(t)) + \tau_M^i\int_{t-\eta_i(t)}^{t} d\bar{X}_i(s)ds] = 0
\end{aligned}
\tag{5.47}
$$

By (5.46) and (5.47), (5.45) can be rewritten as the following inequality to satisfy the requirement of event-triggered mechanism and meet the definite property of LMI,

$$
\begin{aligned}
& E\{\int_0^{t_f} \bar{X}_i^T(t)\bar{Q}_i\bar{X}_i(t)dt\} \\
& \leq E\{\bar{V}_i(\bar{X}_i(0))\}E\{\int_0^{t_f} [\sum_{k=1}^{\infty}\sum_{w,s=1}^{S} h_s(z_i(t))h_w(z_i(t_k h)) \\
& \times \Theta_k(t)\{\bar{X}_i^T(t)[\bar{Q}_i + P_i(\bar{A}_{is} + \bar{B}_{is}\bar{K}_{iw})
\end{aligned}
$$

$$
\begin{aligned}
&+(\bar{A}_{is}+\bar{B}_{is}\bar{K}_{iw})^T P_i]\bar{X}_i(t)-[\bar{B}_{is}\bar{K}_{iw}\int_{t-\eta_i(t)}^{t} d\bar{X}_i(s)ds]^T P_i\bar{X}_i(t)\\
&-\bar{X}_i^T(t)P_i[\bar{B}_{is}\bar{K}_{iw}\int_{t-\eta_i(t)}^{t} d\bar{X}_i(s)ds]-[\bar{B}_{is}\bar{K}_{iw}\bar{e}_i(t)]^T P_i\bar{X}_i(t)\\
&-\bar{X}_i^T(t)P_i[\bar{B}_{is}\bar{K}_{iw}\bar{e}_i(t)]+\rho_i^2\bar{X}_{-i}^T(t-\tau_{-i}(t))\bar{X}_{-i}(t-\tau_{-i}(t))\\
&+\frac{1}{\rho_i^2}\bar{X}_i^T(t)P_i\bar{F}_{-is}\bar{F}_{-is}{}^T P_i\bar{X}_i(t)+\rho_i^2\bar{v}_i^T(t)\bar{v}_i(t)\\
&+\frac{1}{\rho_i^2}\bar{X}_i^T(t)P_iHH^T P_i\bar{X}_i(t)+(\bar{C}_{is}\bar{X}_i(t))^T P_i\bar{C}_{is}\bar{X}_i(t)\\
&+2[\bar{X}_i(t)-\bar{X}_i(t-\eta_i(t))-\int_{t-\eta_i\%(t)}^{t} d\bar{X}_i(s)ds]^T\\
&\times P_i[\bar{\delta}_i\bar{X}_i(t-\eta_i(t))+\tau_M^i\int_{t-\eta_i(t)}^{t} d\bar{X}_i(s)ds]\\
&+\sigma_i\bar{X}_i^T(t-\eta_i(t))\bar{\Phi}_i\bar{X}_i(t-\eta_i(t))-\sigma_i\bar{X}_i^T(t-\eta_i(t))\bar{\Phi}_i\bar{e}_i(t)\\
&-\sigma_i\bar{e}_i^T(t)\bar{\Phi}_i\bar{X}_i(t-\eta_i(t))+(\sigma_i-1)\bar{e}_i^T(t)\bar{\Phi}_i\bar{e}_i(t)\}dt\}
\end{aligned}
\tag{5.48}
$$

$$
\begin{aligned}
&=E\{\bar{V}_i(\bar{X}_i(0))+E\{\int_0^{t_f}[\sum_{k=1}^{\infty}\sum_{s,w=1}^{S}h_s(z_i(t))h_w(z_i(t_kh))\times\Theta_k(t)(\zeta_i^T(t)\Psi_i^{s,w}\zeta_i(t)\\
&\quad+\rho_i{}^2\bar{v}_i^T(t)\bar{v}_i(t))+\rho_i^2(\bar{X}_{-i}^T(t-\tau_{-i}(t))\bar{X}_{-i}(t-\tau_{-i}(t)))]dt\}
\end{aligned}
$$

where,

$$
\zeta_i^T(t)=[\bar{X}_i^T(t),\ \int_{t-\eta_i(t)}^{t} d\bar{X}_i^T(s)ds,\ \bar{e}_i^T(t),\ \bar{X}_i^T(t-\eta_i(t))]
$$

$$
\Psi_i^{s,w}=\begin{bmatrix}\Psi_{i,11}^{s,w} & \Psi_{i,12}^{s,w} & \Psi_{i,13}^{s,w} & \bar{\delta}_iP_i\\ * & -2\tau_M^iP_i & 0_{24\times24} & -(\tau_M^i+\bar{\delta}_i)P_i\\ * & * & \Psi_{i,33}^{s,w} & -\sigma_i\bar{\Phi}_i\\ * & * & * & \Psi_{i,44}^{s,w}\end{bmatrix}
$$

and

$$
\begin{aligned}
\Psi_{i,11}^{s,w}&=\bar{Q}_i+P_i\bar{A}_{is}+\bar{A}_{is}^TP_i+P_i\bar{B}_{is}\bar{K}_{iw}+\bar{K}_{iw}^T\bar{B}_{is}^TP_i\\
&\quad+\frac{1}{\rho_i^2}P_i\bar{F}_{-is}\bar{F}_{-is}^TP_i+\frac{1}{\rho_i^2}P_iHH^TP_i+\bar{C}_{is}^TP_i\bar{C}_{is}\\
\Psi_{i,12}^{s,w}&=-P_i\bar{B}_{is}\bar{K}_{iw}+\tau_M^iP_i,\ \Psi_{i,13}^{s,w}=-P_i\bar{B}_{is}\bar{K}_{iw}\\
\Psi_{i,33}^{s,w}&=(\sigma_i-1)\bar{\Phi}_i,\ \Psi_{i,44}{}^{s,w}=\sigma_i\bar{\Phi}_i-2\bar{\delta}_iP_i
\end{aligned}
$$

It is clear that if $\Psi_i^{s,w} \leq 0$, then the augmented H_∞ decentralized team formation tracking performance in (5.20) can be guaranteed. We define a positive matrix $W_i = P_i^{-1}$. Then pre- and post-multiplying both sides of $\Psi_i^{s,w}$ with $diag\{W_i, W_i, W_i, W_i\}$ and its transpose, respectively, to obtain the responding matrix inequality

$$\Psi_i^{s,w} = \begin{bmatrix} \Psi_{i,11}^{\prime\prime s,w} & \Psi_{i,12}^{\prime s,w} & \Psi_{i,13}^{\prime s,w} & \bar{\delta}_i W_i \\ * & -2\tau_M^i W_i & 0 & \Psi_{i,24}^{\prime s,w} \\ * & * & \Psi_{i,33}^{\prime s,w} & \Psi_{i,34}^{\prime s,w} \\ * & * & * & \Psi_{i,44}^{\prime s,w} \end{bmatrix} \leq 0 \tag{5.49}$$

where,

$$\Psi_{i,11}^{\prime s,w} = W_i \bar{Q}_i W_i + \bar{A}_{is} W_i + W_i \bar{A}_{is}^T + \bar{B}_{is} \bar{K}_{iw} W_i$$
$$+ W_i \bar{K}_{iw}^T \bar{B}_{is}^T + \frac{1}{\rho_i^2} \bar{F}_{-is} \bar{F}_{-is}^T + \frac{1}{\rho_i^2} H H^T + W_i \bar{C}_{is}^T W_i^{-1} \bar{C}_{is} W_i$$
$$\Psi_{i,12}^{\prime s,w} = -\bar{B}_{is} \bar{K}_{iw} W_i + \tau_M^i W_i,\ \Psi_{i,13}^{\prime s,w} = -\bar{B}_{is} \bar{K}_{iw} W_i$$
$$\Psi_{i,33}^{\prime s,w} = (\sigma_i - 1) W_i \bar{\Phi}_i W_i, \Psi_{i,34}^{\prime s,w} = -\sigma_i W_i \bar{\Phi}_i W_i$$
$$\Psi_{i,44}^{\prime s,w} = \sigma_i W_i \bar{\Phi}_i W_i - 2\bar{\delta}_i W_i, \Psi_{i,24}^{\prime s,w} = -((\tau_M^i + \bar{\delta}_i)) W_i$$

The inequality in (5.49) can be transformed to (5.50) by using Schur complement [50].

$$\Psi_i^{s,w} = \begin{bmatrix} \Psi_{i,11}^{\prime\prime s,w} & \Psi_{i,12}^{\prime s,w} & \Psi_{i,13}^{s,w} & \bar{\delta}_i W_i & \\ * & -2\tau_M^i P_i & 0 & \Psi_{i,24}^{\prime s,w} & \\ * & * & \Psi_{i,33}^{\prime s,w} & \Psi_{i,34}^{\prime s,w} & M_{1S} \\ * & * & * & \Psi_{i,44}^{\prime s,w} & \\ & & * & & M_2 \end{bmatrix} \leq 0 \tag{5.50}$$

where

$$\Psi_i^{\prime\prime s,w} = \bar{A}_{is} W_i + W_i \bar{A}_{is}^{\ T} + \bar{B}_{is} \bar{K}_{iw} W_i + W_i \bar{K}_{iw}^T \bar{B}_{is}^T$$

$$M_{1,s} = \begin{bmatrix} W_i \bar{Q}_i^{\frac{1}{2}} & \frac{1}{\rho_i} \bar{F}_{-is} & \frac{1}{\rho_i} H & W_i \bar{C}_{is}^T \\ 0 & 0 & 0 & 0 \\ 0 & 0 & 0 & 0 \\ 0 & 0 & 0 & 0 \end{bmatrix}$$

$$M_2 = \begin{bmatrix} -I_{24} & 0 & 0 & 0 \\ * & -I_{12N_i} & 0 & 0 \\ * & * & -I_{24} & 0 \\ * & * & * & -W_i \end{bmatrix}$$

Considering $\bar{K}_{iw} = \begin{bmatrix} K_{v,iw} & K_{iw} \end{bmatrix}$ and $\bar{\Phi}_i = diag\{\bar{\Phi}_{i1}, \bar{\Phi}_{i2}\}$, we replace $W_i = diag\{W_{i1}, W_{i2}\}$, $\breve{\Phi}_{i1} = W_{i1}\bar{\Phi}_{i1}W_{i1}$, $\breve{\Phi}_{i2} = W_{i2}\bar{\Phi}_{i2}W_{i2}$, $Y_{v,iw} = K_{v,iw}W_{i1}$, $Y_{iw} = K_{iw}W_{i2}$ into the (5.50) to obtain the LMIs in (5.31). As a result, it is obvious that if the condition (5.31) holds for $s = 1,2,\cdots,S$, $w = 1,2,\cdots,S$, $i = 1,2,\cdots,N$, then the decentralized team formation tracking control of the ith quadrotor UAV network subsystem for large-scale quadrotor UAVs team formation NCS with discrete event-triggered mechanism in (5.3) under time-varying network-induced delay and packet dropout caused by transmission in the network configuration can be guaranteed to achieve the robust decentralized H_∞ team formation tracking performance in (5.20) for an arbitrary reference input, external disturbance and coupling from the other quadrotor UAVs in neighborhood. The decentralized fuzzy control gains of the large-scale quadrotor UAV team formation NCS can be obtained from $K_{v,iw} = Y_{v,iw}W_{i1}^{-1}$, $K_{iw} = Y_{iw}W_{i2}^{-1}$, and the event-triggered weighting matrix $\bar{\Phi}_i = diag\{\bar{\Phi}_{i1}, \bar{\Phi}_{i2}\}$ can be obtained from $\bar{\Phi}_{i1} = W_{i1}^{-1}\breve{\Phi}_{i1}W_{i1}^{-1}$, $\bar{\Phi}_{i2} = W_{i2}^{-1}\breve{\Phi}_{i2}W_{i2}^{-1}$. The proof is completed.

Appendix 5C
Figures and Tables of Simulation

TABLE 5A.1
The Shifted Distance of Each Quadrotor UAV in the Large-Scale Quadrotor UAV Team Formation NCS

Shifted Distance	Layer of the Formation
$d_1 = [6,0,0,0,0,0,0,0,0,0,0,0]^T$	4th
$d_2 = [4,0,0,0,0,0,0,0,0,0,0,0]^T$	4th
$d_3 = [2,0,0,0,0,0,0,0,0,0,0,0]^T$	4th
$d_4 = [0,0,0,0,0,0,0,0,0,0,0,0]^T$	4th
$d_5 = [-2,0,0,0,0,0,0,0,0,0,0,0]^T$	4th
$d_6 = [-4,0,0,0,0,0,0,0,0,0,0,0]^T$	4th
$d_7 = [-6,0,0,0,0,0,0,0,0,0,0,0]^T$	4th
$d_8 = [4,0,1.5,0,0,0,0,0,0,0,0,0]^T$	3rd
$d_9 = [2,0,1.5,0,0,0,0,0,0,0,0,0]^T$	3rd
$d_{10} = [0,0,1.5,0,0,0,0,0,0,0,0,0]^T$	3rd
$d_{11} = [-2,0,1.5,0,0,0,0,0,0,0,0,0]^T$	3rd
$d_{12} = [-4,0,1.5,0,0,0,0,0,0,0,0,0]^T$	3rd
$d_{13} = [2,0,3,0,0,0,0,0,0,0,0,0]^T$	2nd
$d_{14} = [0,0,3,0,0,0,0,0,0,0,0,0]^T$	2nd
$d_{15} = [-2,0,3,0,0,0,0,0,0,0,0,0]^T$	2nd
$d_{16} = [0,0,4.5,0,0,0,0,0,0,0,0,0]^T$	1st
$d_{17} = [4,0,-1.5,0,0,0,0,0,0,0,0,0]^T$	5th
$d_{18} = [2,0,-1.5,0,0,0,0,0,0,0,0,0]^T$	5th
$d_{19} = [0,0,-1.5,0,0,0,0,0,0,0,0,0]^T$	5th
$d_{20} = [-2,0,-1.5,0,0,0,0,0,0,0,0,0]^T$	5th
$d_{21} = [-4,0,-1.5,0,0,0,0,0,0,0,0,0]^T$	5th
$d_{22} = [2,0,-3,0,0,0,0,0,0,0,0,0]^T$	6th
$d_{23} = [0,0,-3,0,0,0,0,0,0,0,0,0]^T$	6th
$d_{24} = [-2,0,-3,0,0,0,0,0,0,0,0,0]^T$	6th
$d_{25} = [0,0,4.5,0,0,0,0,0,0,0,0,0]^T$	7th

TABLE 5A.2
Exact Numbers of Transmitted Signals and Packets Dropouts of Each Quadrotor UAV

The *i* th UAV	UAV1	UAV2	UAV3	UAV4	UAV5
Number of packet dropouts	10	21	23	19	19
Number of transmitted signals	1962	1985	2010	2022	2043
	UAV6	UAV7	UAV8	UAV9	UAV10
Number of packet dropouts		13	17	19	30
Number of transmitted signals		2042	1941	1970	1971
	UAV11	UAV12	UAV13	UAV14	UAV15
Number of packet dropouts		21	18	19	17
Number of transmitted signals		1987	1920	1934	1942
	UAV16	UAV17	UAV18	UAV19	UAV20
Number of packet dropouts		21	18	27	20
Number of transmitted signals		2025	2057	2081	2072
	UAV21	UAV22	UAV23	UAV24	UAV25
Number of packet dropouts	26	18	15	18	22
Number of transmitted signals	2070	2116	2115	2111	2172

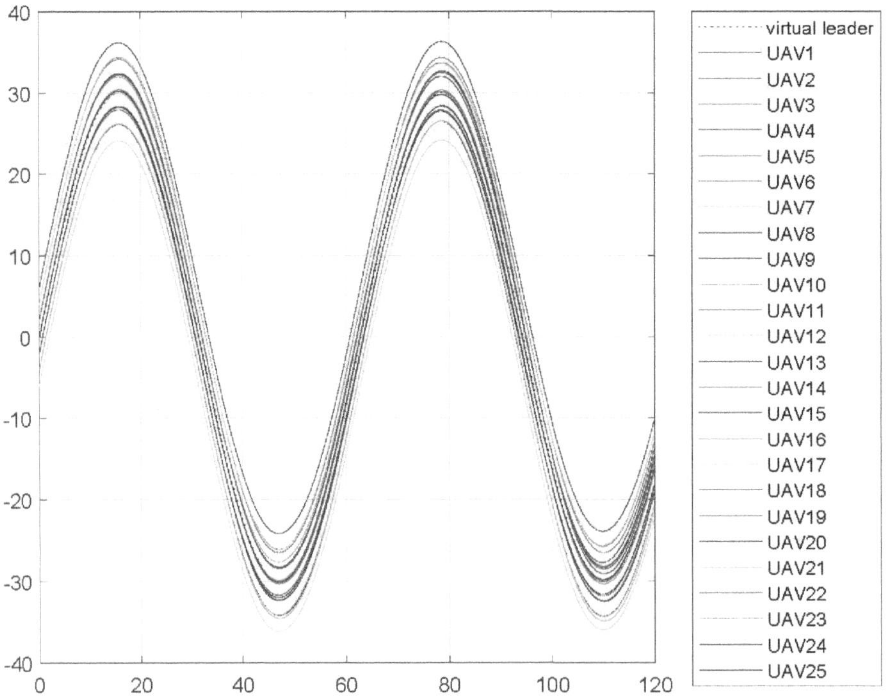

FIGURE 5.A1 Responses of x_1^i of 25 quadrotor UAVs.

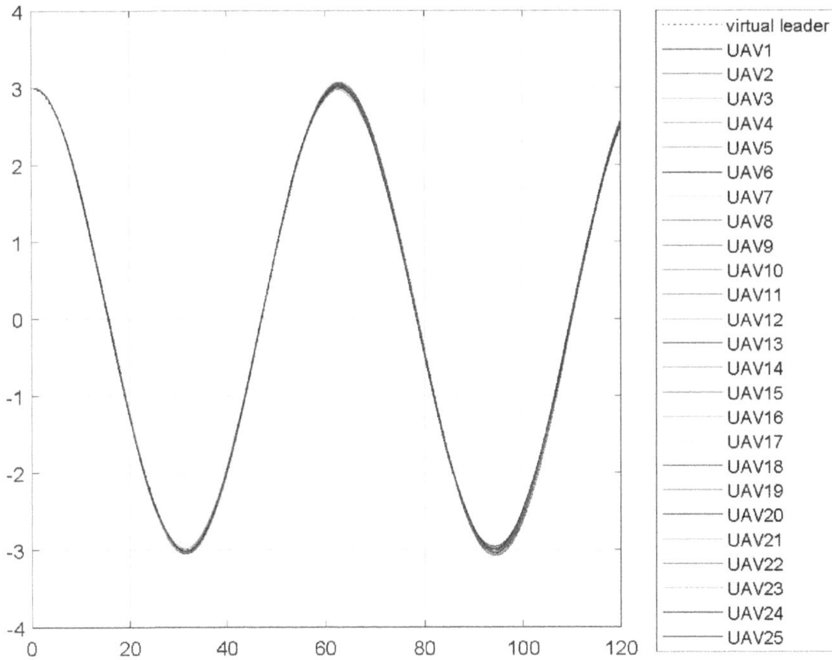

FIGURE 5.A2 Responses of x_2^i of 25 quadrotor UAVs.

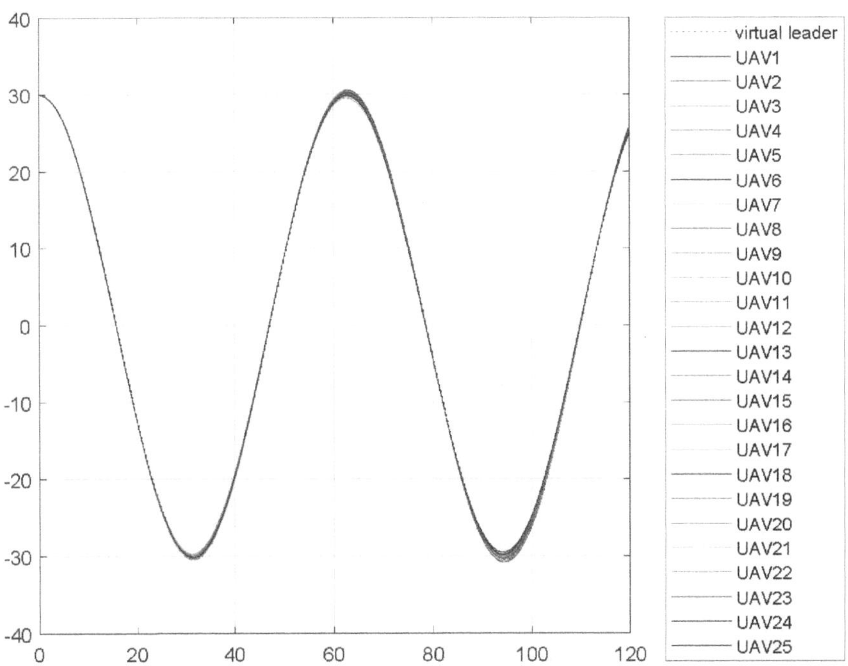

FIGURE 5.A3 Responses of y_1^i of 25 UAVs.

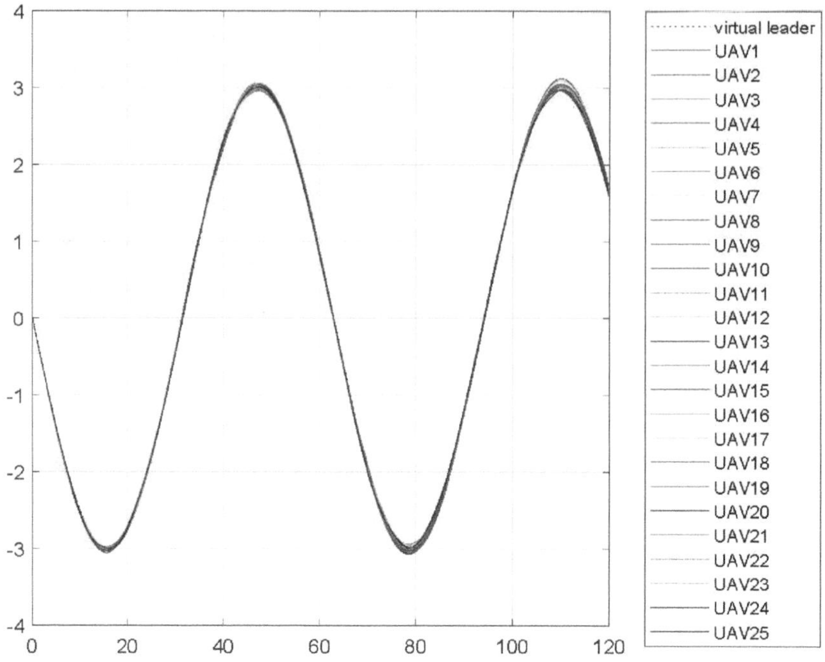

FIGURE 5.A4 Responses of y_2^i of 25 quadrotor UAVs.

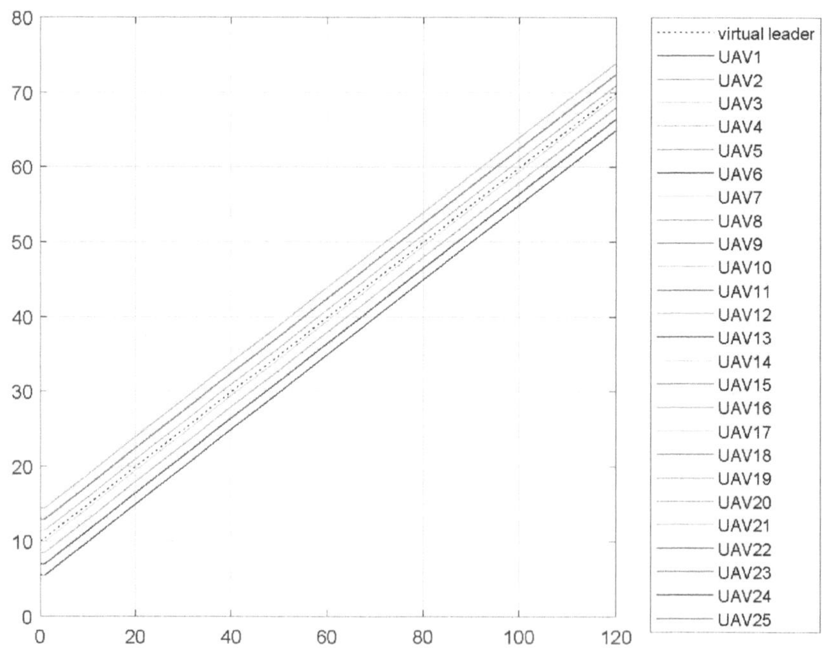

FIGURE 5.A5 Responses of z_1^i of 25 quadrotor UAVs.

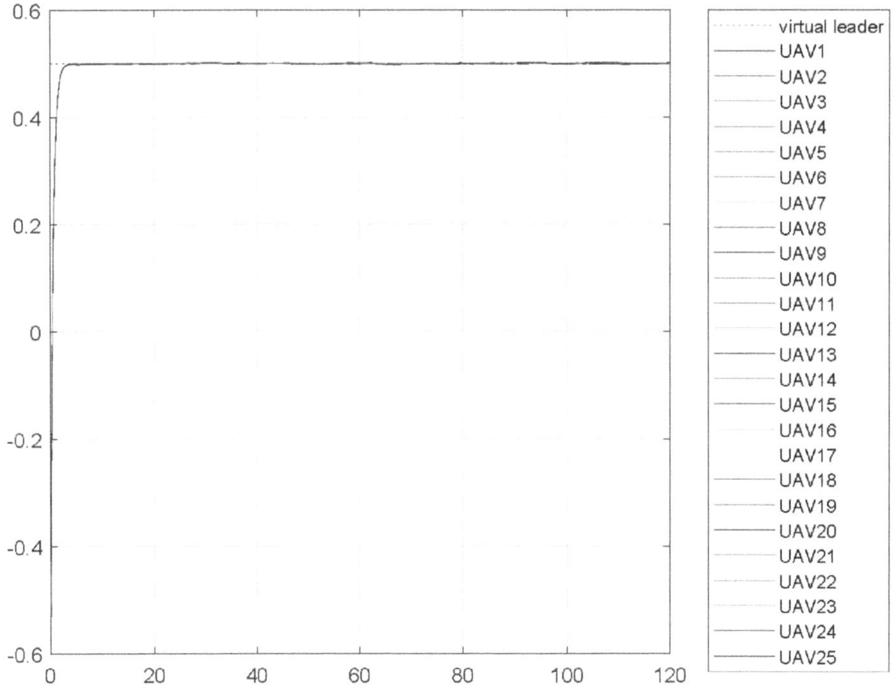

FIGURE 5.A6 Responses of z_2^i of 25 quadrotor UAVs.

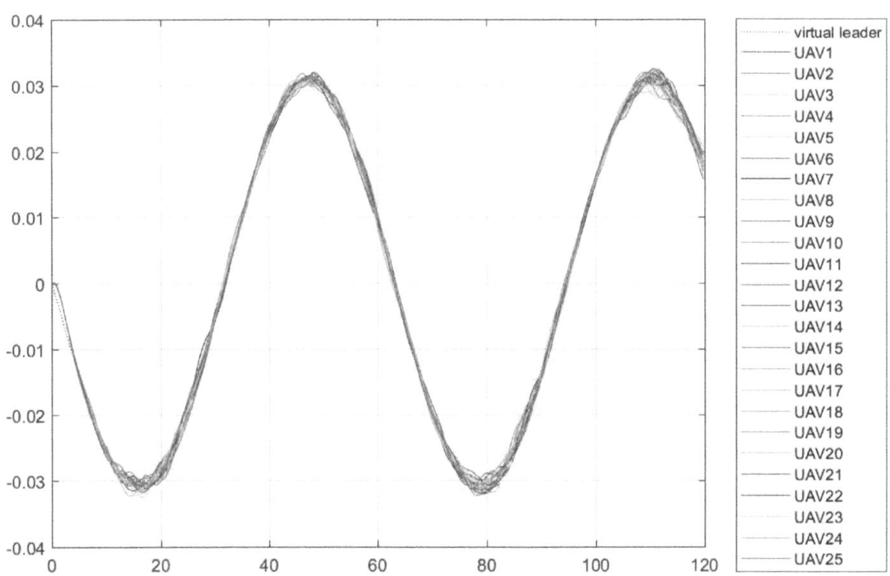

FIGURE 5.A7 Responses of ϕ_1^i of 25 quadrotor UAVs.

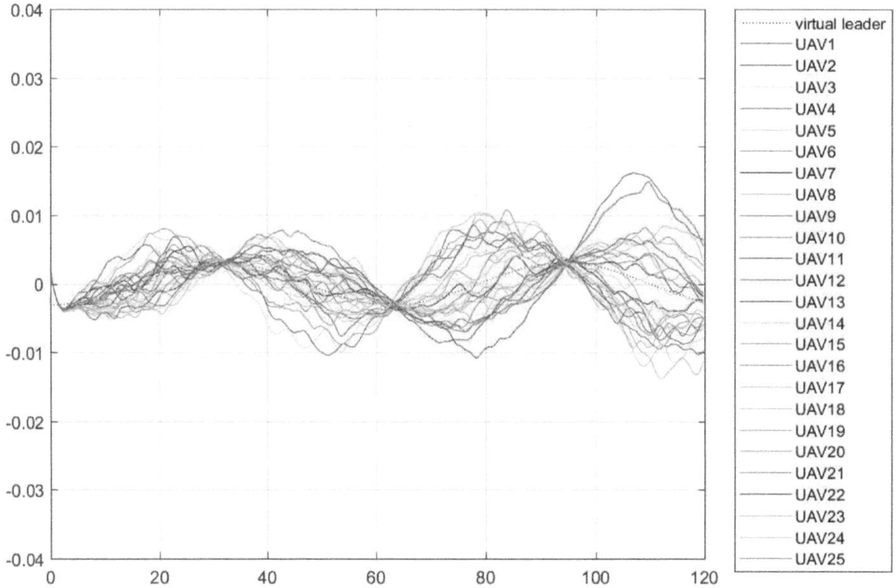

FIGURE 5.A8 Responses of ϕ_2^i of 25 quadrotor UAVs.

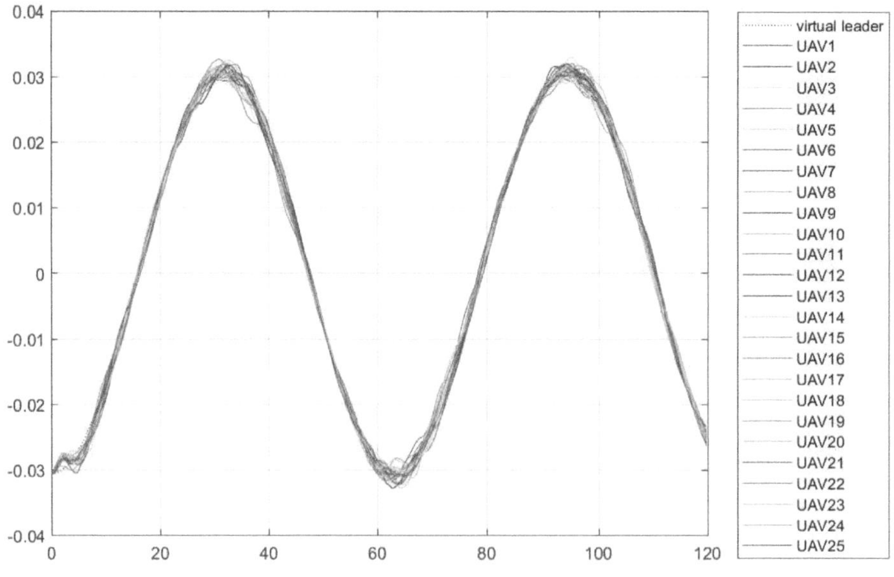

FIGURE 5.A9 Responses of θ_1^i of 25 quadrotor UAVs.

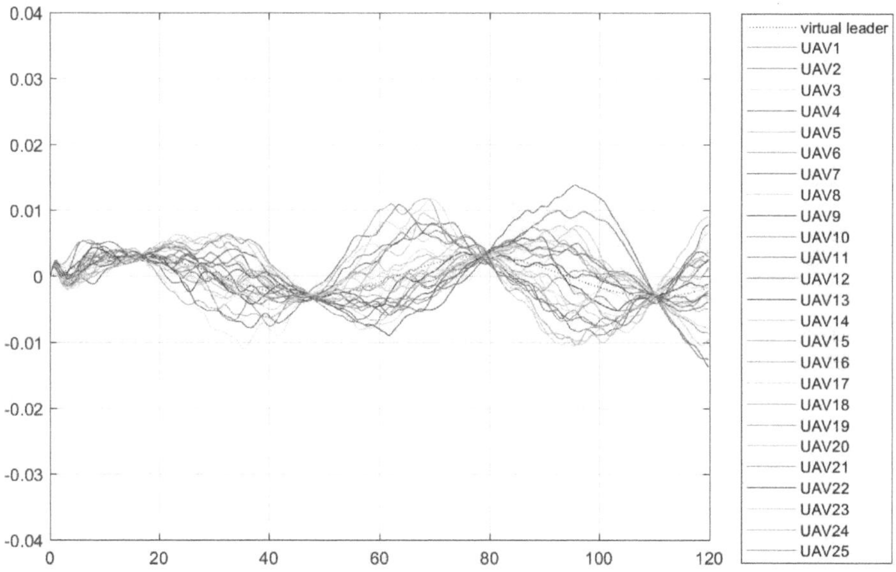

FIGURE 5.A10 Responses of θ_2^i of 25 quadrotor UAVs.

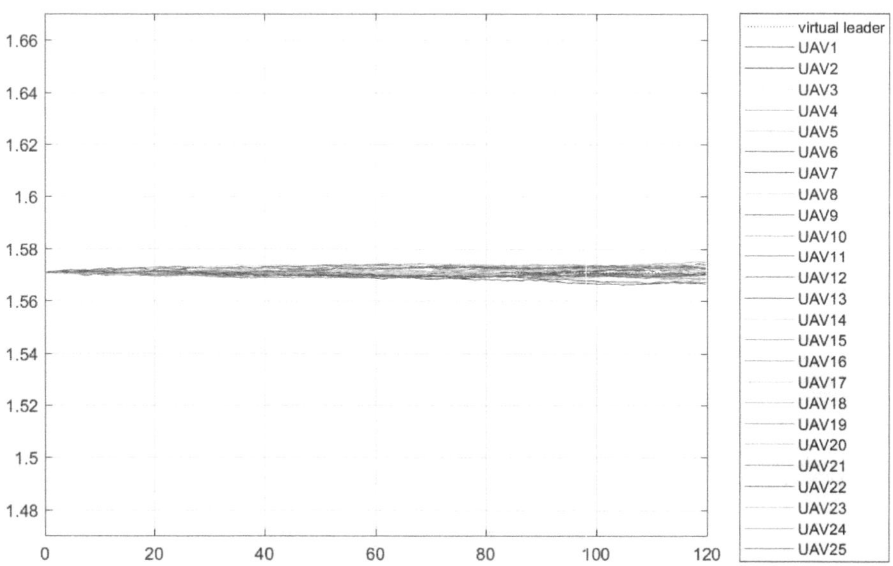

FIGURE 5.A11 Responses of ψ_1^i of 25 quadrotor UAVs.

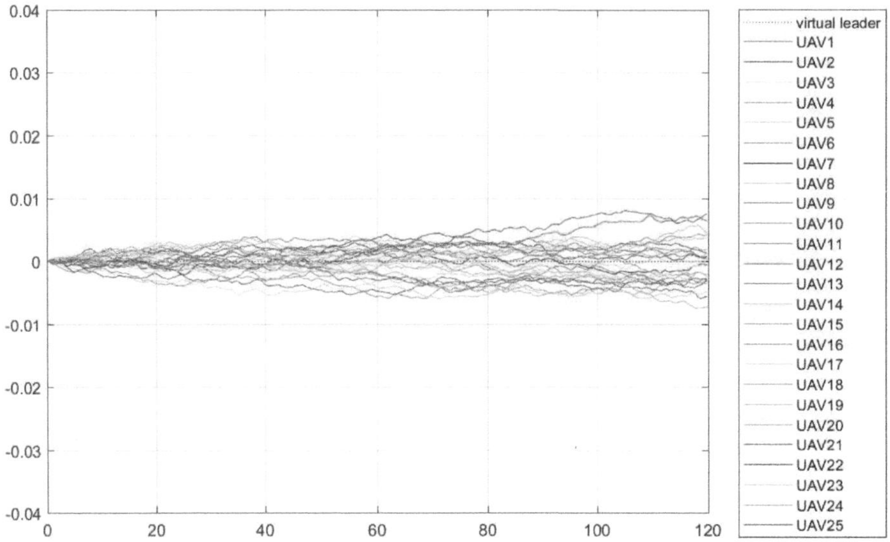

FIGURE 5.A12 Responses of ψ_2^i of 25 quadrotor UAVs.

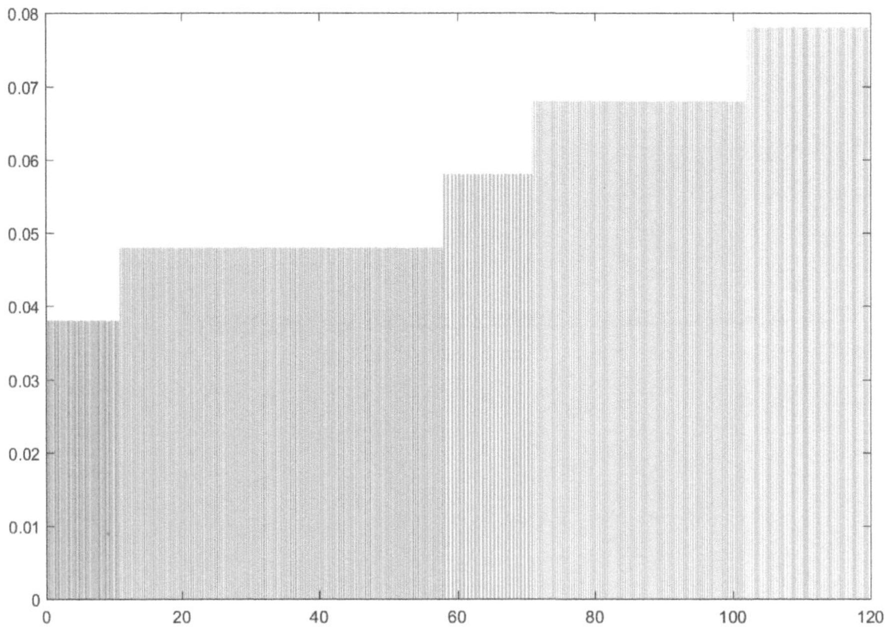

FIGURE 5.A13 Release instants and time intervals at $t \in [0,120]$ of quadrotor UAV1 with the decentralized H_∞ team formation tracking control embedded with the event-triggered mechanism.

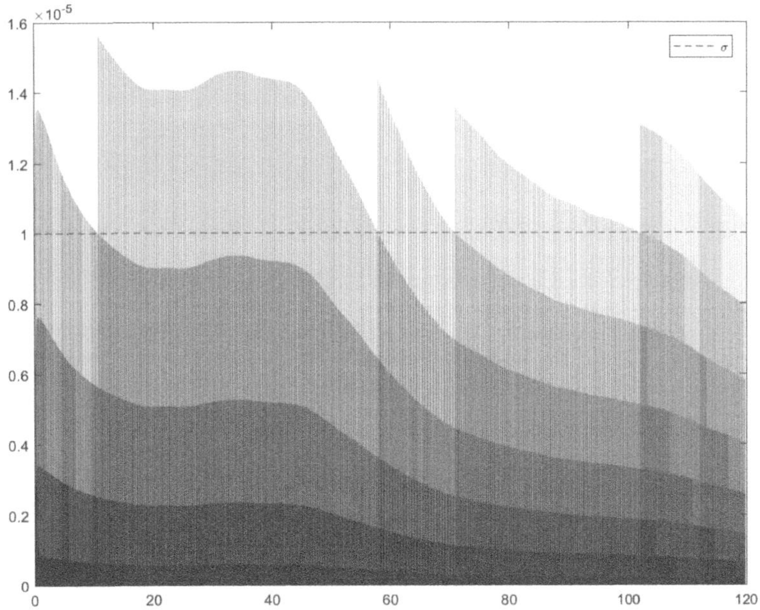

FIGURE 5.A14 The event-triggered process of decentralized H_∞ team formation control of quadrotor UAV1 within $t \in [0,120]$. Blue line represents the value of $\tilde{e}_i^T(t)\Phi_i\tilde{e}_i(t) / X_i^T(t_k{}^ih)\Phi_i X_i(t_k^i h)$. If it is over the level of the event-triggered mechanism (red line), the packet will be transmitted, and vice versa.

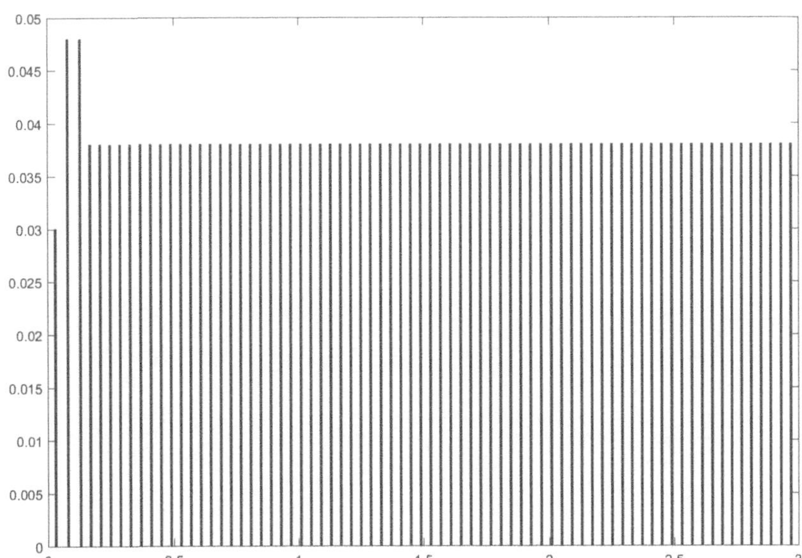

FIGURE 5.A15 Release instants and time intervals within t∈ [0,1] of quadrotor UAV1 with the decentralized control embedded with event-triggered mechanism.

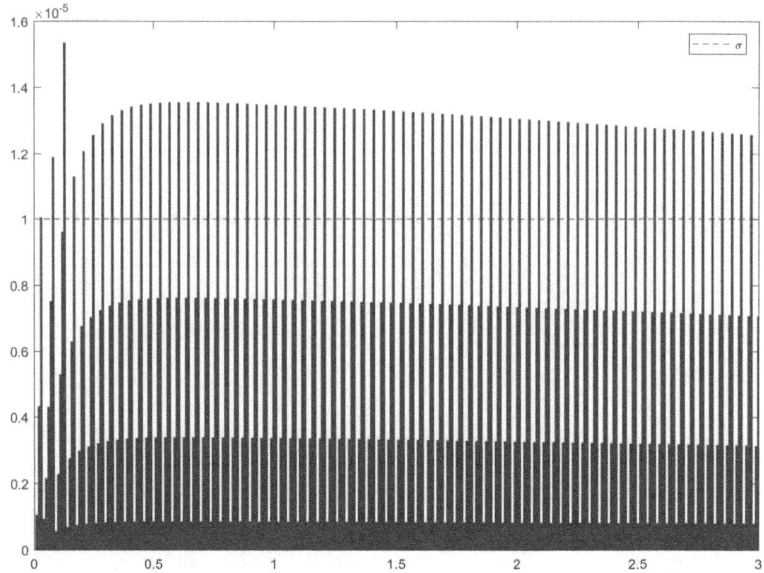

FIGURE 5.A16 The event-triggered process of decentralized H_∞ team formation control of quadrotor UAV1 within $t \in [0,1]$. Blue line represents the value of $\tilde{e}_i^T(t)\Phi_i\tilde{e}_i(t) / X_i^T(t_k{}^i h)$ $\Phi_i X_i(t_k^i h)$. If it is over the level of the event-triggered mechanism (red line), the packet will be transmitted, and vice versa.

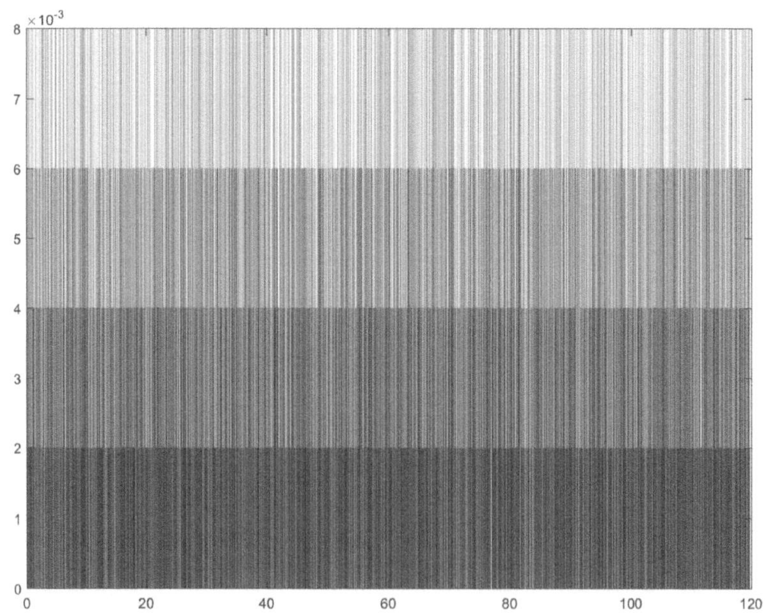

FIGURE 5.A17 The time-varying network-induced delay in the decentralized control process of quadrotor UAV1.

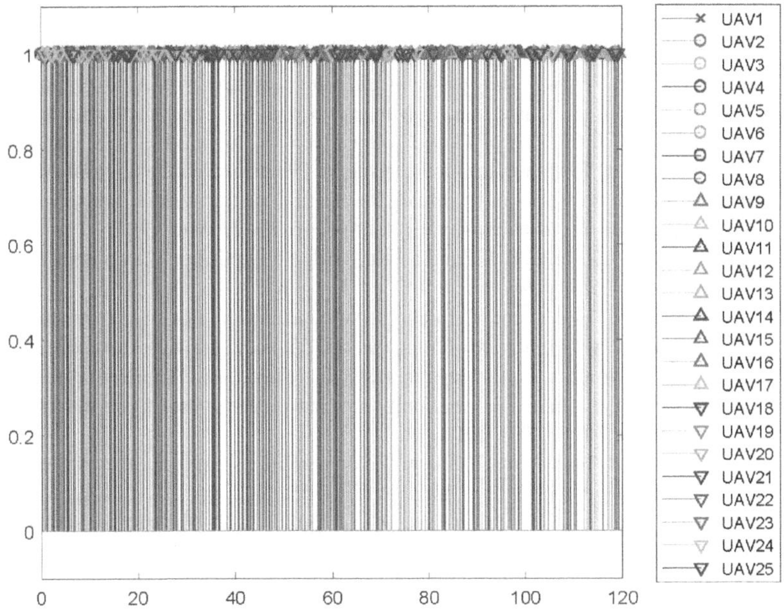

FIGURE 5.A18 The occurrence of packet dropouts of each quadrotor UAV networked control subsystem.

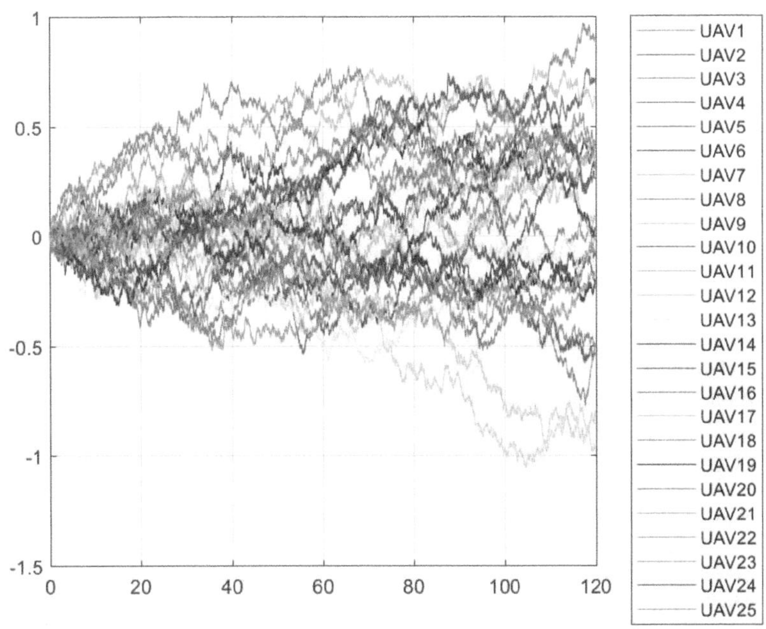

FIGURE 5.A19 The 25 independent Wiener processes of each quadrotor UAV subsystem.

REFERENCES

[1] R. Rajkumar, I. Lee, L. Sha, and J. Stankovic, "Cyber-physical systems: The next computing revolution," *Design Automation Conference*, Anaheim, CA, 2010, pp. 731–736.

[2] A. Zanella, N. Bui, A. Castellani, L. Vangelista, and M. Zorzi, "Internet of things for smart cities," *IEEE Internet of Things Journal*, vol. 1, no. 1, pp. 22–32, Feb. 2014.

[3] M. Wollschlaeger, T. Sauter, and J. Jasperneite, "The future of industrial communication: Automation networks in the era of the internet of things and industry 4.0," *IEEE Industrial Electronics Magazine*, vol. 11, no. 1, pp. 17–27, Mar. 2017.

[4] W. Shi, J. Cao, Q. Zhang, Y. Li, and L. Xu, "Edge computing: Vision and challenges," *IEEE Internet of Things Journal*, vol. 3, no. 5, pp. 637–646, Oct. 2016.

[5] N. H. Motlagh, M. Bagaa, and T. Taleb, "UAV-based IoT platform: A crowd surveillance use case," *IEEE Communications Magazine*, vol. 55, no. 2, pp. 128–134, Feb. 2017.

[6] K. Dorling, J. Heinrichs, G. G. Messier, and S. Magierowski, "Vehicle routing problems for drone delivery," *IEEE Transactions on Systems, Man, and Cybernetics: Systems*, vol. 47, no. 1, pp. 70–85, Jan. 2017.

[7] N. H. Motlagh, M. Bagaa, and T. Taleb, "UAV-based IoT platform: A crowd surveillance use case," *IEEE Communications Magazine*, vol. 55, no. 2, pp. 128–134, Feb. 2017.

[8] N. Hossein Motlagh, T. Taleb, and O. Arouk, "Low-altitude unmanned aerial vehicles-based internet of things services: Comprehensive survey and future perspectives," *IEEE Internet of Things Journal*, vol. 3, no. 6, pp. 899–922, Dec. 2016.

[9] D. Gan, G. Cai, J. Dias, L. Seneviratne, J. Dias, and L. Seneviratne, "Attitude control of quad-rotor UAVs using an intuitive kinematics model," in *2013 IEEE 20th International Conference on Electronics, Circuits, and Systems (ICECS)*, Abu Dhabi, 2013, pp. 597–600.

[10] K. Benzaid, N. Mansouri, and O. Labbani-Igbida, "Robust trajectory tracking control of a quadrotor UAV," in *3rd International Conference on Systems and Control*, Algiers, 2013, pp. 206–211.

[11] M. Nguyen Duc, T. N. Trong, and Y. S. Xuan, "The quadrotor MAV system using PID control, " in *2015 IEEE International Conference on Mechatronics and Automation (ICMA)*, Beijing, 2015, pp. 506–510.

[12] B. Zhao, B. Xian, Y. Zhang, and X. Zhang, "Nonlinear robust adaptive tracking control of a quadrotor UAV via immersion and invariance methodology," *IEEE Transactions on Industrial Electronics*, vol. 62, no. 5, pp. 2891–2902, May 2015.

[13] F. Chen, Q. Wu, B. Jiang, and G. Tao, "A reconfiguration scheme for quadrotor helicopter via simple adaptive control and quantum logic," *IEEE Transactions on Industrial Electronics*, vol. 62, no. 7, pp. 4328–4335, July 2015.

[14] M. Bisheban and T. Lee, "Geometric adaptive control with neural networks for a quadrotor in wind fields," *IEEE Transactions on Control Systems Technology*, pp. 1–16, July 2020.

[15] D. Matouk, O. Gherouat, F. Abdessemed, and A. Hassam, "Quadrotor position and attitude control via backstepping approach," in *2016 8th International Conference on Modelling, Identification and Control (ICMIC)*, Algiers, 2016, pp. 73–79.

[16] D. J. Almakhles, "Robust backstepping sliding mode control for a quadrotor trajectory tracking application," *IEEE Access*, vol. 8, pp. 5515–5525, Dec. 2020.

[17] J. Sun, Y. Wang, Y. Yu, and C. Sun, "Nonlinear robust compensation method for trajectory tracking control of quadrotors," *IEEE Access*, vol. 7, pp. 26766–26776, Mar. 2019.

[18] A. C. Satici, H. Poonawala, and M. W. Spong, "Robust optimal control of quadrotor UAVs," *IEEE Access*, vol. 1, pp. 79–93, 2013.

[19] X. Lin, C. Wu, and B. Chen, "Robust H1 adaptive fuzzy tracking control for MIMO nonlinear stochastic Poisson jump diffusion systems," *IEEE Transactions on Cybernetics*, vol. 49, no. 8, pp. 3116–3130, Aug. 2019.

[20] X. Dong, Y. Zhou, Z. Ren, and Y. Zhong, "Time-varying formation tracking for second-order multi-agent systems subjected to switching topologies with application to quadrotor formation flying," *IEEE Transactions on Industrial Electronics*, vol. 64, no. 6, pp. 5014–5024, June 2017.
[21] H. Du, W. Zhu, G. Wen, Z. Duan, and J. Lü, "Distributed formation control of multiple quadrotor aircraft based on nonsmooth consensus algorithms," *IEEE Transactions on Cybernetics*, vol. 49, no. 1, pp. 342–353, Jan. 2019.
[22] Z. Hou and I. Fantoni, "Interactive leader-follower consensus of multiple quadrotors based on composite nonlinear feedback control," *IEEE Transactions on Control Systems Technology*, vol. 26, no. 5, pp. 1732–1743, Sept. 2018.
[23] D. Zhou, Z. Wang, and M. Schwager, "Agile coordination and assistive collision avoidance for quadrotor swarms using virtual structures," *IEEE Transactions on Robotics*, vol. 34, no. 4, pp. 916–923, Aug. 2018.
[24] C. B. Low, "A dynamic virtual structure formation control for fixed-wing UAVs," in *2011 9th IEEE International Conference on Control and Automation (ICCA)*, Santiago, 2011, pp. 627–632.
[25] B. S. Chen, C. Wang, and M. Lee, "Stochastic robust team tracking control of multi-UAV networked system under wiener and poisson random fluctuations," *IEEE Transactions on Cybernetics*, early access, Jan. 2020. http://doi.org/10.1109/TCYB.2019.2960104.
[26] J. Yang, C. Liu, M. Coombes, Y. Yan, and W. Chen, "Optimal path following for small fixed-wing UAVs under wind disturbances," *IEEE Transactions on Control Systems Technology*, Apr. 2020. http://doi.org/10.1109/TCST.2020.2980727.
[27] D. Shi, Z. Wu, and W. Chou, "Generalized extended state observer based high precision attitude control of quadrotor vehicles subject to wind disturbance," *IEEE Access*, vol. 6, pp. 32349–32359, 2018.
[28] W. Mei and R. Zhang, "Cooperative downlink interference transmission and cancellation for cellular-connected UAV: A divide-and-conquer approach," *IEEE Transactions on Communications*, vol. 68, no. 2, pp. 1297–1311, Feb. 2020.
[29] A. A. Khuwaja, G. Zheng, Y. Chen, and W. Feng, "Optimum deployment of multiple UAVs for coverage area maximization in the presence of co-channel interference," *IEEE Access*, vol. 7, pp. 85203–85212, 2019.
[30] Y. Huang, J. Xu, L. Qiu, and R. Zhang, "Cognitive UAV communication via joint trajectory and power control," in *2018 IEEE 19th International Workshop on Signal Processing Advances in Wireless Communications (SPAWC)*, Kalamata, 2018, pp. 1–5.
[31] C. Shen, T. Chang, J. Gong, Y. Zeng, and R. Zhang, "Multi-UAV interference coordination via joint trajectory and power control," *IEEE Transactions on Signal Processing*, vol. 68, pp. 843–858, Jan. 2020.
[32] W. Mei, Q. Wu, and R. Zhang, "Cellular-connected UAV: Uplink association, power control and interference coordination," *IEEE Transactions on Wireless Communications*, vol. 18, no. 11, pp. 5380–5393, Nov. 2019.
[33] W. Yu, D. Lin, and T. Song, "Adaptive control for UAV close formation flight against disturbances," in *2018 3rd International Conference on Robotics and Automation Engineering (ICRAE)*, Guangzhou, 2018, pp. 196–201.
[34] L. Ding, Q. Han, X. Ge, and X. Zhang, "An overview of recent advances in event-triggered consensus of multiagent systems," *IEEE Transactions on Cybernetics*, vol. 48, no. 4, pp. 1110–1123, Apr. 2018.
[35] X. Zhang and Q. Han, "Event-triggered dynamic output feedback control for networked control systems," *IET Control Theory & Applications*, vol. 8, no. 4, pp. 226–234, Mar. 2014.
[36] S. Yan, S. K. Nguang, M. Shen, and G. Zhang, "Event-triggered H1 control of networked control systems with distributed transmission delay," *IEEE Transactions on Automatic Control*, pp. 4295-4301, Nov. 2019.

[37] M. Lee, B. Chen, Y. Chang, and C. Hwang, "Stochastic robust team formation tracking design of multi-VTOL-UAV networked control system in smart city under time-varying delay and random fluctuation," *IEEE Access*, vol. 8, pp. 131310–131326, 2020.
[38] M. A. S. Teixerira, F. Neves-jr, A. Koubaa, L. Arruda, and A. S. D. Oliveira, "A quadral-fuzzy control approach to flight formation by a fleet of unmanned aerial vehicles," *IEEE Access*, vol. 8, pp. 64366–64381, Apr. 2020.
[39] J. L. Zhong, P. Zhang, and J. G. Yan, "Distributed adaptive finite-time compensation control for UAV swarm with uncertain disturbances," *IEEE Transactions on Circuits and Systems–I*, vol. 68, no. 2, pp. 829–841, Feb. 2021.
[40] H. Y. Luan, Y. T. Xu, D. X. Liu, Z. Y. Du, H. M. Qian, X. D. Liu, and X. B. Tong, "Energy efficient task cooperation for multi-UAV networks: A coalition formation game approach," *IEEE Access*, vol. 8, pp. 149372–149384, Apr. 2020.
[41] J. W. Choi, Y. G. Song, S. G. Lim, C. L. Kwon, and H. D. Oh, "Decentralized multi-subgroup formation control with connectivity preservation and collision avoidance," *IEEE Access*, vol. 8, pp. 71525–71534, Apr. 2020.
[42] C. Jiang, D. X. Zou, and Q. L. Zhang, "Observer-based control for networked control systems with network induced delay and packet dropout," in *Proceedings of the 8th World Congress on Intelligent Control and Automation*, Jinan, China, Jul. 2010, pp. 4376–4379.
[43] Z. H. Pang, G. P. Liu, D. H. Zhou, and D. H. Sun, "Data-based predictive control for networked nonlinear systems with network-induced delay and packet dropout," *IEEE Transactions on Industrial Electronics*, vol. 63, no. 2, pp. 1249–1257, Feb. 2016.
[44] J. N. Li, H. B. Yu, W. Liang, P. Zeng, and M. Z. Yuan, "A survey of packet disordering existing in networked control systems," in *Proceedings of the 5th Annual IEEE International Conference on Cyber Technology in Automation, Control and Intelligent Systems*, Shenyang, China, June 2015, pp. 1797–1801.
[45] J. B. Seo, Y. D. Kim, S. J. Kim, and A. Tsourdos, "Collision avoidance strategies for unmanned aerial vehicles in formation flight," *IEEE Transactions on Aerospace and Electronic Systems*, vol. 53, no. 6, pp. 2718–2734, Dec. 2017.
[46] H. Liu, T. Ma, F. L. Lewis, and Y. Wan, "Robust formation trajectory tracking control for multiple quadrotors with communication delays," *IEEE Transactions on Control Systems Technology*, pp. 2633–2640, Oct. 2019.
[47] W. Zhang and B. S. Chen, "State feedback H1 control for a class of nonlinear stochastic systems," *Siam Journal on Control and Optimization – SIAM*, vol. 44, pp. 1973–1991, 2006.
[48] W. Zhang, L. Xie, and B. S. Chen, *Stochastic H2/H1 Control: A Nash Game Approach*, Boca Raton, FL: CRC Press, 2017.
[49] C. S. Tseng, B. S. Chen, and H. J. Uang, "Fuzzy tracking control design for nonlinear dynamic systems via T-S fuzzy model," *IEEE Transactions on Fuzzy Systems*, vol. 9, no. 3, pp. 381–392, June 2001.
[50] C. S. Tseng and B. S. Chen, "H1 decentralized fuzzy model reference tracking control design for nonlinear interconnected systems," *IEEE Transactions on Fuzzy Systems*, vol. 9, no. 6, pp. 795–809, Dec. 2001.
[51] Y. He, M. Wu, J. H. She, and G. P. Liu, "Parameter-dependent Lyapunov functional for stability of time-delay systems with polytopic-type uncertainties," *IEEE Transactions on Automatic Control*, vol. 49, no. 5, pp. 828–832, May 2004.

6 DNN-Based H_∞ Decentralized Attack-Tolerant Team Formation Tracking Design of Large-Scale UAV Networked Control System under Time-Varying Delay and Interconnected Coupling

6.1 INTRODUCTION

In this generation, along with the technological development, automation system and IoT have attracted more attention in a future smart city. UAV has become a hot topic due to its wide applications such as agriculture, military, rescuing, humanitarian relief, and wireless communication [1–3]. In order to successfully complete the above tasks, a UAV is needed to track the specified trajectory. Among all types of UAVs, quadrotor UAVs have vertical take-off and landing (VTOL) capabilities [4]. In accordance with this maneuverability, quadrotor UAVs can follow a wider variety of trajectories than other types of UAVs. Although quadrotor UAVs have many appealing advantages in practical applications, the power consumption of quadrotor UAVs during flight is a key issue that needs further consideration. Recently, as the need to implement more complex tasks continues to increase, the team formation tracking control problem of multi-quadrotor UAVs is more appealing than the tracking control problem of single quadrotor UAV in the future.

Currently, common team formation methods include the behavior-based control method [5], the leader-follower (L-F) method [6], and the virtual leader structure (V-S) method [7]. Among the above-mentioned methods, the V-S method is considered to be the most recognized strategy for team formation tracking control design problems. Since the leader is virtualized, it avoids the problem of the L-F approach, where the desired team formation shape cannot be maintained once the leader collides during the flight process.

Generally, the control designs of the large-scale team formation system are divided into two kinds: centralized control [8] and decentralized control [9]. The

DOI: 10.1201/9781003545965-6

centralized control strategy is that all quadrotor UAV subsystems are coordinated by a single controller. With an increase in the number of quadrotor UAV subsystems to a large scale, the computational complexity of implementing the centralized formation control strategy will be large and even lead the formation control design to be infeasible. Hence, the decentralized V-S formation control strategy is preferred when considering the formation of large-scale quadrotor UAVs.

Recently, with the advancement of communication technology [10,11], the control signals of physical plants are not calculated by themselves, but by remote computing units via wireless network channels from the ground control system (GCS). In a large-scale quadrotor UAV team formation, the network control system (NCS) is applied to the formation tracking control. Specifically, NCS is divided into two parts: the local side (physical factory) and the remote side (computing unit). When the state information of the quadrotor UAV team is transmitted to the GCS, GCS calculates control commands and then sends the control input back to the quadrotor UAV, which can effectively reduce power consumption and provide greater flexibility and maintainability during tasks. Although the network-based control strategy can optimize the calculation process of control commands, it also creates some problems that need to be dealt with. First, as a result of the uncertainty of quality of service (QoS) in wireless network communication, packet dropout may occur [12–16]. When packets are lost from the transmission, signal information cannot travel over the wireless network communication channel, so the receiver will not be able to receive any signal from the sender [14]. Second, if the signal is affected by the network-induced delay, it cannot be delivered to the receiver on time [16]. According to the above-mentioned, the entire NCS is delay-dependent, and the corresponding control problem is tougher than a delay-free network.

On the flip side, since the rapid development of wireless network-based communication technology and widespread applications in recent years, network security has become a major issue, which has attracted the attention of many researchers and research institutions. Moreover, hackers may attack the network control system to interfere with the transmission of information or even cause the system to crash. Even though various methods have been proposed to resist malicious attacks [17–19], they may still enter the network through security breaches. These malicious attacks interfere with the correctness of signals in wireless network channels, thus disrupting the operation of the entire network system. Due to the fact that these malicious attacks are unobtainable signals to the designer, the impact of malicious attacks is not easily eliminated. To deal with this problem, an attack-tolerant control strategy has been developed to compensate for the impact of unavailable malicious attack signals on the system [20,21]. This strategy constructs a specific observer to simultaneously estimate state variables and malicious attack signals. After that, through the estimated malicious attack signals, the control strategy can be implemented to compensate for the impact of actual malicious attack signals on NCS. Traditionally, to estimate these malicious attack signals, singular descriptor-based observers have been widely used in the field of fault estimation (FE) [22–24]. Nevertheless, the above descriptor-based observers are difficult to implement in most practical applications due to complex algebraic equation constraints.

Except for malicious attacks, during the practice, unexpected external disturbances will inevitably interfere with quadrotor UAVs, for example, a wind gust. Additionally, quadrotor UAVs may suffer unknown coupling effects from other quadrotor UAVs in the large-scale quadrotor UAV team formation system such as trailing vortex coupling from other neighboring quadrotor UAVs. In a nonideal environment, the formation controller design of the quadrotor UAVs should consider the external disturbance, measurement noise, and trailing vortex coupling of other quadrotor UAVs in the team; otherwise, it may lead to instability or even crashes, especially for a large-scale quadrotor UAVs team formation system. Based on the above discussion, the attenuation of these effects should be further considered in the formation tracking control scheme to ameliorate the formation tracking control performance of the NCS of quadrotor UAVs. Hence, we should deal with malicious attack signal estimation and attack-tolerant tracking control problems simultaneously for large-scale team formation quadrotor UAVs NCS under malicious attack, external disturbance, and trailing vortex coupling.

Over several decades, the robust H_∞ observer-based output feedback control strategies have been adopted and widely used to address the attenuation problem of these nonideal effects [25–28]. Through the smoothing model [29], malicious attacks on quadrotor UAV actuators and sensors through network channels can be effectively described and embedded into the augmented state of the augmented system of each quadrotor UAV model and the malicious attack signal smoothing model to avoid the destruction of the state estimate of each quadrotor UAV by the malicious attack signal. Thus, the malicious attack signals and state variables of each quadrotor UAV can be estimated by the traditional Luenberger observer. And then, the estimated state variables and malicious attack signals are adopted for the robust decentralized H_∞ attack-tolerant observer-based team formation tracking control of the NCS of large-scale quadrotor UAVs by compensating for the effect of malicious attacks. Taking advantage of tracking the desired virtual leader team formation shape, the tracking error and estimation error of each quadrotor UAV system can be considered independently by the proposed robust decentralized H_∞ attack-tolerant observer-based tracking control strategy. Therefore, the original robust observer-based team formation tracking control problem of the large-scale quadrotor UAV NCS under malicious attack, external disturbance, measurement noise, and trailing vortex coupling can be converted into an independent robust H_∞ attack-tolerant observer-based tracking control problem for each quadrotor UAV. Furthermore, the robust decentralized H_∞ attack-tolerant observer-based team formation tracking control design problem of large-scale quadrotor UAVs can be converted into the problem of solving an equivalent highly nonlinear partial differential observer/controller-coupled HJIE problem for each quadrotor UAV in the team formation.

However, observer/controller-coupled HJIE is difficult to solve analytically and numerically. Therefore, instead of solving HJIE directly, several interpolation methods (e.g., the global linearization method [30] and the Takagi-Sugeno (T-S) fuzzy interpolation method [31]) have been proposed to interpolate N local linearized systems at different operations to approximate a nonlinear quadrotor UAV system. Then, with the adoption of the quadratic Lyapunov function, observer/controller-coupled HJIE can be replaced by a set of observer/controller-coupled N^2 Riccati-like equations (RLEs). By utilizing Schur's complement technique, N^2 Riccati-like inequalities can be transformed into a set of N^2 observer/controller-coupled LMIs, which can be solved by two-step procedure (one for observer and one for controller of each quadrotor UAV) with the assistance of LMI toolbox in MATLAB. However, these interpolation methods need to solve a large number of LMIs, especially for a highly nonlinear system such as a quadrotor UAV. Further, we need to calculate H_∞ observer-based controller laws for each quadrotor UAV by interpolating N^2 local linearized observer-based control laws via N^2 interpolation functions at every time instant, which makes the design complexity and computational load dramatically increase. Besides, the solution of HJIE is based on a quadratic Lyapunov function, resulting in a more conservative solution for nonlinear HJIE. Lately, an HJIE-embedded DNN learning approach is proposed to solve HJIE for robust H_∞ control of nonlinear time-varying systems [32] and the robust H_∞ observer-based reference tracking control design of nonlinear systems under external disturbance and measurement noise directly through the DNN-based learning algorithm [33].

In this chapter, an HJIE-reinforcement learning DNN observer-based control scheme is proposed to achieve robust decentralized H_∞ attack-tolerant observer-based team formation tracking control of large-scale quadrotor UAV NCS to ensure that the network-controlled quadrotor UAVs can progressively approach the desired formation of the large-scale quadrotor UAV under the effect of external disturbance, malicious attack, trailing vortex coupling, and measurement noise. Instead of conventional methods, a DNN-based observer-based control scheme is trained by HJIE-reinforcement Adam learning algorithm [34] to solve the nonlinear partial differential observer/controller-coupled HJIE directly for the robust decentralized H_∞ attack-tolerant observer-based team formation tracking control law of each quadrotor UAV under the worst-case external disturbance, malicious attack, measurement noise, and coupling, which are employed for generating output signals to produce the state signal, state estimation, estimation error, and tracking error to train DNN for observer gain and control gain of H_∞ observer-based formation control of each quadrotor UAV by HJIE-reinforcement learning-based Adam learning algorithm in the offline training phase as shown in Figure 6.3. This HJIE-reinforcement DNN-based observer-based control scheme will not influence the H_∞ observer-based control tracking performance because the robust decentralized H_∞ attack-tolerant observer-based team formation control design is based on the worst-case effects of external disturbances, malicious attacks, coupling, and measurement noise. In the online operation phase, external disturbances, malicious attacks, coupling, and measurement noise are available, as shown in Figure 6.2. A simulation example with comparison is provided to confirm the performance of the proposed robust decentralized H_∞ attack-tolerant observer-based team formation tracking control scheme of 25 quadrotor UAVs.

The main purposes of this chapter are described as follows:

(1) A novel smoothed dynamic model is employed to model malicious attack signals that are not available on the large-scale quadrotor UAV NCS. Thence, the malicious attack signals are embedded in large-scale quadrotor UAV NCS so that the malicious attack signals and state variables of quadrotor UAVs can be estimated via a traditional Luenberger observer

simultaneously to compensate for the effect of malicious attack signals to achieve attack-tolerant control. The worst-case effects of external disturbance, malicious attacks, measurement noise, and couplings from other quadrotor UAVs on the state estimation and formation tracking control of each quadrotor UAV are minimized to achieve the robust decentralized H_∞ attack-tolerant observer-based team formation network control of large-scale quadrotor UAVs. Further, an observer/controller-coupled HJIE is developed for each quadrotor UAV to accomplish the robust decentralized H_∞ attack-tolerant observer-based team formation tracking control of large-scale quadrotor UAV NCS.

(2) An HJIE-reinforcement DNN scheme is proposed to solve the observer/controller-coupled HJIE directly for the robust decentralized H_∞ attack-tolerant observer-based team formation tracking control design of large-scale quadrotor UAV NCS to simplify the design complexity. Moreover, the proposed HJIE-reinforcement-based DNN learning scheme can reduce a lot of training time and training data compared to traditional big data-driven learning approaches to enlarge the domain of application for DNN via the HJIE-reinforcement learning scheme of decentralized robust H_∞ attack-tolerant observer-based team formation tracking of large-scale UAV NCS.

The structure of the chapter is described as follows. In Section 6.2, a large-scale quadrotor UAV team formation network control system, smoothed model of malicious attack signal description, and problem formulation are introduced. In Section 6.3, the robust decentralized H_∞ attack-tolerant observer-based team formation tracking control strategy is provided. HJIE-reinforcement DNN-based decentralized H_∞ attack-tolerant observer-based team formation tracking control design of large-scale quadrotor UAV NCS is given in Section 6.4. In this section, we also propose the reinforcement learning-based Adam algorithm to train the DNN control scheme. A simulation example with comparison is given in Section 6.5. The conclusion is drawn in Section 6.6.

Notation: M^T: denotes the transpose of the matrix M; $M \geq 0 (M > 0)$: denotes the semi-positive definite symmetric (positive definite symmetric) matrix M, respectively; I_a: denotes the a-dimensional identity matrix; $\mathbb{R}^n$: the set of n-tuple real vectors; $\mathbb{R}^{nxm}$: the set of all real $n \times m$ matrices; $\|\cdot\|_2$: the Euclidean norm; $\pounds_2[0,\infty)$ denotes a set of real stochastic functions $x(t) \in \mathbb{R}^n$ with finite energy, that is, $E\left\{\int_0^\infty x^T(t)x(t)dt\right\} < 0$; $E\{\cdot\}$: the expectation operator; $diag(A_1,...A_n)$: a block diagonal matrix with main diagonal $A_1,...A_n$.

6.2 SYSTEM DESCRIPTION AND PRELIMINARIES

6.2.1 System Model of Quadrotor UAV in Large-Scale Team Formation NCS

In this section, we consider the position and attitude of the quadrotor UAV simultaneously in the dynamic model. The position of the quadrotor UAV with respect to inertial frame is described by Cartesian coordinates (x, y, z) of its mass center, and the

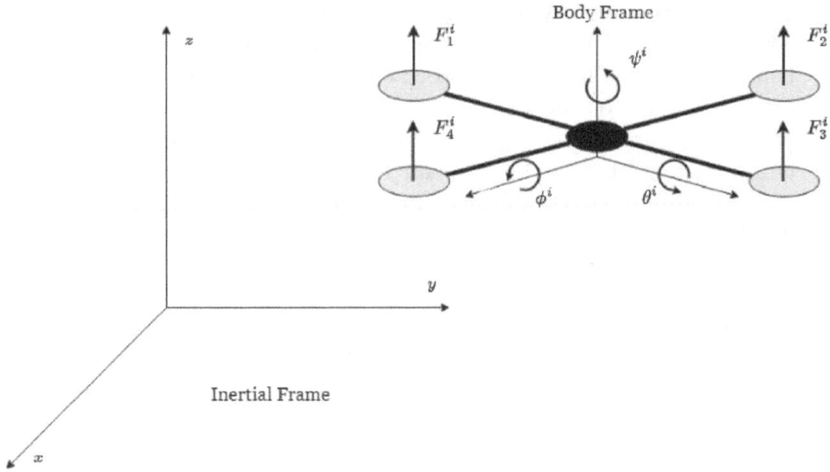

FIGURE 6.1 Structure of the ith quadrotor UAV in the large-scale quadrotor UAV formation team under the inertial frame and body frame.

attitude of the quadrotor UAV in body frame is described by three rotation Euler angles (ϕ, θ, ψ) on the body of the quadrotor UAV, which represent the roll angle $(-\frac{\pi}{2} < \phi < \frac{\pi}{2})$, the pitch angle $(-\frac{\pi}{2} < \theta < \frac{\pi}{2})$, and the yaw angle $(-\pi < \psi < \pi)$ related to the orientation of the quadrotor UAV, respectively, as shown in Figure 6.1.

For a large-scale quadrotor UAV team formation system, it is indicated as a set $S = \{S_1, S_2, ..., S_N\}$, consisting of single quadrotor UAV subsystems $S_i, i = 1,...N$. In practice, each single quadrotor UAV suffers not only from external disturbances but also from the effect of interconnected couplings by nonideal communications or aerodynamics from other nearby quadrotor UAVs. Taking these effects into consideration, the ith team formation quadrotor UAV subsystem can be formulated as follows:

$$\begin{aligned} S_i : \dot{X}_i(t) &= F_i(X_i(t)) + G_i(X_i(t))U_i(t) + D_i(t)v_i(t) \\ &+ \sum_{j \in N_i} F_{ij}(X_i(t))X_j(t - \tau_{ij}(t)) \\ Y_i(t) &= C_i(X_i(t)) + n_i(t), for\ i, j = 1, ..., N; i \neq j \end{aligned} \tag{6.1}$$

where

$X_i(t) = [x_1^i(t), x_2^i(t), y_1^i(t), y_2^i(t), z_1^i(t), z_2^i(t), \phi_1^i(t), \phi_2^i(t), \theta_1^i(t), \theta_2^i(t), \psi_1^i(t), \psi_2^i(t)]^T$ is the state vector of the system, $x_1^i(t), y_1^i(t), z_1^i(t) \in \mathbb{R}$ are the positions of the ith quadrotor UAV, $x_2^i(t), y_2^i(t), z_2^i(t) \in \mathbb{R}$ are the velocities of the ith quadrotor UAV, $\phi_1^i(t), \theta_1^i(t), \psi_1^i(t) \in \mathbb{R}$ are the attitudes of the ith quadrotor UAV, and $\phi_2^i(t), \theta_2^i(t), \psi_2^i(t) \in \mathbb{R}$ are the angular velocities of the ith quadrotor UAV. $U_i(t) = [F^i(t), \tau_\phi^i(t), \tau_\theta^i(t), \tau_\psi^i(t)]^T$ is the control input of the ith quadrotor UAV in the formation team,

$F^i(t) \in \mathbb{R}$ is the total thrust acting on the ith quadrotor UAV, the rotational forces $\tau_\phi^i(t), \tau_\theta^i(t), \tau_\psi^i(t) \in \mathbb{R}$ are generated by four rotors of the ith quadrotor UAV, $v_i(t) = [v_x^i(t), v_y^i(t), v_z^i(t), v_\phi^i(t), v_\theta^i(t), v_\psi^i(t)]^T$ is the external disturbance of the ith quadrotor UAV. $Y_i(t)$ is the measurement output of the ith quadrotor UAV. $n_i(t)$ is the measurement noise. $C_i(X_i(t))$ is the nonlinear output matrix. N denotes the number of quadrotor UAVs in the team formation NCS. The system matrices of the ith quadrotor UAV in (6.1) are given as:

$$F_i(X_i(t)) = \begin{bmatrix} x_2^i(t), \\ -\frac{K_x^i}{m_i} x_2^i(t), \\ y_2{}^i(t), \\ -\frac{K_y^i}{m_i} y_2^i(t), \\ z_2^i(t), \\ -\frac{K_z^i}{m_i} z_2^i(t) - g, \\ \phi_2^i(t), \\ \frac{J_y^i - J_z^i}{J_x^i} \theta_1^i(t)\psi_1^i(t) - \frac{K_\phi^i}{J_x^i} \phi_2^i(t), \\ \theta_2^i(t), \\ \frac{J_z^i - J_x^i}{J_x^i} \phi_1^i(t)\psi_1^i(t) - \frac{K_\theta^i}{J_y^i} \theta_2{}^i(t), \\ \psi_2^i(t), \\ \frac{J_x^i - J_y^i}{Jz} \phi_1^i(t)\theta_1^i(t) - \frac{K_\psi^i}{J_z^i} \psi_2^i(t) \end{bmatrix}$$

$$D_i(X_i(t)) = \begin{bmatrix} 0 & 0 & 0 & 0 & 0 & 0 \\ 1 & 0 & 0 & 0 & 0 & 0 \\ 0 & 0 & 0 & 0 & 0 & 0 \\ 0 & 1 & 0 & 0 & 0 & 0 \\ 0 & 0 & 0 & 0 & 0 & 0 \\ 0 & 0 & 1 & 0 & 0 & 0 \\ 0 & 0 & 0 & 0 & 0 & 0 \\ 0 & 0 & 0 & 1 & 0 & 0 \\ 0 & 0 & 0 & 0 & 0 & 0 \\ 0 & 0 & 0 & 0 & 1 & 0 \\ 0 & 0 & 0 & 0 & 0 & 0 \\ 0 & 0 & 0 & 0 & 0 & 1 \end{bmatrix}$$

$$G_i(X_i(t)) = \begin{bmatrix} 0 & 0 & 0 & 0 \\ \begin{array}{c}(\cos\phi_1^i(t)\sin\theta_1^i(t)\cos\psi_1^i(t) \\ +\sin\phi_1^i(t)\sin\psi_1^i(t))\frac{1}{m_i}\end{array} & 0 & 0 & 0 \\ 0 & 0 & 0 & 0 \\ \begin{array}{c}(\cos\phi_1^i(t)\sin\theta_1^i(t)\sin\psi_1^i(t) \\ -\sin\phi_1^i(t)\sin\psi_1^i(t))\frac{1}{m_i}\end{array} & 0 & 0 & 0 \\ 0 & 0 & 0 & 0 \\ \frac{1}{m_i}\cos\phi_1^i(t)\cos\theta_1^i(t) & 0 & 0 & 0 \\ 0 & 0 & 0 & 0 \\ 0 & \frac{1}{J_x^i} & 0 & 0 \\ 0 & 0 & 0 & 0 \\ 0 & 0 & \frac{1}{J_y^i} & 0 \\ 0 & 0 & 0 & 0 \\ 0 & 0 & 0 & \frac{1}{J_z^i} \end{bmatrix}$$

where $F_i(X_i(t))$ is the nonlinear dynamic vector of the ith quadrotor UAV. g is the gravitational acceleration; J_x^i, J_y^i, and J_z^i are the moments of inertia of the ith quadrotor UAV; $K_x^i, K_y^i, K_z^i, K_\phi^i, K_\theta^i$, and K_ψ^i represent aerodynamic damping coefficient of the ith quadrotor UAV; and m_i is the total mass of the ith quadrotor UAV. $G_i(X_i(t)) \in \mathbb{R}^{12\times4}$ is the input matrix of the ith quadrotor UAV. $F_{ij}(X_i(t)) \in \mathbb{R}^{12\times12}$ is the coupling matrix from the jth quadrotor UAV to the ith quadrotor UAV. $X_j(t-\tau_{ij}(t)) \in \mathbb{R}^{12}$ is the coupling term from the jth quadrotor UAV to the ith quadrotor UAV with a time-varying delay $\tau_{ij}(t)$ N_i denotes the neighboring quadrotor UAVs coupling with the ith quadrotor UAV.

6.2.2 A Large-Scale Quadrotor UAV Team Formation Network Control System

Under the structure of the network control system (NCS) of large-scale quadrotor UAV team formation as shown in Figure 6.2, each quadrotor UAV is controlled through an observer-based controller at the remote side. Through the sensor, output measurement data $Y_i(t)$ of the ith quadrotor UAV is sampled and transmitted to the observer at the remote side. Then, the control signals are computed at each observer-based controller at the remote side and besides transmitted via wireless network channel to the zero-order holder (ZOH) which is set on the quadrotor UAV at the local side of team formation NCS, as shown in Figure 6.2 at every sampling period. Once the sampled data with control commands arrives at the ZOH of the quadrotor UAV in the team formation NCS, the ZOH will convert the sampled data

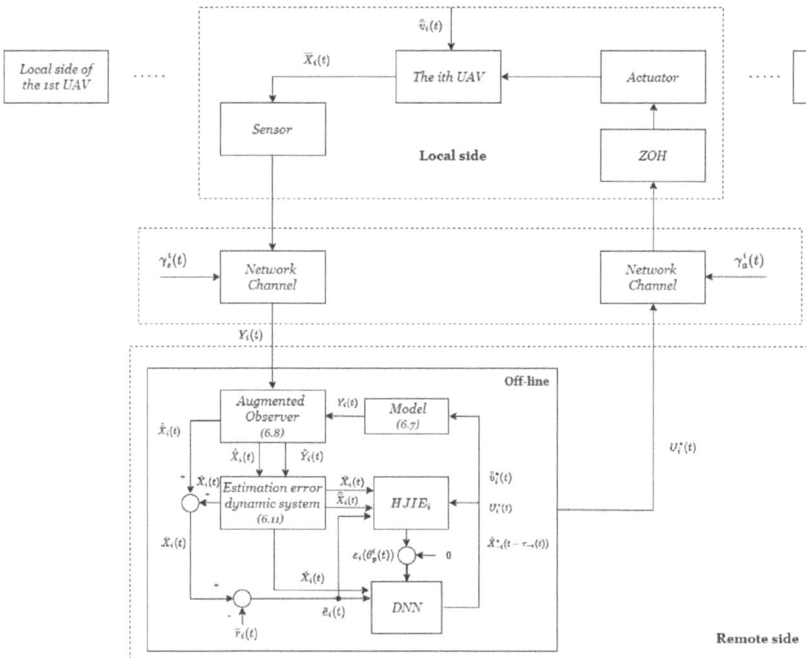

FIGURE 6.2 Structure of team formation NCS of large-scale quadrotor UAVs. In the offline training phase, we need the stochastic augmented model of each quadrotor UAV to produce $Y_i(t)$ with the worst-case external disturbance $\tilde{\bar{v}}_i^*(t)$ and worst-case coupling $\tilde{\bar{X}}_{-i}^*(t-\tau_{-i})$, and then augmented Luenberger observer and estimation error model are employed to produce $\hat{\bar{X}}_i(t)$ and $\tilde{X}_i(t)$ to obtain $\bar{X}_i(t) = \hat{\bar{X}}_i(t) + \tilde{X}_i(t)$ and to get formation tracking error $\bar{e}_i(t) = \bar{X}_i(t) - \bar{r}_i(t)$. Then $\bar{e}_i(t)$ and $\tilde{X}_i(t)$ are transmitted into DNN to train DNN by HJIE-reinforcement learning to output $\frac{\partial V(\hat{\bar{X}}_i(t), \bar{e}_i(t), t)}{\partial[\tilde{X}_i^T(t)\,\bar{e}_i^T(t)\,t]^T}$ to produce H_∞ tracking control $U_i^*(t)$ and observer gain $L_i^*(\widehat{\bar{X}_i}(t))$ until HJIE-reinforcement learning converge as shown in Figure 6.3. In the online operation phase, the output .. can be obtained from the real quadrotor UAV. We don't need the worst-case $\tilde{\bar{v}}_i^*(t)$, $\tilde{\bar{X}}_{-i}^*(t-\tau_{-i})$ to produce $Y_i(t)$ by the augmented stochastic model of quadrotor UAV in (6.7). (See Figure 6.3 in detail)

into continuous control signals and then transmit them to the actuator to control the quadrotor UAV. The control commands are stored in ZOH and kept until new control signals are received to update.

For NCS, when signals are transmitted through the wireless network channels, some hackers may attack NCS, which can even make NCS unstable. Hence, we need to take these malicious attacks into consideration in the NCS of the large-scale quadrotor UAV team formation system via a wireless network. In Figure 6.2, two malicious attack signals $\gamma_a^i(t)$ and $\gamma_s^i(t)$ will interfere with the performance of both the reference tracking and the state estimation of the NCS of the large-scale quadrotor UAV team formation system via wireless network channels, respectively.

In this case, the sensor and actuator would receive erroneous information from the wireless network channel, and their corruption could lead to the entire team formation system of large-scale quadrotor UAVs being out of control. That is, the two malicious attack signals $\gamma_a^i(t)$ and $\gamma_s^i(t)$ can be equivalently referred to as the sensor attack signal and the actuator attack signal in each quadrotor UAV NCS of team formation of large-scale quadrotor UAVs. Then, the ith quadrotor UAV NCS in (6.1) under actuator attack and sensor attack via the wireless network channel should be modified as follows:

$$\begin{aligned}\dot{X}_i(t) &= F_i(X_i(t)) + G_i(X_i(t))(U_i(t) + \gamma_a^i(t)) + D_i(X_i(t))v_i(t) \\ &\quad + \sum_{j\in N_i} F_{ij}(X_i(t))X_j(t-\tau_{ij}(t)) \\ Y_i(t) &= C_i(X_i(t)) + n_i(t) + D_s^i(X_i(t))\gamma_s^i(t), \\ &\quad for\ i,j = 1,\ldots,N; i \neq j\end{aligned} \tag{6.2}$$

where $\gamma_a^i(t)$ denotes the actuator attack signal, $\gamma_s^i(t)$ denotes the sensor attack signal, and $D_s^i(X_i(t))$ denotes the effect matrix of sensor attack signal $\gamma_s^i(t)$.

Assumption 6.1 Before the start of state estimation and team formation tracking control process of quadrotor UAVs, the system state $X_i(t)$, actuator attack signal $\gamma_a^i(t)$, sensor attack signal $\gamma_s^i(t)$, and external disturbance $v_i(t)$ of the ith quadrotor UAV in nonlinear NCS of team formation of quadrotor UAVs are assumed to be zero, that is, $X_i(t)=0$, $\gamma_a^i(t)=0$, $\gamma_s^i(t)=0$, and $v_i(t)=0$, $\forall t<0$, for $i=1,\ldots,N$.

As the actuator attack $\gamma_a^i(t)$ and the sensor attack $\gamma_s^i(t)$ are not available, they cannot be estimated directly by the traditional Luenberger observer. In order to estimate attack signals $\gamma_a^i(t)$ and $\gamma_s^i(t)$, the novel dynamic smoothed models of the attack signals $\gamma_a^i(t)$ and $\gamma_s^i(t)$ in Reference [29] are employed for avoiding their corruption on state estimation and tracking control of quadrotor UAVs in the team formation NCS. By the novel dynamic smoothed models of the attack signals $\gamma_a^i(t)$ and $\gamma_s^i(t)$, the state of the system and the attack signals can be estimated by the traditional Luenberger observer simultaneously. Similar to Reference [29], the dynamic smoothed model is proposed for malicious attack signals $\gamma_a^i(t)$ and $\gamma_s^i(t)$. Initially, by the right derivative of the actuator attack signal $\dot{\gamma}_a^i(t) = \lim_{\tau\to 0} \frac{\gamma_a^i(t+\tau)-\gamma_a^i(t)}{\tau}$, we can construct the following approximations:

$$\begin{aligned}\dot{\gamma}_a^i(t) &= \tfrac{1}{\tau}(\gamma_a^i(t+\tau) - \gamma_a^i(t)) + \epsilon_{1,a}^i(t), \\ \dot{\gamma}_a^i(t-\tau) &= \tfrac{1}{\tau}(\gamma_a{}^i(t) - \gamma_a^i(t-\tau)) + \epsilon_{2,a}^i(t), \\ &\vdots \\ \dot{\gamma}_a^i(t-k\tau) &= \tfrac{1}{\tau}(\gamma_a{}^i(t-(k-1)\tau) - \gamma_a^i(t-k\tau)) + \epsilon_{k,a}^i(t)\end{aligned} \tag{6.3}$$

where $\epsilon_{1,a}^i(t), \epsilon_{2,a}^i(t), ..., \epsilon_{k,a}^i(t)$ denote the derivative approximation errors of $\dot{\gamma}_a^i(t), \dot{\gamma}_a^i(t-\tau), ...\dot{\gamma}_a^i(t-k\tau)$, respectively, and $\tau > 0$ denotes a small time interval and $k \in \mathbb{N}$ denotes the number of delay samples. Moreover, the future attack signal $\gamma_a^i(t+\tau)$ could be extrapolated as follows:

$$\gamma_a^i(t+\tau) = \sum_{n=0}^{k} a_n \gamma_a{}^i(t-n\tau) + \delta_a^i(t), \tag{6.4}$$

where $\{a_n\}_{n=0}^k$ are extrapolation coefficients such as the Lagrange extrapolation approach [35] or other extrapolation approaches with $\sum_{n=0}^{k} a_n = 1$, $\delta_a^i(t)$ is the extrapolation error of $\gamma_a^i(t+\tau)$. By combing (6.3) and (6.4), the smoothed model of the actuator attack signal $\gamma_a^i(t)$ can be constructed as

$$\dot{\Gamma}_a^i(t) = A_{\gamma_a} \Gamma_a^i(t) + \epsilon_a^i(t) \tag{6.5}$$

where $\Gamma_a^i(t) = [\gamma_a^{iT}(t), \gamma_a^{iT}(t-\tau), ..., \gamma_a^{iT}(t-k\tau)]^T$, $\epsilon_a^i(t) = [(\epsilon_{1,a}^i(t) + \delta_a^i(t)/\tau)^T$, $\epsilon_{2,a}^{i\ T}(t), ..., \epsilon_{k,a}^{i\ T}(t)]^T$ denotes the derivative approximation error, and

$$A_{\gamma_a} = \begin{bmatrix} \frac{\bar{a}_0}{\tau} I_{n_a} & \frac{a_1}{\tau} I_{n_a} & \frac{a_2}{\tau} I_{n_a} & \cdots & \frac{a_k}{\tau} I_{n_a} \\ \frac{1}{\tau} I_{n_a} & -\frac{1}{\tau} I_{n_a} & & \cdots & 0 \\ 0 & \frac{1}{\tau} I_{n_a} & -\frac{1}{\tau} I_{n_a} & \cdots & 0 \\ \vdots & \vdots & \ddots & \ddots & \vdots \\ 0 & \cdots & 0 & \frac{1}{\tau} I_{n_a} & -\frac{1}{\tau} I_{n_a} \end{bmatrix},$$

with $\bar{a}_0 = -1 + a_0$. Similar to the above, the smoothed model of the sensor attack signal $\gamma_s^i(t)$ can be constructed as:

$$\dot{\Gamma}_s^i(t) = A_{\gamma_s} \Gamma_s^i(t) + \epsilon_s^i(t) \tag{6.6}$$

where $\Gamma_s^i(t) = [\gamma_s^{iT}(t), \gamma_s^{iT}(t-\tau), ..., \gamma_s^{iT}(t-k\tau)]^T$, $\epsilon_s^i(t) = [(\epsilon_{1,s}^i(t) + \delta_s^i(t)/\tau)^T$, $\epsilon_{2,s}^{i\ T}(t), ..., \epsilon_{k,s}^{i\ T}(t)]^T$, and

$$A_{\gamma s} = \begin{bmatrix} \frac{\bar{b}_0}{\tau} I_{n_a} & \frac{b_1}{\tau} I_{n_a} & \frac{b_2}{\tau} I_{n_a} & \cdots & \frac{b_k}{\tau} I_{n_a} \\ \frac{1}{\tau} I_{n_a} & -\frac{1}{\tau} I_{n_a} & & \cdots & 0 \\ 0 & \frac{1}{\tau} I_{n_a} & -\frac{1}{\tau} I_{n_a} & \cdots & 0 \\ \cdot & \cdot & & & \cdot \\ \cdot & & \cdot & & \cdot \\ \cdot & & & \cdot & \cdot \\ 0 & \cdots & 0 & \frac{1}{\tau} I_{n_a} & -\frac{1}{\tau} I_{n_a} \end{bmatrix},$$

with $\bar{b}_0 = -1 + b_0$ and $b_i, i = 0, \ldots, k$ are the extrapolation coefficient with $\sum_{i=0}^{k} b_i = 1$. Then, to estimate $X_i(t), \gamma_a^i(t), \gamma_s^i(t)$ simultaneously, we can augment these states as $\bar{X}_i(t) = \left[X_i^T(t) \quad \Gamma_a^{iT}(t) \quad \Gamma_s^{iT}(t) \right]^T$, and the corresponding ith augmented quadrotor UAV team formation NCS is given as follows:

$$\begin{aligned} \dot{\bar{X}}_i(t) &= \bar{F}_i(\bar{X}_i(t)) + \bar{G}_i(\bar{X}_i(t))U_i(t) + \bar{D}_i(\bar{X}_i(t))\bar{v}_i(t) \\ &\quad + \sum_{j \in N_i} \bar{F}_{ij}(\bar{X}_i(t))\bar{X}_j(t - \tau_{ij}(t)) \\ Y_i(t) &= \bar{C}_i(\bar{X}_i(t)) + n_i(t), \\ &\quad for\ i, j = 1, \ldots, N; i \neq j \end{aligned} \tag{6.7}$$

where,

$$\bar{F}_i(\bar{X}_i(t)) = \begin{bmatrix} F_i(X_i(t)) + \bar{G}_i(\bar{X}_i(t))S_a\Gamma_a^i(t) \\ A_{\gamma a}^i \Gamma_a^i(t) \\ A_{\gamma s}^i \Gamma_s^i(t) \end{bmatrix}, \ S_a = [I_{na}, 0, \ldots, 0],$$

$$\bar{G}_i(\bar{X}_i(t)) = \begin{bmatrix} G_i(X_i(t)) \\ 0 \\ 0 \end{bmatrix}, \ \bar{D}_i(\bar{X}_i(t)) = \begin{bmatrix} D_i(X_i(t)) & 0 & 0 \\ 0 & I & 0 \\ 0 & 0 & I \end{bmatrix},$$

$$\bar{v}_i(t) = \begin{bmatrix} v_i(t) \\ \epsilon_a^i(t) \\ \epsilon_s^i(t) \end{bmatrix}, \bar{F}_{ij}(\bar{X}_i(t)) = \begin{bmatrix} F_{ij}(X_i(t)) & 0 & 0 \\ 0 & I & 0 \\ 0 & 0 & I \end{bmatrix},$$

$$\bar{X}_j(t - \tau_{ij}(t)) = \left[X_j^T(t - \tau_{ij}(t)), 0, 0 \right]^T,$$

$$\bar{C}_i(\bar{X}_i(t)) = C_i(X_i(t)) + D_s^i S_s \Gamma_s^i(t), S_s = [I_{ns}, 0, \ldots, 0].$$

In the ith augmented quadrotor UAV NCS of the team formation of large-scale quadrotor UAVs in (6.7), the attack signals $\gamma_a^i(t)$ and $\gamma_s^i(t)$ are embedded in the augmented state $\bar{X}_i(t)$ so that their corruption on state estimation and control can be avoided by the decentralized H_∞ attack-tolerant observer-based team formation tracking control strategy of the large-scale UAV NCS in the sequel. Generally, it is not easy to ensure the observability of the ith augmented nonlinear quadrotor UAV system in (6.7). To facilitate observer design in the sequel, the Assumption 6.2 is made as follows.

Assumption 6.2 Each augmented quadrotor UAV NCS in (6.7) is observable.

Then, the Luenberger observer-based control law for each augmented quadrotor UAV NCS in (6.7) is proposed as:

$$\begin{aligned}
&\dot{\hat{\bar{X}}}_i(t)=\bar{F}_i(\widehat{\bar{X}_i}(t))+\bar{G}_i(\widehat{\bar{X}_i}(t))U_i(t)+L_i(\widehat{\bar{X}_i}(t))\left(Y_i(t)-\hat{Y}_i(t)\right)\\
&\hat{Y}_i(t)=\bar{C}_i(\widehat{\bar{X}_i}(t))
\end{aligned} \tag{6.8}$$

where $\widehat{\bar{X}_i}(t)$ is the estimated state of the ith augmented quadrotor UAV NCS in (6.7), $\hat{Y}_i(t)$ denotes the estimated measurement output and $L_i(\widehat{\bar{X}_i}(t))$ is the nonlinear observer gain.

In a practical application, we would like to control the state $X_i(t)$ of the ith quadrotor UAV to follow a specified reference $r_i(t)$ to achieve some tasks, where $[r_1(t),...,r_i(t),...,r_N(t)]^T$ is specified as a desired time-varying formation shape in the team formation of N quadrotor UAVs by designer. Then the reference tracking error dynamic of the ith quadrotor UAV in the team formation NCS is indicated as follows:

$$\begin{aligned}
\dot{\bar{e}}_i(t) &=\dot{\bar{X}}_i(t)-\dot{\bar{r}}_i(t)\\
&=F_{e,i}(\bar{e}_i(t),t)+G_{e,i}(\bar{e}_i(t),t)U_i(t)+D_{e,i}(\bar{e}_i(t),t)\times\bar{v}_i(t)\\
&\quad+\sum_{j\in N_i}F_{e,ij}(\bar{e}_i(t),t)\bar{X}_j(t-\tau_{ij}(t))
\end{aligned} \tag{6.9}$$

where $\bar{r}_i(t)=[r_i^T(t)\quad 0\quad 0]^T, F_{e,i}(\bar{e}_i(t),t)=\bar{F}_i(\bar{e}_i(t)+\bar{r}_i(t))-\dot{\bar{r}}_i(t),$
$G_{e,i}(\bar{e}_i(t),t)=\bar{G}_i(\bar{e}_i(t)+\bar{r}_i(t)), D_{e,i}(\bar{e}_i(t),t)=\bar{D}_i(\bar{e}_i(t)+\bar{r}_i(t)),$
$F_{e,ij}(\bar{e}_i(t),t)=\bar{F}_{ij}(\bar{e}_i(t)+\bar{r}_i(t))$

6.2.3 Problem Description

In practical applications under some situations, certain tasks need to be achieved by a team formation of multi-quadrotor UAVs. Thence, the team formation problem of quadrotor UAVs should be solved to achieve a certain task. In this chapter, we employed a virtual leader-based team formation structure to address the large-scale quadrotor UAVs team formation tracking control design problem. In order to construct the desired time-varying formation shape $[r_1(t),...,r_i(t),...,r_N(t)]^T$ of large-scale quadrotor UAVs, each quadrotor UAV should move a specific formation offset $r_i(t)$ from the virtual leader, respectively, and then the ith quadrotor UAV will follow the trajectory of the virtual leader and move the formation offset $r_i(t)$ to maintain the desired formation shape. Hence, the problem of team formation tracking control for large-scale quadrotor UAVs can be converted into N decentralized observer-based reference tracking control problems of each quadrotor UAV in the team formation. In practice, malicious attacks may appear in the team formation network control system. Then, we employed the decentralized nonlinear Luenberger observer in (6.8) to estimate each quadrotor UAV's state $X_i(t)$ as well as malicious attacks $\gamma_a^i(t)$ and $\gamma_s^i(t)$ on actuator and sensor, respectively. If malicious attack signals on the actuator and sensor of each quadrotor UAV can be effectively estimated, their effects on the quadrotor UAV NCS can be reduced or even eliminated by the compensation of observer-based reference tracking control. On the other hand, intending to make each quadrotor UAV track the desired trajectory $r_i(t)$ efficiently, an attack-tolerant observer-based tracking control $U_i(t) = K_i(\hat{\bar{X}}_i(t), \bar{e}_i(t))$ is needed for each UAV based on $\hat{\bar{X}}_i(t)$ and $\bar{e}_i(t)$. The external disturbances such as coupling effects from neighboring quadrotor UAVs and measurement noise during the team flight process are unavoidable, too. The robust decentralized H_∞ attack-tolerant observer-based team formation tracking control strategy of large-scale quadrotor UAVs NCS is proposed to effectively attenuate these undesired effects on the observer-based reference tracking control performance of each quadrotor UAV in the team formation NCS:

$$\min_{\substack{K_i(\hat{\bar{X}}_i(t),\bar{e}_i(t)),\\ L_i(\hat{\bar{X}}_i(t))}} \max_{\substack{\bar{v}_i(t),\\ n_i(t),\\ \bar{X}_j(t-\tau_{ij}(t))\\ \in \pounds_2[0,t_f]}} \frac{\begin{array}{l} E\{\int_0^{t_f}[(\bar{X}_i(t)-\widehat{\bar{X}}_i(t))^T Q_{1,i}(\bar{X}_i(t)\\ -\widehat{\bar{X}}_i(t)) + \bar{e}_i^T(t) Q_{2,i}\bar{e}_i(t)\\ +U_i^T(t)R_iU_i(t) - V(\widetilde{X}_i(0),\bar{e}_i(0),0)]dt\} \end{array}}{\begin{array}{l} E\{\int_0^{t_f}[\bar{v}_i^T(t)\bar{v}_i(t) + n_i^T(t)n_i(t)\\ +\sum\limits_{j\in N_j}\bar{X}_j^T(t-\tau_{ij}(t))\bar{X}_j(t-\tau_{ij}(t))]dt\} \end{array}} \le \rho_i^2, \tag{6.10}$$

$$for\ i = 1,\dots,N$$

where t_f denotes the terminal time; $Q_{1,i} \ge 0, Q_{2,i} \ge 0$ and $R_i = R_i^T \ge 0$ are the weighting matrices to trade-off between the estimation error, team formation tracking error and control input. ρ_i denotes the prescribed attenuation level; $V(\overline{X_i}(0) - \hat{\bar{X}}_i(0), \bar{e}_i(0), 0)$ represents the effect of nonzero initial condition of the augmented system in (6.7) to be extracted from the decentralized H_∞ attack-tolerant observer-based team formation tracking strategy in (6.10). The physical meaning of the robust decentralized H_∞ attack-tolerant observer-based team formation tracking control strategy is how to attenuate the worst-case effect of external disturbance, measurement noise, and couplings from other quadrotor UAVs on the state estimation error of observer, team formation tracking error, and control efforts below a prescribed level ρ_i^2 by specifying the observer-based control gain $K_i(\hat{\bar{X}}_i(t)$ and observer gain $L_i(\widehat{\bar{X}_i}(t))$ for each quadrotor UAV in the team formation from the mean energy perspective. Since the worst-case effect of all possible finite-energy interconnected couplings from other quadrotor UAVs is minimized, the decentralized observer and controller of each quadrotor UAV in the team formation NCS will be achieved from the mean energy perspective, which will be confirmed in the sequel.

6.3 DECENTRALIZED H_∞ ATTACK-TOLERANT OBSERVER-BASED TEAM FORMATION TRACKING CONTROL OF LARGE-SCALE QUADROTOR UAV NCS

In this section, the decentralized H_∞ attack-tolerant observer-based team formation tracking control design of quadrotor UAV NCS is investigated. First, we formulate the state estimation error as $\widetilde{X_i}(t) = \overline{X_i}(t) - \widehat{\overline{X_i}}(t)$ and the corresponding estimation error dynamic equation is derived as follows:

$$\begin{aligned}\dot{\widetilde{X_i}}(t) &= \widetilde{F_i}(\widetilde{X_i}(t)) + \widetilde{G_i}(\widetilde{X_i}(t))U_i(t) - L_i(\widehat{\overline{X_i}}(t))\widetilde{C_i}(\widetilde{X_i}(t)) \\ &\quad + \sum_{j\in N_i} \bar{F}_{ij}(\bar{X}_i(t))\bar{X}_j(t-\tau_{ij}(t)) + \left[\bar{D}_i(\bar{X}_i(t)) - L_i(\widehat{\overline{X_i}}(t))\right]\bar{\tilde{v}}_i(t)\end{aligned} \tag{6.11}$$

where $\widetilde{F_i}(\widetilde{X_i}(t)) = \overline{F_i}(\bar{X}_i(t)) - \overline{F_i}(\widehat{\overline{X_i}}(t)), \widetilde{G_i}(\widetilde{X_i}(t)) = \bar{G}_i(\bar{X}_i(t)) - \overline{G_i}(\widehat{\overline{X_i}}(t))$ $\widetilde{C_i}(\widetilde{X_i}(t)) = \bar{C}_i(\bar{X}_i(t)) - \bar{C}_i(\widehat{\overline{X_i}}(t))$, $\bar{\tilde{v}}_i(t) = \begin{bmatrix}\bar{v}_i(t) \\ n_i(t)\end{bmatrix}$. Moreover, we can augment the state estimation error $\tilde{X}_i(t)$ in (6.11) and team formation tracking error $\bar{e}_i(t)$ in (6.9), then the robust decentralized H_∞ attack-tolerant observer-based team formation tracking control strategy for large-scale quadrotor UAV NCS can be rewritten as follows:

$$\min_{\substack{K_i(\hat{\bar{X}}_i(t),\bar{e}_i(t)) \\ L_i(\hat{\bar{X}}_i(t))}} \max_{\substack{\tilde{\bar{v}}_i(t), \\ \tilde{\bar{X}}_{-i}(t-\tau_{-i}(t)) \\ \in \mathcal{L}_2[0,tf]}} \frac{\begin{aligned}&E\{\int_0^{t_f}[\tilde{\bar{X}}_i^T(t)\bar{Q}_i\tilde{\bar{X}}_i(t) \\ &+U_i^T(t)R_iU_i(t) - V(\tilde{\bar{X}}_i(0),\bar{e}_i(0),0)]dt\}\end{aligned}}{\begin{aligned}&E\{\int_0^{t_f}[\tilde{\bar{v}}_i^T(t)\tilde{\bar{v}}_i(t) \\ &+\tilde{\bar{X}}_{-i}^T(t-\tau_{-i}(t))\tilde{\bar{X}}_{-i}(t-\tau_{-i}(t))]dt\}\end{aligned}} \le \rho_i^2, \tag{6.12}$$

$for\ i = 1,\ldots,N$

where $\tilde{\bar{X}}_i(t) = \begin{bmatrix}\tilde{X}_i^T(t) & \bar{e}_i^T(t)\end{bmatrix}^T, \bar{Q}_i = diag\{Q_{1,i}, Q_{2,i}\}, \tilde{\bar{X}}_{-i}(t-\tau_{-i}(t)) = \begin{bmatrix}\ldots\bar{X}_j(t-\tau_{ij}(t))\ldots \\ \ldots\bar{X}_j(t-\tau_{ij}(t))\ldots\end{bmatrix}$

Thus, the robust decentralized H_∞ attack-tolerant observer-based team formation tracking control strategy for large-scale quadrotor UAVs NCS in (6.10) turns into the robust decentralized H_∞ stabilization strategy in (6.12) of the following augmented time-varying error system of large-scale quadrotor UAV team formation NCS:

$$\begin{aligned}\dot{\tilde{\bar{X}}}_i(t) &= \begin{bmatrix}\widetilde{F}_i(\widetilde{X}_i(t)) - L_i(\widehat{\widetilde{X}}_i(t))\widetilde{C}_i(\widetilde{X}_i(t)) \\ F_{e,i}(\overline{e}_i(t),t)\end{bmatrix} + \begin{bmatrix}\widetilde{G}_i(\widetilde{X}_i(t)) \\ G_{e,i}(\overline{e}_i(t),t)\end{bmatrix} U_i(t) \\ &+ \begin{bmatrix}\overline{D}_i(\overline{X}_i(t)) & -L_i(\widehat{\widetilde{X}}_i(t)) \\ D_{e,i}(\overline{e}_i(t),t) & 0\end{bmatrix} \tilde{\bar{v}}_i(t) \\ &+ \begin{bmatrix}\overline{F}_{-i}(\overline{X}_i(t)) \\ F_{e,-i}(\overline{e}_i(t),t)\end{bmatrix} \tilde{\bar{X}}_{-i}(t-\tau_{-i}(t))\end{aligned} \tag{6.13}$$

where

$$F_{-i}(\overline{X}_i(t)) = \left[\ldots\overline{F}_{ij}(\overline{X}_i(t))\ldots\right],\; F_{e,-i}(\overline{e}_i(t),t) = \left[\ldots F_{e,ij}(\overline{e}_i(t),t)\ldots\right]$$

After the above operation, the complicated robust decentralized H_∞ attack-tolerant observer-based team formation control design problem of large-scale quadrotor UAV NCS in (6.10) is converted into how to solve min-max decentralized H_∞ stabilization design problem in (6.12) for the nonlinear augmented system in (6.13) to simplify the design procedure.

Nevertheless, it is still difficult to directly solve the min-max game problem of the fractional payoff function in (6.12). Since $\tilde{\bar{v}}_i(t)$ and $\tilde{\bar{X}}_{-i}(t-\tau_{-i}(t))$ seek to maximize the payoff function while other players $K_i(\hat{\bar{X}}_i(t),\overline{e}_i(t))$ and $L_i(\widehat{\widetilde{X}}_i(t))$ seek to minimize the payoff function. Since the players $\tilde{\bar{v}}_i(t)$ and $\tilde{\bar{X}}_{-i}(t-\tau_{-i}(t))$ in the denominator $\tilde{\bar{v}}_i(t)$ are independent of other players $K_i(\hat{\bar{X}}_i(t),\overline{e}_i(t))$ and $L_i(\widehat{\widetilde{X}}_i(t))$. We employed the indirect two-step method [25] in the following to solve the min-max Nash game problem in (6.12). Thus, the min-max H_∞ game problem in (6.12) is equivalent to the following nonzero sum Nash min-max quadratic game problem [25, 28]:

$$\begin{aligned}&\min_{\substack{K_i(\hat{\bar{X}}_i(t),\overline{e}_i(t)) \\ L_i(\hat{\bar{X}}_i(t))}} \; \max_{\substack{\tilde{\bar{v}}_i(t), \\ \tilde{\bar{X}}_{-i}(t-\tau_{-i}(t)) \\ \in \mathcal{L}_2[0,tf]}} \; E\{\textstyle\int_0^{t_f} \tilde{\bar{X}}_i^T(t)\overline{Q}_i\tilde{\bar{X}}_i(t) + U_i^T(t)R_iU_i(t) \\ &-\rho_i^2(\tilde{\bar{v}}_i^T(t)\tilde{\bar{v}}_i(t) + \tilde{\bar{X}}_{-i}^T(t-\tau_{-i}(t)) \times \tilde{\bar{X}}_{-i}(t-\tau_{-i}(t)))dt\} \\ &\leq E\{V(\tilde{\bar{X}}_i(0),\overline{e}_i(0),0)\},\; for\; i = 1,\ldots,N\end{aligned} \tag{6.14}$$

Then, the constrained min-max Nash quadratic game problem in (6.14) could be solved by a two-step method as follows: (i) at first, we solve the following Nash min-max quadratic game problem:

$$J_i = \min_{\substack{K_i(\hat{\tilde{X}}_i(t),\bar{e}_i(t)) \\ L_i(\hat{\tilde{X}}_i(t))}} \max_{\substack{\tilde{\bar{v}}_i(t), \\ \tilde{\bar{X}}_{-i}(t-\tau_{-i}(t)) \\ \in \pounds_2[0,tf]}} E\{\int_0^{t_f} \tilde{\bar{X}}_i^T(t)\bar{Q}_i\widetilde{\tilde{X}}_i(t)+U_i^T(t)R_iU_i(t) - \rho_i^2(\tilde{\bar{v}}_i^T(t)\tilde{\bar{v}}_i(t)+\tilde{\bar{X}}_{-i}^T(t-\tau_{-i}(t))\times\tilde{\bar{X}}_{-i}(t-\tau_{-i}(t)))dt\}, \quad for\ i=1,\ldots,N \tag{6.15}$$

(ii) In the second step, $J_i \le E\{V(\widetilde{\tilde{X}}_i(0),\bar{e}_i(0),0)\}$ needs to be guaranteed.

According to the above two-step method, the following theorem can solve the robust decentralized H_∞ attack-tolerant observer-based team formation tracking control strategy of large-scale quadrotor UAV NCS in (6.12).

Theorem 6.1 (a) The robust decentralized H_∞ attack-tolerant observer-based team formation tracking control strategy for large-scale quadrotor UAV NCS in (6.12) can be solved by the following H_∞ control $U_i^*(t)$, observer gain $L_i^*(\widehat{\tilde{X}_i}(t))$, the worst-case external disturbances $\tilde{\bar{v}}_i^*(t)$, and interconnected coupling effects $\tilde{\bar{X}}_{-i}^*(t-\tau\_i(t))$:

$$\tilde{\bar{v}}_i^*(t)=\frac{1}{2\rho_i^2}\begin{bmatrix}\bar{D}_i(\bar{X}_i(t)) & -L_i(\hat{\tilde{X}}_i(t))\\ D_{e,i}(\bar{e}_i(t),t) & 0\end{bmatrix}^T\times(\frac{\partial V(\tilde{X}_i(t),\bar{e}_i(t),t)}{\partial[\tilde{X}_i^T(t)\,\bar{e}_i^T(t)]^T}) \tag{6.16}$$

$$U_i^*(t)=K_i^*(\hat{\tilde{X}}_i(t),\bar{e}_i(t))=-\frac{1}{2}R_i^{-1}\begin{bmatrix}\tilde{G}_i(\tilde{X}_i(t))\\ G_{e,i}(\bar{e}_i(t),t)\end{bmatrix}^T\times(\frac{\partial V(\tilde{X}_i(t),\bar{e}_i(t),t)}{\partial[\tilde{X}_i^T(t)\,\bar{e}_i^T(t)]^T}) \tag{6.17}$$

$$\tilde{\bar{X}}_{-i}^*(t-\tau_{-i}(t))=\frac{1}{2\rho_i^2}\begin{bmatrix}\bar{F}_{-i}(\overline{X_i}(t))\\ F_{e,-i}(\bar{e}_i(t),t)\end{bmatrix}^T\times(\frac{\partial V(\tilde{X}_i(t),\bar{e}_i(t),t)}{\partial[\tilde{X}_i^T(t)\,\bar{e}_i^T(t)]^T}) \tag{6.18}$$

$$L_i^*(\widehat{\tilde{X}_i}(t))=\frac{1}{2}\frac{\frac{\partial V(\tilde{X}_i(t),\bar{e}_i(t),t)}{\partial \tilde{X}_i(t)}}{\left\|\frac{\partial V(\tilde{X}_i(t),\bar{e}_i(t),t)}{\partial \tilde{X}_i(t)}\right\|^2}\tilde{C}_i^T(\widetilde{X}_i(t)),\ for\ i=1,\ldots,N \tag{6.19}$$

where the Lyapunov function $V(\tilde{X}_i(t),\bar{e}_i(t),t)\ge 0$ with $V(0,0,t)=0$ is the solution of the observer/controller-coupled HJIE_i as follows:

$$\begin{aligned}HJIE_i = &\frac{\partial V(\tilde{X}_i(t),\bar{e}_i(t),t)}{\partial t}+\tilde{\bar{X}}_i^T(t)\bar{Q}_i\tilde{\bar{X}}_i(t)\\ &-\frac{1}{4}(\frac{\partial V(\tilde{X}_i(t),\bar{e}_i(t),t)}{\partial[\tilde{X}_i^T(t)\,\bar{e}_i^T(t)]^T})^T\tilde{\bar{G}}_i(\tilde{X}_i(t),\bar{e}_i(t),t)R_i^{-1}\\ &\times\tilde{\bar{G}}_i^T(\tilde{X}_i(t),\bar{e}_i(t),t)(\frac{\partial V(\tilde{X}_i(t),\bar{e}_i(t),t)}{\partial[\tilde{X}_i^T(t)\,\bar{e}_i^T(t)]^T})\\ &+(\frac{\partial V(\tilde{X}_i(t),\bar{e}_i(t),t)}{\partial[\tilde{X}_i^T(t)\,\bar{e}_i^T(t)]^T})^T\tilde{\bar{F}}_i(\tilde{X}_i(t),\bar{e}_i(t),t)\\ &+\frac{1}{4\rho_i^2}(\frac{\partial V(\tilde{X}_i(t),\bar{e}_i(t),t)}{\partial[\tilde{X}_i^T(t)\,\bar{e}_i^T(t)]^T})^T\tilde{\bar{D}}_i(\tilde{X}_i(t),\bar{e}_i(t),t)\end{aligned} \tag{6.20}$$

$$\times\left(\frac{\partial V(\tilde{X}_i(t),\bar{e}_i(t),t)}{\partial[\tilde{X}_i^T(t)\ \bar{e}_i^T(t)]^T}\right)+\frac{1}{4\rho_i^2}\left(\frac{\partial V(\tilde{X}_i(t),\bar{e}_i(t),t)}{\partial[\tilde{X}_i^T(t)\ \bar{e}_i^T(t)]^T}\right)^T$$
$$\times\bar{F}_{-i}(\tilde{\bar{X}}_i(t))\left(\frac{\partial V(\tilde{X}_i(t),\bar{e}_i(t),t)}{\partial[\tilde{X}_i^T(t)\ \bar{e}_i^T(t)]^T}\right)-\rho_i^2\tilde{C}_i^T(\tilde{X}_i(t))\tilde{C}_i(\tilde{X}_i(t))=0$$

where,

$$\frac{\partial V(\tilde{X}_i(t),\bar{e}_i(t),t)}{\partial[\tilde{X}_i^T(t)\ \bar{e}_i^T(t)\ ]^T}=\begin{bmatrix}\frac{\partial V(\tilde{X}_i(t),\bar{e}_i(t),t)}{\partial\tilde{X}_i^T(t)}\\ \frac{\partial V(\tilde{X}_i(t),\bar{e}_i(t),t)}{\partial\bar{e}_i^T(t)}\end{bmatrix},\ \tilde{\bar{G}}_i(\tilde{X}_i(t),\bar{e}_i(t),t)=\begin{bmatrix}\tilde{G}_i(\tilde{X}_i(t))\\ G_{e,i}(\bar{e}_i(t),t)\end{bmatrix},$$

$$\tilde{\tilde{F}}_i(\tilde{X}_i(t),\bar{e}_i(t),t)=\begin{bmatrix}\tilde{F}_i(\tilde{X}_i(t))\\ F_{e,i}(\bar{e}_i(t),t)\end{bmatrix},$$

$$\tilde{\bar{D}}_i(\tilde{X}_i(t),\bar{e}_i(t),t)$$
$$=\begin{bmatrix}\bar{D}_i(\bar{X}_i(t))\bar{D}_i^T(\bar{X}_i(t)) & \bar{D}_i(\bar{X}_i(t))D_{e,i}^T(\bar{e}_i(t),t)\\ D_{e,i}(\bar{e}_i(t),t)\bar{D}_i^T(\bar{X}_i(t)) & D_{e,i}(\bar{e}_i(t),t)D_{e,i}^T(\bar{e}_i(t),t)\end{bmatrix},$$

$$\bar{F}_{-i}(\tilde{X}_i(t),\bar{e}_i(t),t)$$
$$=\begin{bmatrix}\bar{F}_{-i}(\overline{X_i}(t))\bar{F}_{-i}^T(\overline{X_i}(t)) & \bar{F}_{-i}(\overline{X_i}(t))F_{e,-i}^T(\bar{e}_i(t),t)\\ F_{e,-i}(\bar{e}_i(t),t)\bar{F}_{-i}^T(\overline{X_i}(t)) & F_{e,-i}(\bar{e}_i(t),t)F_{e,-i}^T(\bar{e}_i(t),t)\end{bmatrix}$$

(b) For each quadrotor UAV in the team formation NCS, if external disturbances $\bar{v}_i(t)\in\pounds_2[0,\infty)$, measurement noise $n_i(t)\in\pounds_2[0,\infty)$, and couplings $\tilde{\bar{X}}_{-i}(t-\tau_{ij}(t))\in\pounds_2[0,\infty)$, then the proposed robust decentralized H_∞ attack-tolerant observer-based team formation tracking control strategy of each quadrotor UAV NCS will achieve the asymptotical mean square team formation tracking and estimation ability, that is, $E\{U_i^T(t)U_i(t)\}\to 0$, $E\{\tilde{X}_i^T(t)\tilde{X}_i(t)\}\to 0$ and $E\{\bar{e}_i^T(t)\bar{e}_i(t)\}\to 0$ as $t\to\infty$, for each UAV in the team.

Proof. Please refer to Appendix 6A.

Remark 6.1 In the conventional H_∞ observer-based control problems, the H_∞ observer design and the H_∞ controller design are always separated based on their corresponding strategies [28,38], that is,

$$\min_{L_i(\hat{\bar{X}}_i(t))}\ \max_{\substack{\bar{\bar{v}}_i(t)\\ \in\pounds_2[0,t_f]}}\frac{E\{\int_0^{t_f}\tilde{X}_i^T(t)Q_{1,i}\tilde{X}_i(t)dt\}}{E\{\int_0^{t_f}\bar{\bar{v}}_i^T(t)\bar{\bar{v}}_i(t)dt\}}$$

and

$$\min_{K_i(\hat{\bar{X}}_i(t),\bar{e}_i(t))}\ \max_{\substack{\bar{\bar{v}}_i(t)\\ \in\pounds_2[0,t_f]}}\frac{E\{\int_0^{t_f}\bar{e}_i^T(t)Q_{2,i}\bar{e}_i(t)+U_i^T(t)R_iU_i(t)]dt\}}{E\{\int_0^{t_f}\bar{\bar{v}}_i^T(t)\bar{\bar{v}}_i(t)dt\}}$$

respectively. In this situation, two coupled HJIEs are needed to be solved, which will be more complicated than HJIE_i in (6.20).

Remark 6.2 From $\text{HJIE}_i = 0, i = 1,\ldots,N$ in (6.20), it is seen that each $\text{HJIE}_i = 0$ can be solved without the information of other quadrotor UAVs, that is, the robust decentralized H_∞ attack-tolerant observer-based team formation tracking control is achieved for each quadrotor UAV in the team formation NCS by the proposed method in Theorem 6.1.

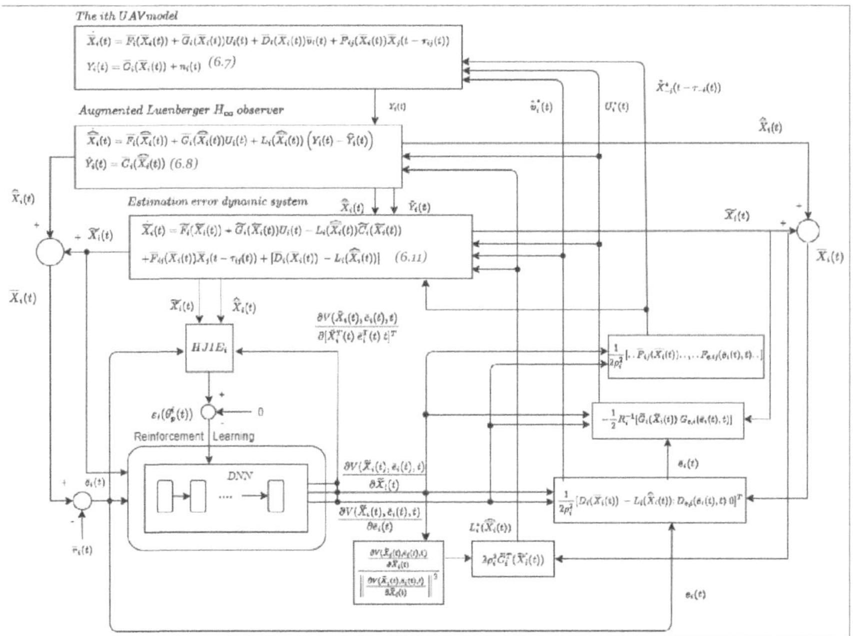

FIGURE 6.3 The flow chart of the offline training phase for the robust decentralized H_∞ attack-tolerant observer-based team formation DNN-based reference tracking control scheme of the large-scale quadrotor UAV team formation NCS via HJIE-reinforcement learning algorithm in the remote side as shown in Figure 6.2. The worst-case $U_i^*(t)$, $\tilde{\bar{v}}_i^*(t)$, and $\tilde{\bar{X}}_{-i}^*(t-\tau_{-i}(t))$ are fed back to the model of the ith quadrotor UAV in the team formation to produce $Y_i(t)$. Then $Y_i(t)$ is inputted to the augmented Luenberger observer in (6.8) to produce $\hat{\bar{X}}_i(t)$ with observer gain $L_i^*(\widehat{\bar{X}_i}(t))$. Further $U_i^*(t)$, $\tilde{\bar{v}}_i^*(t)$, $L_i^*(\widehat{\bar{X}_i}(t))$, and $\tilde{\bar{X}}_{-i}^*(t-\tau_{ij}(t))$ are also inputted to the estimation error system in (6.11) to produce $\widetilde{\bar{X}}_i(t)$. After that, we can acquire $\bar{X}_i(t) = \hat{\bar{X}}_i(t) + \widetilde{\bar{X}}_i(t)$ to get the tracking error by $\bar{e}_i(t) = \bar{X}_i(t) - \bar{r}_i(t)$. Eventually, the estimation error $\widetilde{\bar{X}}_i(t)$ and reference tracking error $\bar{e}_i(t)$ are both inputted to DNN to output $\frac{\partial V(\tilde{\bar{X}}_i(t),\bar{e}_i(t),t)}{\partial[\tilde{\bar{X}}_i^T(t)\ \bar{e}_i^T(t)\ t]^T}$ to solve the HJIE_i. The error of HJIEi_i, $\varepsilon_i(\theta_p^i(t))$, will provide feedback to train DNN by HJIE_i-reinforcement Adam learning algorithms (6.24)–(6.27). In online operation phase, $\tilde{\bar{v}}_i^*(t)$ and $\tilde{\bar{X}}_{-i}^*(t-\tau_{-i}(t))$ are replaced by the real external disturbance $\tilde{\bar{v}}_i(t)$ and coupling $\tilde{\bar{X}}_{-i}(t-\tau_{-i}(t))$ of the real ith quadrotor UAV. The remote side only needs to receive $Y_i(t)$ from the ith quadrotor UAV at the local side and send $U_i^*(t)$ to the ith quadrotor UAV, as shown in Figure 6.2. In general, the reinforcement learning-based Adam algorithm is stopped during the online operation phase. However, when $\left|\varepsilon_i(\theta_p^i(t))\right| > \kappa$ for prescribed small κ, the reinforcement learning-based Adam algorithm can be performed to enhance DNN performance without effect on the observer and tracking controller in the online operation phase.

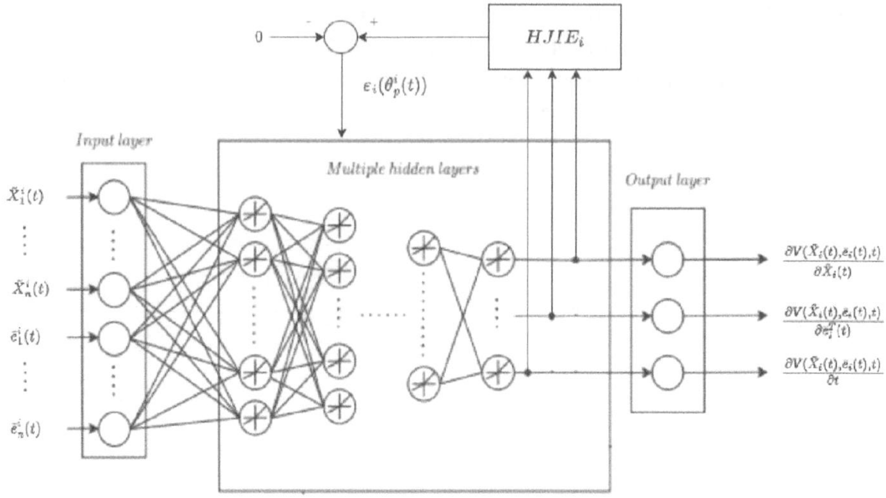

FIGURE 6.4 HJIE-reinforcement learning-based DNN architecture of robust H_∞ attack-tolerant observer-based team formation control scheme of Figure 6.3 with input $\tilde{X}_i(t)$ and $\bar{e}_i(t)$ to output $\frac{\partial V(\tilde{X}_i(t),\bar{e}_i(t),t)}{\partial[\tilde{X}_i^T(t)\ \bar{e}_i^T(t)\ ]^T}$ and $\frac{\partial V(\tilde{X}_i(t),\bar{e}_i(t),t)}{\partial t}$ to solve HJIE$_i$ in (6.20) after the training by Adam learning algorithm in (6.24)–(6.27) according to the error of HJIE$_i$ $= \varepsilon_i(\theta_p^i(t))$.

6.4 HJIE-REINFORCEMENT DNN-BASED DECENTRALIZED H_∞ ATTACK-TOLERANT OBSERVER-BASED TEAM FORMATION TRACKING CONTROL DESIGN OF LARGE-SCALE QUADROTOR UAV NCS

For the nonlinear team formation NCS under external disturbance, malicious attacks, coupling effects, and output measurement noise in (6.2), the decentralized H_∞ attack-tolerant observer-based team formation tracking control strategy in (6.12) is used to minimize the worst-case effect of all possible external disturbance $\bar{v}_i(t) \in \pounds_2[0,\infty)$, coupling effect $\tilde{\bar{X}}_{-i}(t-\tau_{-i}(t)) \in \pounds_2[0,\infty)$, and measurement noise $n_i(t) \in \pounds_2[0,\infty)$ on the estimation error $\tilde{X}_i(t)$, tracking error $\bar{e}_i(t)$, and control input $U_i(t)$. From Theorem 6.1, in order to complete the decentralized H_∞ attack-tolerant observer-based formation tracking control design for each quadrotor UAV of large-scale UAV NCS, we have to solve the $\frac{\partial V(\tilde{X}_i(t),\bar{e}_i(t),t)}{\partial[\tilde{X}_i^T(t)\ \bar{e}_i^T(t)\ ]^T}$ and $\frac{\partial V(\tilde{X}_i(t),\bar{e}_i(t),t)}{\partial t}$ from complicated time-varying partial differential HJIE$_i$ in (6.20) to obtain the optimal control input gain $K_i^*(\hat{\bar{X}}_i(t),\bar{e}_i(t))$ in (6.17), observer gain $L_i^*(\widehat{X_i}(t))$ in (6.19), and the worst-case external disturbances $\bar{v}_i^*(t)$, measurement noise $n_i^*(t)$ in (6.16), and interconnected coupling effects $\tilde{\bar{X}}_{-i}^*(t-\tau_{-i}(t))$ in (6.18) for each quadrotor UAV in the team formation. In this design, in order to solve HJIE$_i$ in (6.20) directly and more efficiently for the decentralized H_∞ attack-tolerant observer-based team formation tracking control strategy of each quadrotor UAV, an HJIE-reinforcement learning DNN method is adopted to approach the theoretical solution $\frac{\partial V(\tilde{X}_i(t),\bar{e}_i(t),t)}{\partial[\tilde{X}_i^T(t)\ \bar{e}_i^T(t)\ ]^T}$ and $\frac{\partial V(\tilde{X}_i(t),\bar{e}_i(t),t)}{\partial t}$ in (6.20). The reason is that it is very hard to calculate $\frac{\partial V(\tilde{X}_i(t),\bar{e}_i(t),t)}{\partial[\tilde{X}_i^T(t)\ \bar{e}_i^T(t)\ ]^T}$ and $\frac{\partial V(\tilde{X}_i(t),\bar{e}_i(t),t)}{\partial t}$ in (6.20) for $K_i^*(\hat{\bar{X}}_i(t),\bar{e}_i(t))$, $L_i^*(\widehat{\bar{X}_i}(t))$, $\bar{v}_i^*(t)$, $n_i^*(t)$, and $\tilde{\bar{X}}^*_{-i}(t-\tau_{-i}(t))$ in (6.16)–(6.19) in the real-time decentralized H_∞ attack-tolerant observer-based team formation control process even if the Lyapunov function $V(\tilde{X}_i(t),\bar{e}_i(t),t)$ of HJIE$_i$ in (6.20) is solved [27].

To simplify the design, we denote:

$$\frac{\partial V(\tilde{X}_i(t),\bar{e}_i(t),t)}{\partial[\tilde{X}_i^T(t)\ \bar{e}_i^T(t)\ t]^T}=\begin{bmatrix}\frac{\partial V(\tilde{X}_i(t),\bar{e}_i(t),t)}{\partial \tilde{X}_i(t)}\\ \frac{\partial V(\tilde{X}_i(t),\bar{e}_i(t),t)}{\partial \bar{e}_i(t)}\\ \frac{\partial V(\tilde{X}_i(t),\bar{e}_i(t),t)}{\partial t}\end{bmatrix} \tag{6.21}$$

Then, according to (6.21), HJIE$_i$ = 0 in (6.20) can be reformulated as:

$$\begin{aligned}
HJIE_i(t)&=\tilde{\bar{X}}_i^T(t)\bar{Q}_i\tilde{\bar{X}}_i^T(t)-\frac{1}{4}\Big(\frac{\partial V(\tilde{X}_i(t),\bar{e}_i(t),t)}{\partial[\tilde{X}_i^T(t)\ \bar{e}_i^T(t)\ t]^T}\Big)^T\\
&\times\begin{bmatrix}\tilde{\bar{G}}_i(\tilde{X}_i(t),\bar{e}_i(t),t)\\ 0\end{bmatrix}R_i^{-1}\begin{bmatrix}\tilde{\bar{G}}_i(\tilde{X}_i(t),\bar{e}_i(t),t)\\ 0\end{bmatrix}^T\\
&\times\Big(\frac{\partial V(\tilde{X}_i(t),\bar{e}_i(t),t)}{\partial[\tilde{X}_i^T(t)\ \bar{e}_i^T(t)\ t]^T}\Big)+\Big(\frac{\partial V(\tilde{X}_i(t),\bar{e}_i(t),t)}{\partial[\tilde{X}_i^T(t)\ \bar{e}_i^T(t)\ t]^T}\Big)^T\\
&\times\begin{bmatrix}\tilde{\bar{F}}_i(\tilde{X}_i(t),\bar{e}_i(t),t)\\ 1\end{bmatrix}+\frac{1}{4\rho_i^2}\Big(\frac{\partial V(\tilde{X}_i(t),\bar{e}_i(t),t)}{\partial[\tilde{X}_i^T(t)\ \bar{e}_i^T(t)\ t]^T}\Big)^T\\
&\times\begin{bmatrix}\tilde{\bar{D}}_i(\tilde{X}_i(t),\bar{e}_i(t),t) & 0\\ 0 & 0\end{bmatrix}\Big(\frac{\partial V(\tilde{X}_i(t),\bar{e}_i(t),t)}{\partial[\tilde{X}_i^T(t)\ \bar{e}_i^T(t)\ t]^T}\Big)\\
&+\frac{1}{4\rho_i^2}\Big(\frac{\partial V(\tilde{X}_i(t),\bar{e}_i(t),t)}{\partial[\tilde{X}_i^T(t)\ \bar{e}_i^T(t)\ t]^T}\Big)^T\begin{bmatrix}\tilde{\bar{F}}_i(\tilde{X}_i(t),\bar{e}_i(t),t) & 0\\ 0 & 0\end{bmatrix}\\
&\times\Big(\frac{\partial V(\tilde{X}_i(t),\bar{e}_i(t),t)}{\partial[\tilde{X}_i^T(t)\ \bar{e}_i^T(t)\ t]^T}\Big)-\rho_i^2\tilde{C}_i^T(\tilde{X}_i(t))\tilde{C}_i(\tilde{X}_i(t))\\
&=0, for\ i=1,\dots,N
\end{aligned} \tag{6.22}$$

Thus, in this chapter, we employ a reinforcement learning-based DNN scheme to solve the partial differential $\frac{\partial V(\tilde{X}_i(t),\bar{e}_i(t),t)}{\partial[\tilde{X}_i^T(t)\ \bar{e}_i^T(t)\ t]^T}$ in (6.22) to obtain control input $K_i^*(\hat{\bar{X}}_i(t),\bar{e}_i(t))$, observer gain $L_i^*(\widehat{\bar{X}_i}(t))$, and the worst-case external disturbances $\bar{v}_i^*(t)$, coupling effects $\tilde{\bar{X}}^*_{-i}(t-\tau_{-i}(t))$, and measurement noise $n_i^*(t)$. The training process of the HJIE-reinforcement deep learning approach is separated into an offline training phase and an online operation phase. In the offline training period as shown in Figure 6.3, since the actual state, external disturbance, measurement noise, coupling effect, and malicious attack signals are unavailable, we substitute the worst-case $\bar{v}_i^*(t)$, $n_i^*(t)$, and $\bar{\tilde{X}}_{-i}^*(t-\tau_{-i}(t))$ for the actual $\bar{v}_i(t)$, $n_i(t)$, and $\bar{\tilde{X}}_{-i}(t-\tau_{-i}(t))$. This does not have any effect on the decentralized H_∞ attack-tolerant observer-based team formation tracking control strategy of large-scale UAV team formation NCS in (6.10) because this strategy is designed on the basis of the worst-case $\bar{v}_i^*(t)$, $n_i^*(t)$, and $\bar{\tilde{X}}_{-i}^*(t-\tau_{-i}(t))$. Then, the Luenberger observer is proposed to produce $\hat{\bar{X}}_i(t)$, and use the augmented state estimation error model in (6.11) to produce the estimation error $\tilde{X}_i(t)$ to calculate $\bar{X}_i(t) = \hat{\bar{X}}_i(t) + \tilde{X}_i(t)$. Therefore, we can also get the tracking error $\bar{e}_i(t)$ by $\bar{e}_i(t) = \bar{X}_i(t) - \bar{r}_i(t)$. After that, $\hat{\bar{X}}_i(t)$ and $\bar{e}_i(t)$ are employed as input of DNN to produce $\frac{\partial V(\tilde{X}_i(t),\bar{e}_i(t),t)}{\partial[\tilde{X}_i^T(t)\ \bar{e}_i^T(t)\ t]^T}$ at the output of DNN. Then the output $\frac{\partial V(\tilde{X}_i(t),\bar{e}_i(t),t)}{\partial[\tilde{X}_i^T(t)\ \bar{e}_i^T(t)\ t]^T}$ of DNN will be sent back to calculate the value of HJIE_i as follows:

$$\begin{aligned}
HJIE_{i,\varepsilon}(t) &= \bar{\tilde{X}}_i^T(t)\bar{Q}_i\bar{\tilde{X}}_i^T(t) - \frac{1}{4}\left(\frac{\partial V(\tilde{X}_i(t),\bar{e}_i(t),t)}{\partial[\tilde{X}_i^T(t)\ \bar{e}_i^T(t)\ t]^T}\right)_\varepsilon^T \\
&\times \begin{bmatrix} \bar{\tilde{G}}_i(\tilde{X}_i(t),\bar{e}_i(t),t) \\ 0 \end{bmatrix} R_i^{-1} \begin{bmatrix} \bar{\tilde{G}}_i(\tilde{X}_i(t),\bar{e}_i(t),t) \\ 0 \end{bmatrix}^T \\
&\times \left(\frac{\partial V(\tilde{X}_i(t),\bar{e}_i(t),t)}{\partial[\tilde{X}_i^T(t)\ \bar{e}_i^T(t)\ t]^T}\right)_\varepsilon + \left(\frac{\partial V(\tilde{X}_i(t),\bar{e}_i(t),t)}{\partial[\tilde{X}_i^T(t)\ \bar{e}_i^T(t)\ t]^T}\right)_\varepsilon^T \\
&\times \begin{bmatrix} \bar{\tilde{F}}_i(\tilde{X}_i(t),\bar{e}_i(t),t) \\ 1 \end{bmatrix} + \frac{1}{4\rho_i^2}\left(\frac{\partial V(\tilde{X}_i(t),\bar{e}_i(t),t)}{\partial[\tilde{X}_i^T(t)\ \bar{e}_i^T(t)\ t]^T}\right)_\varepsilon^T \\
&\times \begin{bmatrix} \bar{\tilde{D}}_i(\tilde{X}_i(t),\bar{e}_i(t),t) & 0 \\ 0 & 0 \end{bmatrix} \left(\frac{\partial V(\tilde{X}_i(t),\bar{e}_i(t),t)}{\partial[\tilde{X}_i^T(t)\ \bar{e}_i^T(t)\ t]^T}\right)_\varepsilon \\
&+ \frac{1}{4\rho_i^2}\left(\frac{\partial V(\tilde{X}_i(t),\bar{e}_i(t),t)}{\partial[\tilde{X}_i^T(t)\ \bar{e}_i^T(t)\ t]^T}\right)_\varepsilon^T \begin{bmatrix} \bar{\tilde{F}}_i(\tilde{X}_i(t),\bar{e}_i(t),t) & 0 \\ 0 & 0 \end{bmatrix} \\
&\times \left(\frac{\partial V(\tilde{X}_i(t),\bar{e}_i(t),t)}{\partial[\tilde{X}_i^T(t)\ \bar{e}_i^T(t)\ t]^T}\right)_\varepsilon - \rho_i^2\tilde{C}_i^T(\tilde{X}_i(t))\tilde{C}_i(\tilde{X}_i(t)) \\
&= \varepsilon_i(\theta_p^i(t)),\ for\ i = 1,\ldots,N
\end{aligned} \tag{6.23}$$

Next, $\varepsilon_i(\theta_p^i(t))$ will be sent back to train the weighting parameters in the hidden layer of DNN until $\varepsilon_i(\theta_p^i(t))$ converges to zero or below a small threshold. The output $\left(\frac{\partial V(\tilde{X}_i(t),\bar{e}_i(t),t)}{\partial[\tilde{X}_i^T(t)\ \bar{e}_i^T(t)\ t]^T}\right)_\varepsilon$ of DNN is taken to calculate $K_i^*(\hat{\bar{X}}_i(t),\bar{e}_i(t))$, $L_i^*(\widehat{\bar{X}_i}(t))$, $\bar{\tilde{v}}_i^*(t)$, and $\bar{\tilde{X}}_{-i}^*(t-\tau_{-i}(t))$ in (6.17)–(6.19) at the same time. Based on the above information, we can again obtain $\tilde{X}_i(t)$ and $\bar{e}_i(t)$ as input into DNN as shown in Figure 6.3. Consequently, repeat the above DNN reinforcement learning procedure until $\varepsilon_i(\theta_p^i(t)) = 0$ and $(\frac{\partial V(\tilde{X}_i(t),\bar{e}_i(t),t)}{\partial[\tilde{X}_i^T(t)\ \bar{e}_i^T(t)\ t]^T})_\varepsilon$ approach the theoretical solution $(\frac{\partial V(\tilde{X}_i(t),\bar{e}_i(t),t)}{\partial[\tilde{X}_i^T(t)\ \bar{e}_i^T(t)\ t]^T})$, then the offline training is over, and we can switch the offline training phase to the online operation phase.

In the online operation phase, as shown in Figure 6.2, the observer can receive the output $Y_i(t)$ from each quadrotor UAV on the local side with authentic external disturbance, malicious attacks, coupling effects, measurement noise, and $U_i^*(t)$. Hence, we do not need the worst-case $\bar{v}_i^*(t)$, $n_i^*(t)$, and $\bar{\bar{X}}_{-i}^*(t-\tau_{-i}(t))$ to produce $Y_i(t)$ from the stochastic model of the i th quadrotor UAV in the team formation. The other procedure to output $(\frac{\partial V(\tilde{X}_i(t),\bar{e}_i(t),t)}{\partial[\tilde{X}_i^T(t)\ \bar{e}_i^T(t)\ t]^T})$ by DNN to produce control $K_i(\hat{\bar{X}}_i(t),\bar{e}_i(t))$ and observer gain $L_i(\widehat{X_i}(t))$ is in a similar way as the offline training phase. In general, we don't have to train DNN by the reinforcement learning scheme again as in the online operation phase. Nevertheless, in some situations, if $\varepsilon_i(\theta_p^i(t)) > \kappa$ for a specified little value κ, we can start the reinforcement learning-based Adam algorithm to continue training without affecting the online operation process.

The DNN architecture based on HJIE-reinforcement learning in the robust decentralized H_∞ attack-tolerant observer-based team formation tracking control scheme is composed of an input layer, multiple hidden layers, an output layer, and HJIE-reinforcement learning, as shown in Figure 6.4. The DNN inputs are the estimation error $\tilde{X}_i(t)$ and the tracking error $\bar{e}_i(t)$. In hidden layers, we adopted LeakyReLU instead of ReLU as an activation function to avoid the dead neuron problem. If the input to a ReLU neuron is negative, the output will be zero, which means those neurons may never get updated. This is not the case with LeakyReLU, which has a constant gradient when the input is negative, instead of zero. The LeakyReLU function is formulated as follows:

$$l(X(t)) = \begin{cases} \mu_1 X(t), if\ X(t) > 0 \\ \mu_2 X(t), if\ X(t) \leq 0 \end{cases}$$

where μ_1 and μ_2 are some constants with $\mu_1, \mu_2 \in (0,1)$.

As shown in the offline training phase in Figure 6.3 at the remote side, the error $\varepsilon_i(\theta_p^i(t))$ of HJIE-reinforcement learning will be sent back to update the weighting parameters of DNN through Adam learning algorithm as follows [34,36]:

$$\theta_p^i(t) = \theta_{p-1}^i(t) - \frac{\iota}{\sqrt{\hat{\nu}_p^i(t)+\varsigma}}\hat{m}_p^i(t), p = 1,\ldots,P \tag{6.24}$$

$$\hat{m}_p^i(t) = \frac{m_p^i(t)}{1-\alpha_1^p},\ \hat{\nu}_p^i(t) = \frac{\nu_p^i(t)}{1-\alpha_2^p} \tag{6.25}$$

$$\begin{aligned} m_p^i(t) &= \alpha_1 m_{p-1}^i(t) + (1-\alpha_1)\nabla_p^i(t) \\ \nu_p^i(t) &= \alpha_2 \nu_{p-1}^i(t) + (1-\alpha_2)\nabla_p^{i\,2}(t) \end{aligned} \tag{6.26}$$

$$\nabla_p^i(t) = \frac{\partial}{\partial \theta_p^i(t)} \sqrt{\frac{1}{B}\sum_{p=1}^{B} \varepsilon_i^2(\theta_p^i(t))} \tag{6.27}$$

where $\theta_p^i(t)$ represents the weighting parameter vector in the hidden layers at time t, ι represents the learning rate of the learning algorithm, and P represents the number of training time steps at time t. ζ denotes a small number to prevent a zero denominator. $\hat{\nu}_p^i(t)$ and $\hat{m}_p^i(t)$ denote bias-corrected estimators defined in (6.25). $\nabla_p^i(t)$ is the gradient, that is, the vector of the derivative of the root mean square error of HJIE$_i$, $\varepsilon_i(\theta_p^i(t))$, in (6.23) with respect to $\theta_p^i(t)$ in time step p at time t. B is batch size of the input data. $\alpha_1, \alpha_2 \in [0,1]$ in (6.26) denotes the degree of influence of the previous impact on the current direction. The designer specifies these two parameters to prevent DNN parameters from being captured by local minima [34,36]. Moreover, if the current gradient $\nabla_p^i(t)$ direction is the same as the accumulated gradient, then the gradient will conduct a larger update, Conversely, it will conduct a smaller update. α_1^p and α_2^p are the pth power of α_1 and α_2, respectively. $m_p^i(t)$ and $\nu_p^i(t)$ in (6.26) are the moving average of the gradient and the squared gradient of $\nabla_p^i(t)$ at time t. The HJIE-reinforcement learning-based Adam algorithm combines the advantages of momentum and RMSProp [36] and is an adaptive gradient descent algorithm. This method is easy to use and does not require a lot of memory. Overall, it is a robust optimizer and suitable for non-convex optimization problems in the field of machine learning and deep learning [37].

Remark 6.3 If the time steps and the number of hidden neurons are large enough, then the Adam learning algorithm, which is employed for updating the weighted parameter vector $\theta_p^i(t)$ of DNN, can converge to a globally optimal $\theta_p^{i*}(t)$ with a linear convergence rate as $p \to \infty$ in (6.24)–(6.27). It has been proven in Reference [34].

When the error $\varepsilon_i(\theta_p^i(t))$ approaches to zero using the HJIE-reinforcement-based Adam learning algorithm in (6.24)–(6.27), it can be proven that the output $(\frac{\partial V(\tilde{X}_i(t),\bar{e}_i(t),t)}{\partial[\tilde{X}_i^T(t)\ \bar{e}_i^T(t)\ t]^T})_\varepsilon$ in (6.23) of DNN can approach $(\frac{\partial V(\tilde{X}_i(t),\bar{e}_i(t),t)}{\partial[\tilde{X}_i^T(t)\ \bar{e}_i^T(t)\ t]^T})$ of HJIE$_i$ in (6.22) using the following theorem.

Theorem 6.2 If $\varepsilon_i(\theta_p^i(t)) \to 0$ in (6.23) by HJIE-reinforcement learning-based Adam algorithm in (6.24)–(6.27), then $(\frac{\partial V(\tilde{X}_i(t),\bar{e}_i(t),t)}{\partial[\tilde{X}_i^T(t)\ \bar{e}_i^T(t)\ t]^T})_\varepsilon$ will approach to $\frac{\partial V(\tilde{X}_i(t),\bar{e}_i(t),t)}{\partial[\tilde{X}_i^T(t)\ \bar{e}_i^T(t)\ t]^T}$ in (6.22), that is, the proposed HJIE-reinforcement learning DNN decentralized $H\infty$ attack-tolerant observer-based team formation tracking control scheme in Figure 6.2 via Adam learning algorithm (6.24)–(6.27) will approach to the theoretical decentralized $H\infty$ attack-tolerant observer-based team formation tracking control design (6.16)–(6.19) of large-scale quadrotor UAV NCS in Theorem 6.1.

Proof. Please refer to Appendix 6B.

Remark 6.4 In Theorem 6.2, we can obtain the output $\frac{\partial V(\tilde{X}_i(t),\bar{e}_i(t),t)}{\partial[\tilde{X}_i^T(t)\ \bar{e}_i^T(t)\ t]^T}$ through training DNN by HJIE-reinforcement Adam learning algorithm (6.24)–(6.27) to calculate H_∞ control $U_i^*(t)$ and H_∞ observer gain $L_i^*(\hat{\tilde{X}}_i(t))$ if the error $\varepsilon_i(\theta_p^i(t))$ of HJIE$_{\varepsilon,i}$

in (6.23) approaches to zero in the offline training phase, as shown in Figure 6.3. Nevertheless, for the practical design, if $\left|\varepsilon(\theta_p^i(t))\right| \leq \kappa$ for a little prescribed κ, we will stop the reinforcement learning-based Adam algorithm in the off-line training phase and switch to the online operation phase.

Remark 6.5 $Y_i(t)$ can be obtained through the actual quadrotor UAV system in (6.2) with real $v_i(t)$, coupling effect $X_j(t-\tau_{ij}(t))$ and malicious attacks by the decentralized H_∞ attack-tolerant observer-based team formation tracking control $U_i^*(t)$ in the online operation phase, as shown in Figure 6.2. We take $\tilde{X}_i(t)$ and $\bar{e}_i(t)$ as input into well-trained DNN to obtain $\frac{\partial V(\tilde{X}_i(t),\bar{e}_i(t),t)}{\partial[\tilde{X}_i^T(t)\ \bar{e}_i^T(t)\ t]^T}$ to calculate $U_i^*(t)$ and $L_i^*(\hat{\bar{X}}_i(t))$ for the robust decentralized H_∞ attack-tolerant observer-based team formation tracking control of the large-scale quadrotor UAV NCS. However, the weighting parameter $\theta_p^i(t)$ of the DNN can still be updated by the HJIE-reinforced Adam learning algorithm in (6.24)–(6.27) if $\left|\varepsilon(\theta_p^i(t))\right| > \kappa$ during the online operation phase, without any impact on the HJIE-reinforcement learning-based DNN observer-based team formation tracking control scheme of large-scale quadrotor UAV NCS.

The state information is transmitted through the wireless network channel in the form of data packets in practical applications. Therefore, the quadrotor UAV formation NCS in (6.2) can be reformulated by the nonlinear sampled-data system as follows:

$$
\begin{aligned}
S_i : \frac{X_i(t+h)-X_i(t)}{h} &= F_i(X_i(t)) + G_i(X_i(t))(U_i(t)+\gamma_a^i(t)) + D_i(X_i(t))v_i(t) \\
&\quad + \sum_{j\in N_i} F_{ij}(X_i(t))X_j(t-\tau_{ij}(t)) \\
Y_i(t) &= C_i(X_i(t)) + n_i(t) + D_s^i(t)\gamma_s^i(t)
\end{aligned}
\tag{6.28}
$$

where h is the sampling time. Or we can rewrite (6.28) as follows:

$$
\begin{aligned}
S_i : X_i(t+h) &= (X_i(t) + hF_i(X_i(t))) + hG_i(X_i(t))(U_i(t) \\
&\quad + \gamma_a^i(t)) + hD_i(X_i(t))v_i(t) \\
&\quad + h\sum_{j\in N_i} F_{ij}(X_i(t))X_j(t-\tau_{ij}(t)) \\
Y_i(t) &= C_i(X_i(t)) + n_i(t) + D_s^i(t)\gamma_s^i(t)
\end{aligned}
\tag{6.29}
$$

From smoothed models of actuator malicious attacks in (6.5) and the sensor malicious attacks in (6.6), the augmented system (6.7) can be reformulated as follows:

$$
\begin{aligned}
\bar{X}_i(t+h) &= (\bar{X}_i(t) + h\bar{F}_i(\bar{X}_i(t))) + h\bar{G}_i(\bar{X}_i(t))U_i(t) \\
&\quad + h\bar{D}_i(\bar{X}_i(t))\bar{v}_i(t) \\
&\quad + h\sum_{j\in N_i} \bar{F}_{ij}(\bar{X}_i(t))\bar{X}_j(t-\tau_{ij}(t)) \\
Y_i(t) &= \bar{C}_i(\bar{X}_i(t)) + n_i(t)
\end{aligned}
\tag{6.30}
$$

In this case, the observer dynamic model in (6.8) is reformulated as:

$$\begin{aligned}\hat{\bar{X}}_i(t+h) &= (\hat{\bar{X}}_i(t)+h\bar{F}_i(\widehat{\bar{X}}_i(t))+h\bar{G}_i(\widehat{\bar{X}}_i(t))U_i(t)\\ &\quad +hL_i(\widehat{\bar{X}}_i(t))\left(Y_i(t)-\hat{Y}_i(t)\right)\\ \hat{Y}_i(t) &= \bar{C}_i(\widehat{\bar{X}}_i(t))\end{aligned} \tag{6.31}$$

Correspondingly, the tracking error dynamic equation in (6.9) and the state estimation error dynamic equation in (6.11) are reformulated as follows:

$$\begin{aligned}\bar{e}_i(t+h) &= (e(t)+hF_{e,i}(\bar{e}_i(t),t))+hG_{e,i}(\bar{e}_i(t),t)U_i(t)\\ &\quad +hD_{e,i}(\bar{e}_i(t),t)\bar{v}_i(t)+h\sum_{j\in N_i}F_{e,ij}(\bar{e}_i(t),t)\bar{X}_j(t-\tau_{ij}(t))\end{aligned} \tag{6.32}$$

and

$$\begin{aligned}\widetilde{X}_i(t+h) &= (\widetilde{X}_i(t)+h\widetilde{F}_i(\widetilde{X}_i(t)))+h\widetilde{G}_i(\widetilde{X}_i(t))U_i(t)\\ &\quad -hL_i(\widehat{\bar{X}}_i(t))\widetilde{C}_i(\widetilde{X}_i(t))+h\sum_{j\in N_i}\bar{F}_{ij}(\bar{X}_i(t))\bar{X}_j(t-\tau_{ij}(t))\\ &\quad +h\left[\bar{D}_i(\bar{X}_i(t))-L_i(\widehat{\bar{X}}_i(t))\right]\tilde{\bar{v}}_i(t)\end{aligned} \tag{6.33}$$

respectively.

The above sampled-data quadrotor UAV systems will be employed for the simulation of the reinforcement learning-based DNN robust decentralized H_∞ attack-tolerant observer-based team formation tracking control strategy of large-scale quadrotor UAVs NCS in the following section.

6.5 SIMULATION RESULTS

In order to validate the effectiveness of the proposed reinforcement learning-based DNN robust decentralized H_∞ attack-tolerant observer-based team formation tracking control scheme of large-scale quadrotor UAV NCS under external disturbances, malicious attacks, measurement noises, and couplings, a simulation example of a team formation NCS of 25 quadrotor UAVs under external disturbances, malicious attack, measurement noises, and couplings is provided in this section. The parameters of the ith quadrotor UAV in NCS are given as [33]:

$$\begin{aligned}&m=2\text{ (kg)}, g=9.8\text{ (m/s}^2), J_x=J_y=J_z=0.1(\text{Ns}^2/\text{rad}),\\ &K_x=K_y=K_z=0.01(\text{Ns/m}), K_\phi=K_\theta=K_\psi=0.01\text{ (Ns/m)}\end{aligned}$$

The weighting matrices are set as $Q_{1,i}=0.02diag\{I_{20}\}, Q_{2,i}=0.01diag\{I_{20}\}$ and $R=10^{-4}I_4$ in the H_∞ attack-tolerant observer-based team formation tracking control strategy in (6.10). The sampling time h is set as $0.01\,\text{s}$ and the terminal time t_f is $30\,s$. The quadrotor UAV NCS may suffer interference from external

TABLE 6.1
The Reference $r_i(t)$ of 25 Quadrotor UAVs for Team Formation NCS

Quadrotor UAV	The Reference $r_i(t)$
UAV 1	$x_1 = 15\sin(0.5t), y_1 = 15\cos(0.5t), z_1 = t, \psi_1 = 0$
UAV 2	$x_2 = 17\sin(0.5t), y_2 = 17\cos(0.5t), z_2 = t, \psi_2 = 0$
UAV 3	$x_3 = 19\sin(0.5t), y_3 = 19\cos(0.5t), z_3 = t, \psi_3 = 0$
UAV 4	$x_4 = 21\sin(0.5t), y_4 = 21\cos(0.5t), z_4 = t, \psi_4 = 0$
UAV 5	$x_5 = 13\sin(0.5t), y_5 = 13\cos(0.5t), z_5 = t, \psi_5 = 0$
UAV 6	$x_6 = 15\sin(0.5t), y_6 = 15\cos(0.5t), z_6 = t+2, \psi_6 = 0$
UAV 7	$x_7 = 17\sin(0.5t), y_7 = 17\cos(0.5t), z_7 = t+2, \psi_7 = 0$
UAV 8	$x_8 = 19\sin(0.5t), y_8 = 19\cos(0.5t), z_8 = t+2, \psi_8 = 0$
UAV 9	$x_9 = 21\sin(0.5t), y_9 = 21\cos(0.5t), z_9 = t+2, \psi_9 = 0$
UAV 10	$x_{10} = 13\sin(0.5t), y_{10} = 13\cos(0.5t), z_{10} = t+2, \psi_{10} = 0$
UAV 11	$x_{11} = 15\sin(0.5t), y_{11} = 15\cos(0.5t), z_{11} = t+1, \psi_{11} = 0$
UAV 12	$x_{12} = 17\sin(0.5t), y_{12} = 17\cos(0.5t), z_{12} = t+1, \psi_{12} = 0$
UAV 13	$x_{13} = 19\sin(0.5t), y_{13} = 19\cos(0.5t), z_{13} = t+1, \psi_{13} = 0$
UAV 14	$x_{14} = 21\sin(0.5t), y_{14} = 21\cos(0.5t), z_{14} = t+1, \psi_{14} = 0$
UAV 15	$x_{15} = 13\sin(0.5t), y_{15} = 13\cos(0.5t), z_{15} = t+1, \psi_{15} = 0$
UAV 16	$x_{16} = 15\sin(0.5t), y_{16} = 15\cos(0.5t), z_{16} = t+3, \psi_{16} = 0$
UAV 17	$x_{17} = 17\sin(0.5t), y_{17} = 17\cos(0.5t), z_{17} = t+3, \psi_{17} = 0$
UAV 18	$x_{18} = 19\sin(0.5t), y_{18} = 19\cos(0.5t), z_{18} = t+3, \psi_{18} = 0$
UAV 19	$x_{19} = 21\sin(0.5t), y_{19} = 21\cos(0.5t), z_{19} = t+3, \psi_{19} = 0$
UAV 20	$x_{20} = 13\sin(0.5t), y_{20} = 13\cos(0.5t), z_{20} = t+3, \psi_{20} = 0$
UAV 21	$x_{21} = 15\sin(0.5t), y_{21} = 15\cos(0.5t), z_{21} = t+4, \psi_{21} = 0$
UAV 22	$x_{22} = 17\sin(0.5t), y_{22} = 17\cos(0.5t), z_{22} = t+4, \psi_{22} = 0$
UAV 23	$x_{23} = 19\sin(0.5t), y_{23} = 19\cos(0.5t), z_{23} = t+4, \psi_{23} = 0$
UAV 24	$x_{24} = 21\sin(0.5t), y_{24} = 21\cos(0.5t), z_{24} = t+4, \psi_{24} = 0$
UAV 25	$x_{25} = 13\sin(0.5t), y_{25} = 13\cos(0.5t), z_{25} = t+4, \psi_{25} = 0$

disturbances in the system plant and measurement noise in the measurement output. Therefore, the external disturbances are given in the following:

$$v_i(t) = 0.1[\sin(0.5t), \sin(0.5t), \cos(0.5t), \cos(0.5t), \sin(0.5t), \sin(0.5t)]^T$$

and the measurement noise $n(t) \in N(0, 0.1I_{20})$. Since there are only four control inputs in $U_i(t)$ for each quadrotor UAV in (6.1), the desired position and the yaw angle for the reference input $r_i(t)$ of each quadrotor UAV in the team are given in Table 6.1.

Furthermore, the attitude of the quadrotor UAV depends on its corresponding position information. Therefore, the reference attitudes of ϕ_i and θ_i are given as follows:

$$\phi_i = \sin^{-1}\left(\frac{m}{F(t)}(\ddot{x}_i \sin\psi_i - \ddot{y}_i \cos\psi_i)\right),$$

$$\theta_i = \tan^{-1}\left(\frac{1}{\ddot{z}_i + g}(\ddot{x}_i \cos\psi_i + \ddot{y}_i \sin\psi_i)\right)$$

where $\ddot{x}_i, \ddot{y}_i, \ddot{z}_i$ are the double derivative of x_i, y_i, z_i representing acceleration, respectively.

From the perspective of the wireless network, the observation of the position will be affected by the sensor attack $\gamma_s^i(t)$ via the wireless network channel. On the other hand, the actuator attack signal $\gamma_a^i(t)$ will be transmitted to the actuator with control signals to destroy the tracking control of the quadrotor UAV. To construct the smoothed model in (6.5) and (6.6) for the estimation of the actuator attack signal $\gamma_a^i(t)$ and the sensor attack signal $\gamma_s^i(t)$ as shown in Figure 6.5, the extrapolation coefficients are specified, respectively, as follows:

$$a_1 = 0.9, a_2 = 0.06, a_3 = 0.03, a_4 = 0.01$$
$$b_1 = 0.96, b_2 = 0.02, b_3 = 0.01, b_4 = 0.01$$

Due to a large-scale team formation of quadrotor UAVs among 25-quadrotor-UAV NCS, some interconnected coupling effects will occur, such as trailing vortex couplings of neighboring quadrotor UAVs. The strength of the trailing vortex couplings are given below to affect the shape of the quadrotor UAV:

$$\begin{aligned} & F_{ij}(X_i(t))X_j(t-\tau_{ij}(t)) \\ & = [0, 0.01x_2^i(t)x_2^j(t), 0, 0, 0, 0.01z_2^i(t)z_2^j(t), 0, 0, 0, 0, 0, 0]^T, j \in N_i \end{aligned}$$

where N_i denotes the neighboring quadrotor UAVs of the ith quadrotor UAV.

According to Theorem 6.1, the design of the decentralized H_∞ attack-tolerant observer-based team formation tracking control must solve the following HJIE$_i$ to obtain the decentralized H_∞ team formation control $U_i^*(t)$ in (6.17) and observer gain $L_i^*(\widehat{X}_i(t))$ in (6.19) for each quadrotor UAV in the team formation NCS to achieve robust decentralized H_∞ attack-tolerant observer-based team formation tracking control strategy.

$$\begin{aligned} HJIE_i = & \frac{\partial V(\tilde{X}_i(t),\bar{e}_i(t),t)}{\partial t} + \tilde{\bar{X}}_i^T(t)\bar{Q}_i\tilde{\bar{X}}_i(t) \\ & -\frac{1}{4}\left(\frac{\partial V(\tilde{X}_i(t),\bar{e}_i(t),t)}{\partial[\tilde{X}_i^T(t)\ \bar{e}_i^T(t)]^T}\right)^T \tilde{\bar{G}}_i(\tilde{X}_i(t),\bar{e}_i(t),t)R_i^{-1} \\ & \times\tilde{\bar{G}}_i^T(\tilde{X}_i(t),\bar{e}_i(t),t)\left(\frac{\partial V(\tilde{X}_i(t),\bar{e}_i(t),t)}{\partial[\tilde{X}_i^T(t)\ \bar{e}_i^T(t)]^T}\right) \\ & +\left(\frac{\partial V(\tilde{X}_i(t),\bar{e}_i(t),t)}{\partial[\tilde{X}_i^T(t)\ \bar{e}_i^T(t)]^T}\right)^T \widetilde{\widetilde{F}}_i(\tilde{X}_i(t),\bar{e}_i(t),t) \\ & +\frac{1}{4\rho_i^2}\left(\frac{\partial V(\tilde{X}_i(t),\bar{e}_i(t),t)}{\partial[\tilde{X}_i^T(t)\ \bar{e}_i^T(t)]^T}\right)^T \tilde{\bar{D}}_i(\tilde{X}_i(t),\bar{e}_i(t),t) \\ & \times\left(\frac{\partial V(\tilde{X}_i(t),\bar{e}_i(t),t)}{\partial[\tilde{X}_i^T(t)\ \bar{e}_i^T(t)]^T}\right) + \frac{1}{4\rho_i^2}\left(\frac{\partial V(\tilde{X}_i(t),\bar{e}_i(t),t)}{\partial[\tilde{X}_i^T(t)\ \bar{e}_i^T(t)]^T}\right)^T \\ & \times\bar{F}_{-i}(\overline{\widetilde{\widetilde{X}}}_i(t))\left(\frac{\partial V(\tilde{X}_i(t),\bar{e}_i(t),t)}{\partial[\tilde{X}_i^T(t)\ \bar{e}_i^T(t)]^T}\right) \\ & -\rho_i^2\widetilde{C}_i^T(\widetilde{X}_i(t))\widetilde{C}_i(\widetilde{X}_i(t)) = 0 \end{aligned} \tag{6.34}$$

It is nearly impossible to solve $\frac{\partial V(\tilde{X}_i(t),\bar{e}_i(t),t)}{\partial[\tilde{X}_i^T(t)\ \bar{e}_i^T(t)]^T}$ analytically or numerically for $U_i^*(t)$ in (6.17) and $L_i^*(\widehat{X_i(t)})$ in (6.19) directly, since (6.34) is a highly nonlinear time-varying partial differential equation. Thus, the proposed HJIE-reinforcement DNN-based H_∞ observer-based tracking control scheme in Figure 6.2 derived from quadrotor UAV sample data systems in (6.28)–(6.33) with sampling time $h = 0.01$ in Figure 6.3 is adopted to solve $\frac{\partial V(\tilde{X}_i(t),\bar{e}_i(t),t)}{\partial[\tilde{X}_i^T(t)\ \bar{e}_i^T(t)\ t]^T}$ from (6.34) for $U_i^*(t)$, $\tilde{\bar{v}}_i^*(t)$ and $\tilde{\bar{X}}_{-i}^*(t-\tau_{-i}(t))$. In the simulation example, the HJIE-reinforcement DNN architecture in Figure 6.4 contains an input layer with input $\tilde{X}_i(t)$, $\bar{e}_i(t)$, four hidden layers, an output layer, and a feedback HJIE loop for reinforcement learning. Each hidden layer has 256,128,32 and 12 neurons sequentially. The parameters of the Adam learning algorithm in (6.24)–(6.27) are specified as $\alpha_1 = 0.9$, $\alpha_2 = 0.999$, $\zeta = 10^{-7}$, and $\iota = 10^{-3}$. The training steps are specified as $P = 30$ and batch size B is 800. In the offline training phase as shown in Figure 6.4, we randomly select a set of training inputs, which are 20000 initial tracking errors $\bar{e}_i(0)$ and estimation errors $\tilde{X}_i(0)$ around the origin $X(0)$.

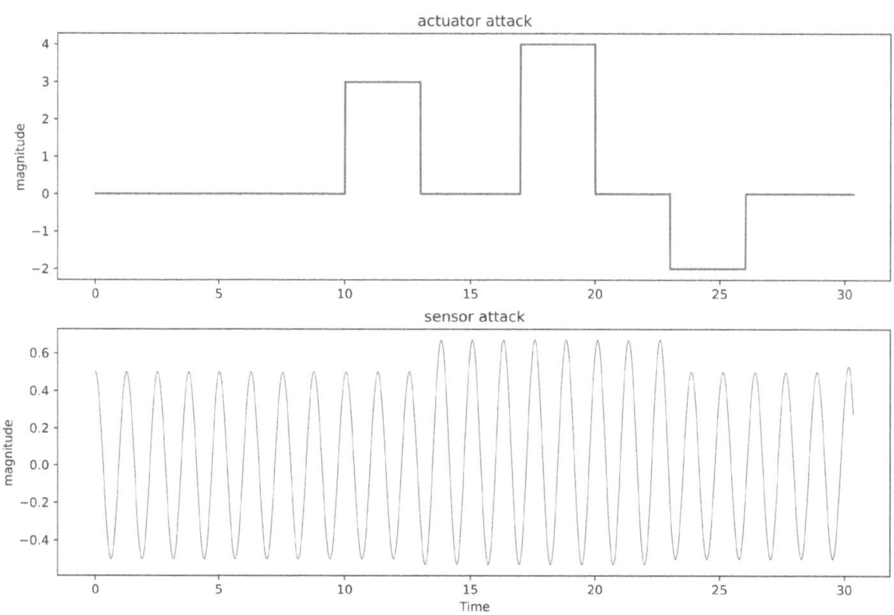

FIGURE 6.5 (a) The square actuator attack signal on each quadrotor UAV in the team formation NCS. (b) The cosine-type sensor attack signal on each quadrotor UAV in the team formation NCS.

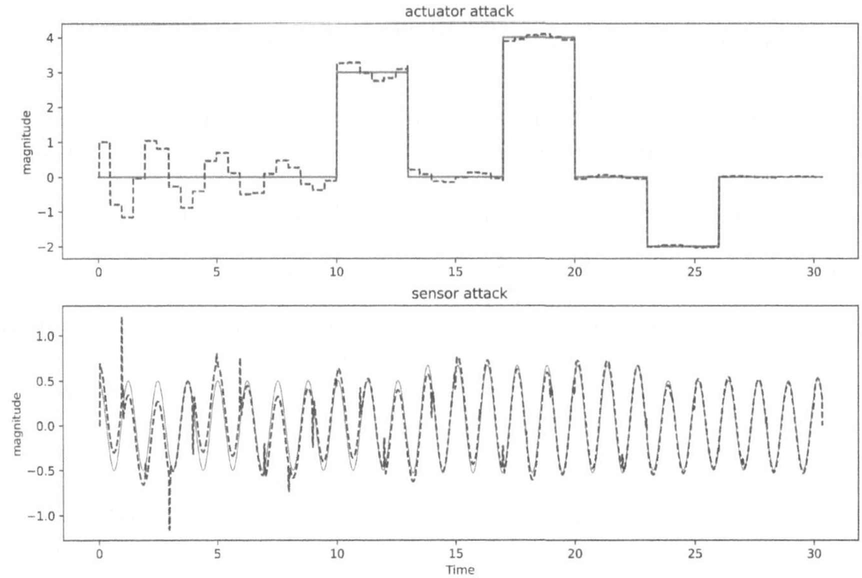

FIGURE 6.6 (a) The square actuator attack signal and its estimation of each quadrotor UAV in the team formation NCS. (b) The cosine-type sensor attack signal and its estimation of each quadrotor UAV in the team formation NCS.

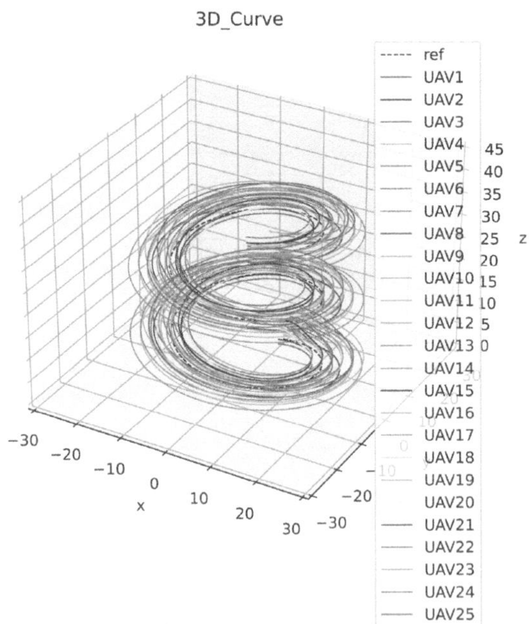

FIGURE 6.7 The flight trajectories of NCS team formation of 25 quadrotor UAVs using the proposed HJIE-reinforcement DNN-based H_∞ attack-tolerant observer-based team formation tracking control scheme.

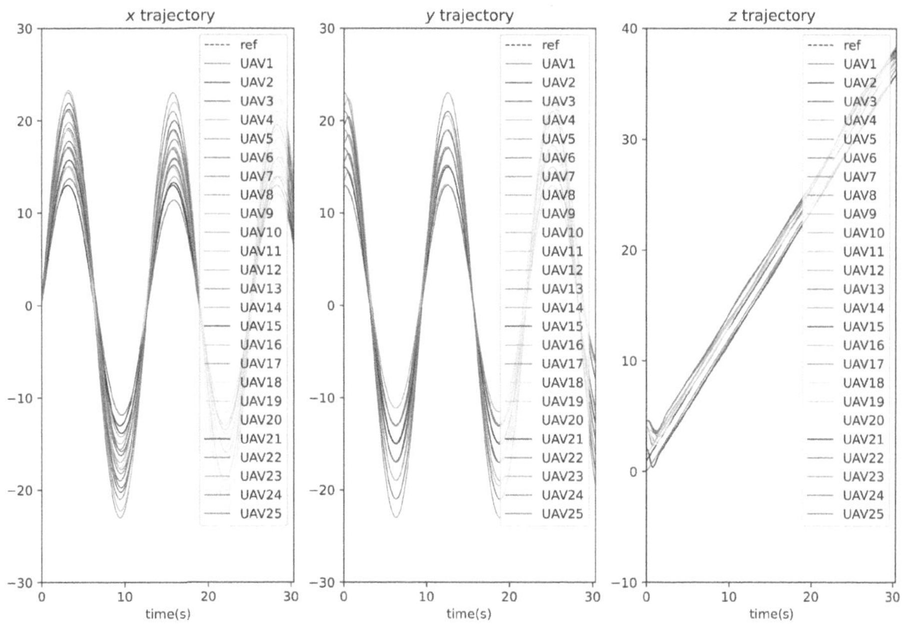

FIGURE 6.8 The position trajectories of 25 quadrotor UAVs in the team formation NCS using the proposed robust decentralized H_∞ attack-tolerant observer-based team formation control strategy via HJIE-reinforcement learning-based DNN approach.

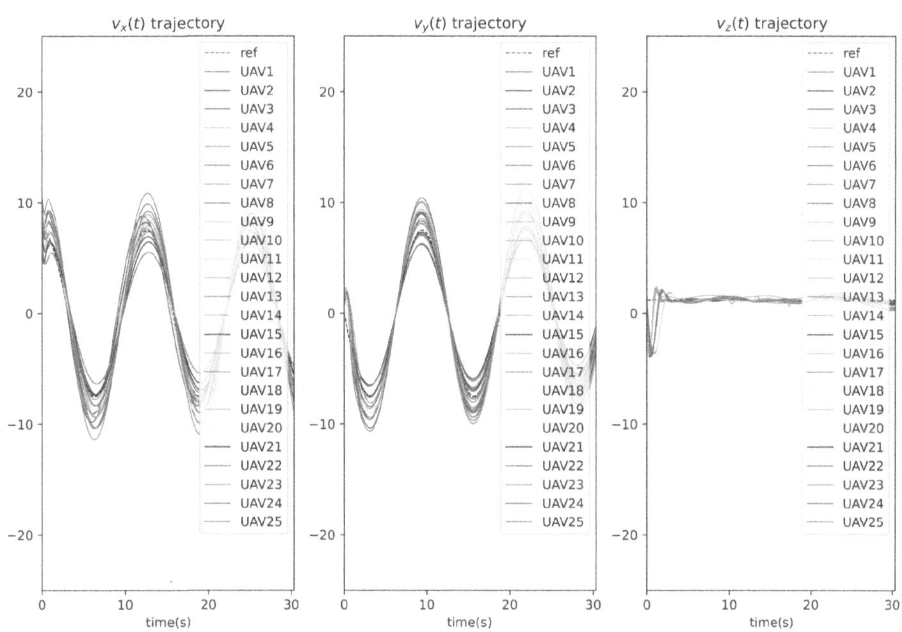

FIGURE 6.9 The velocity trajectories of 25 quadrotor UAVs in the team formation NCS using the proposed robust decentralized H_∞ attack-tolerant observer-based team formation control strategy via HJIE-reinforcement learning-based DNN approach.

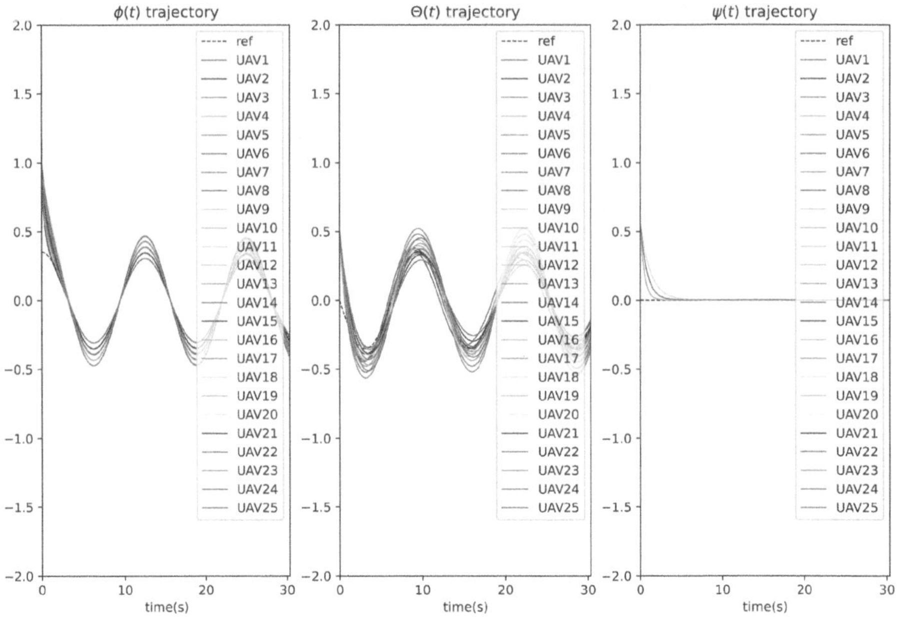

FIGURE 6.10 The angular trajectories of 25 quadrotor UAVs in the team formation NCS using the proposed robust decentralized H_∞ attack-tolerant observer-based team formation control strategy via HJIE-reinforcement learning-based DNN approach.

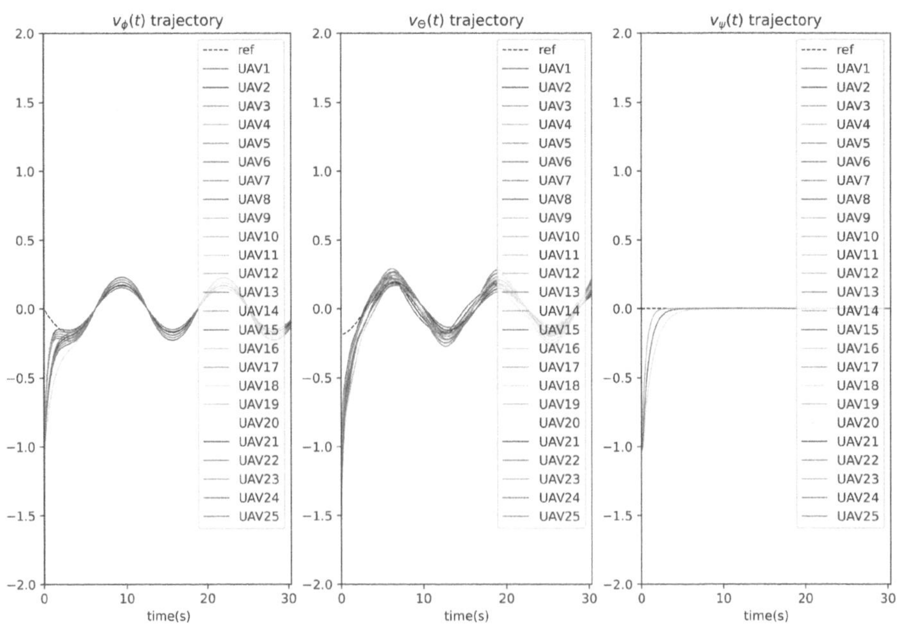

FIGURE 6.11 The angular velocity trajectories of 25 quadrotor UAVs in the team formation NCS using the proposed robust decentralized H_∞ attack-tolerant observer-based team formation control strategy via HJIE-reinforcement learning-based DNN approach.

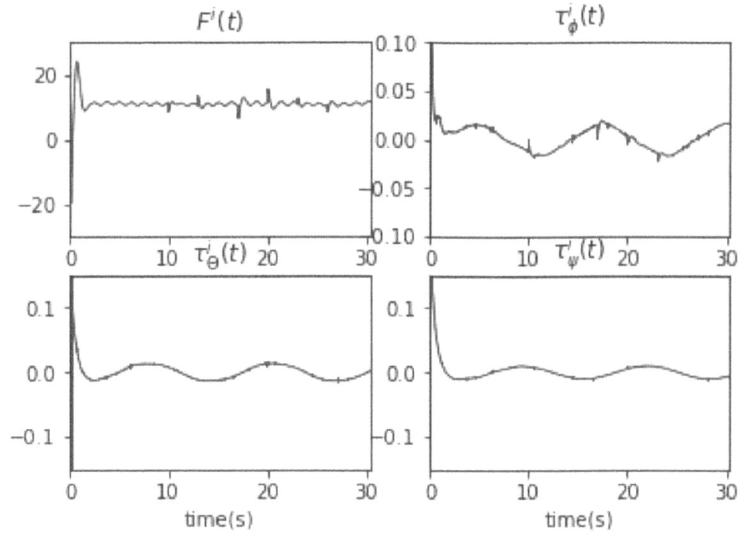

FIGURE 6.12 The control inputs of the first quadrotor UAV in the team formation NCS using the proposed robust decentralized H_∞ attack-tolerant observer-based team formation control strategy via HJIE-reinforcement learning-based DNN approach.

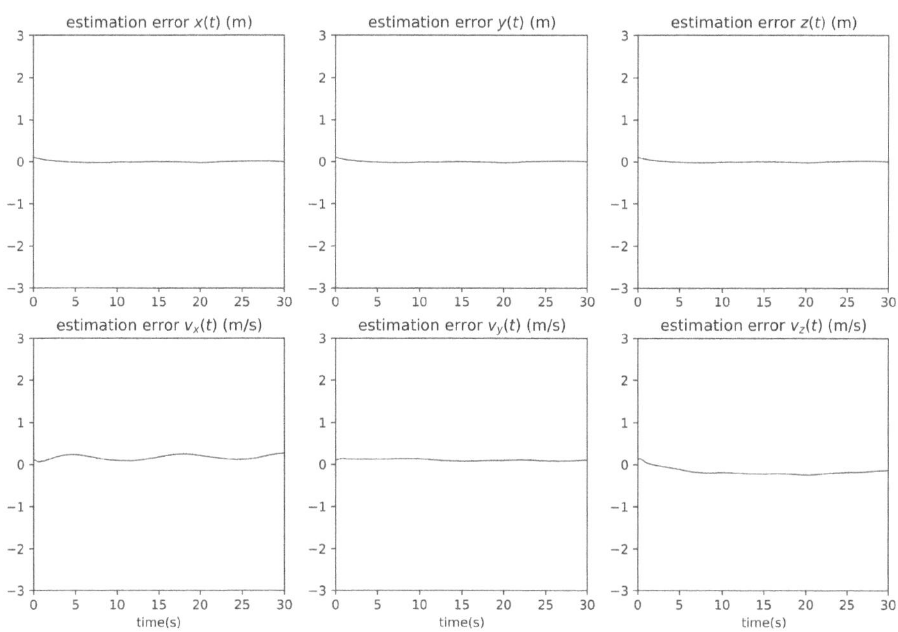

FIGURE 6.13 Estimation errors of the position/velocity of the first quadrotor UAV in the team formation NCS using the proposed robust decentralized H_∞ attack-tolerant observer-based team formation control strategy via HJIE-reinforcement learning-based DNN approach.

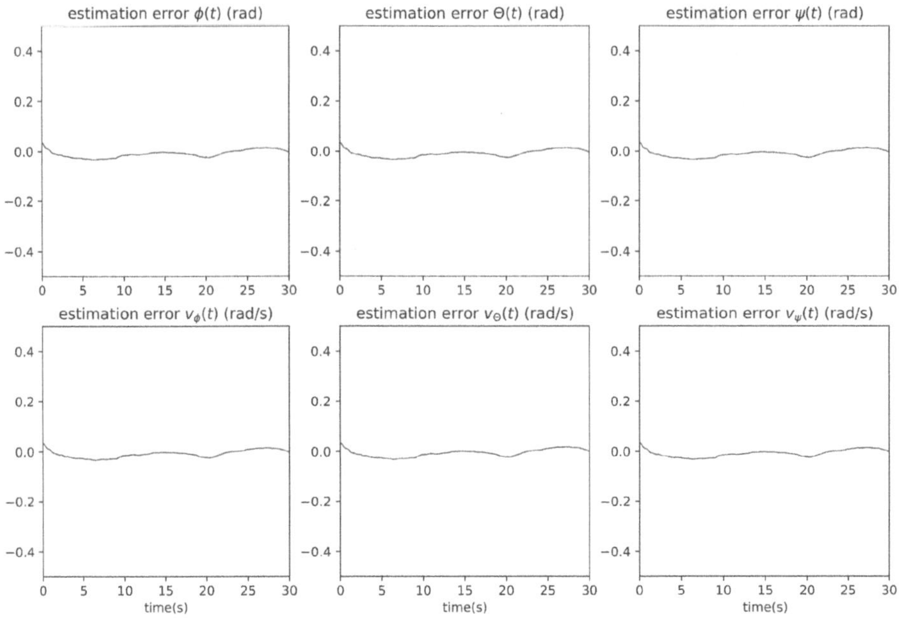

FIGURE 6.14 Estimation errors of the attitude/angular velocity of the 1st quadrotor UAV in the team formation NCS using the proposed robust decentralized H_∞ attack-tolerant observer-based team formation control strategy via HJIE-reinforcement learning-based DNN approach.

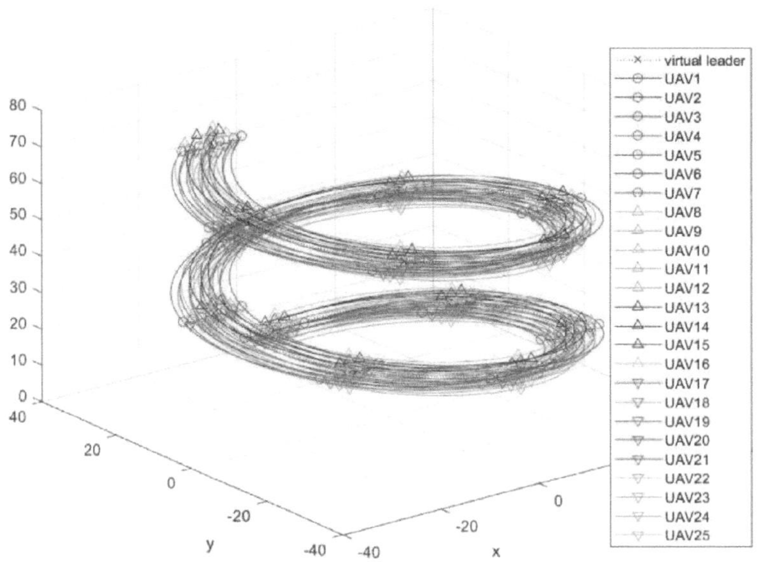

FIGURE 6.15 The flight trajectories of NCS formation of 25 quadrotor UAVs in the team formation NCS by the decentralized H_∞ T-S fuzzy team formation tracking control scheme in [9].

The simulation results are shown in Figures 6.7–6.14. The malicious attack signals $\gamma_a^i(t)$ and $\gamma_s^i(t)$ on the ith quadrotor UAV and their estimation are shown in Figure 6.6. It can be shown in Figure 6.7 that the team of 25-quadrotor UAV NCS is capable of tracking a spiral trajectory formation along the virtual leader reference trajectory $r_i(t)$. In Figures 6.8–6.11, the trajectories of the large-scale quadrotor UAV NCS can track the reference trajectories by the HJIE-reinforcement DNN-based decentralized H_∞ attack-tolerant observer-based team formation tracking control approach. The control inputs of the first quadrotor UAV in the team formation NCS are shown in Figure 6.12. Additionally, using the estimation of malicious attack signals, the total thrust $F^i(t)$ automatically changes its control amplitude to compensate for malicious attacks on the NCS actuator of the first quadrotor UAV in the team formation NCS, as shown in Figure 6.12. On the other hand, since output measurement $Y_i(t)$ of the ith quadrotor UAV is transmitted to the observer in (6.8) at the remote side directly via the wireless network channel, it will be affected by the sensor attack signal $\gamma_s^i(t)$. Then, the sensor attack signal $\gamma_s^i(t)$ will affect the observer state $\tilde{X}_i(t)$ which in turn affects $U_i(t)$. In Figures 6.13 and 6.14, the estimation error of the first quadrotor UAV quickly approaches to zero, which means that accurate estimation can be efficiently achieved by the proposed HJIE-reinforcement DNN-based decentralized H_∞ attack-tolerant observer-based team formation tracking control approach under external disturbance, malicious attacks, coupling effects, and measurement noise. The robust decentralized H_∞ team formation tracking control of the large-scale UAV networked control system by the T-S fuzzy controller in Reference [9] is carried out for the performance comparison with the proposed method and the results are also shown in Figure 6.15. However, the T-S fuzzy method needs to interpolate 128 local linearized systems to approximate a nonlinear system. In order to improve the team formation tracking performance, the number of fuzzy rules must be increased. For a highly nonlinear system such as a quadrotor UAV, a large number of 128 × 128 locally linearized observer-based controllers need to be interpolated by fuzzy bases and need to be computed at every time instant for the observer-based control signal, which will increase the computational complexity dramatically. Furthermore, the T-S fuzzy interpolation solution of the HJIE is based on a quadratic Lyapunov function, resulting in a more conservative solution of the nonlinear HJIE. The average attenuation level of the 25-quadrotor-UAV $\rho \approx 2.3112$ by the decentralized H_∞ team formation control in Reference [9] is poor than the proposed method which is calculated as $\rho \approx 2$.

6.6 CONCLUSION

In this chapter, an HJIE-reinforcement learning-based DNN robust decentralized H_∞ attack-tolerant observer-based team formation tracking control strategy is proposed to deal with a large-scale quadrotor UAV NCS under external disturbance, measurement noise, malicious attack, and coupling effect. At first, the dynamic model of the position and attitude of quadrotor UAV NCS under malicious attack, external disturbance, measurement noise, and coupling has been constructed. Then, with the aid of the virtual leader team formation structure, the large-scale quadrotor UAV team formation tracking control problem can be reformulated as the decentralized H_∞ attack-tolerant observer-based team formation tracking control design problem of large-scale quadrotor UAVs. To estimate the state variables of the system and the malicious attack signals on the sensors and actuators in the large-scale quadrotor UAV NCS, these malicious attack signals are augmented with each UAV NCS through a smooth dynamic model to simplify the estimation of attack signals and state of the system. Through the smooth dynamic model embedded in the system model of quadrotor UAV NCS, we can avoid its corruption and efficiently estimate the augmented system state by a traditional nonlinear Luenberger observer. Furthermore, by the proposed robust decentralized H_∞ attack-tolerant observer-based team formation tracking control strategy of large-scale UAV NCS, the decentralized observer-based controller of each UAV can effectively attenuate the worst-case effect of external disturbance, measurement noise, malicious attacks, and couplings from other UAVs on the state estimation and reference team formation control of large-scale quadrotor UAVs NCS to achieve robust H_∞ state estimation performance and reference tracking performance of each quadrotor UAV in the team formation NCS simultaneously. To address the difficulties in solving the HJIE_i of each quadrotor UAV for robust decentralized H_∞ attack-tolerant observer-based team formation tracking control, an HJIE-reinforcement learning DNN observer-based control scheme has been proposed in Figure 6.3. Using the output $\frac{\partial V(\tilde{X}_i(t),\bar{e}_i(t),t)}{\partial[\tilde{X}_i^T(t)\ \bar{e}_i^T(t)\ t]^T}$ of DNN, H_∞ control input, the worst-case augmented external disturbance, and coupling are sent back to the ith quadrotor UAV model to produce the next system output to start a new training period without affecting H_∞ team formation tracking performance. Thence, a DNN based on HJIE-reinforcement learning is used as a dynamic model-based learning scheme with a great potential to solve nonlinear partial differential HJIE_i for the complicated decentralized H_∞ attack-tolerant observer-based team formation tracking control design problems of large-scale quadrotor UAVs. We have also proven that if the reinforcement learning-based DNN Adam algorithm makes the HJIE_i error $\varepsilon_i(\theta_p^i(t))$ approach zero, the proposed HJIE-reinforcement learning-based DNN robust decentralized H_∞ attack-tolerant observer-based team formation tracking control scheme can approach to the theoretical robust decentralized H_∞ attack-tolerant observer-based team formation tracking control performance of large-scale quadrotor UAVs NCS. Eventually, the simulation example of decentralized H_∞ attack-tolerant observer-based team formation tracking control of 25-quadrotor UAV NCS is also given to validate the effectiveness of the proposed method with comparison.

Appendix 6A
Proof of Theorem 6.1

(a) Through the two-step procedure of the indirect method, we should first solve the Nash min-max quadratic game problem in (6.15) in step (i) to guarantee $J_i \leq E\{V(\tilde{X}_i(0), \overline{e}_i(0), 0)\}$ in step (ii). Thus, Theorem 6.1 can be proved as follows:

From (6.15), we can obtain:

$$
\begin{aligned}
J_i &= \min_{\substack{K_i(\hat{\bar{X}}_i(t), \overline{e}_i(t)),\\ L_i(\hat{\bar{x}}_i(t))}} \max_{\substack{\tilde{\bar{v}}_i(t),\\ \tilde{\bar{X}}_{-i}(t-\tau_{-i}(t))\\ \in \pounds_2[0,t_f]}} E\{-V(\tilde{X}_i(t_f), \overline{e}_i(t_f), t_f) \\
&\quad +V(\tilde{X}_i(0), \overline{e}_i(0), 0) + \int_0^{t_f} (\tilde{\bar{X}}_i^T(t) \bar{Q}_i \tilde{\bar{X}}_i(t) \\
&\quad +U_i(t) R_i U_i(t) - \rho_i^2 (\tilde{\bar{X}}_{-i}^T(t-\tau_{-i}(t)) \tilde{\bar{X}}_{-i}(t-\tau_{-i}(t))) \\
&\quad +\tilde{\bar{v}}_i(t) \tilde{\bar{v}}_i(t)) + \tfrac{d}{dt} V(\tilde{X}_i(t), \overline{e}_i(t), t)) dt\}
\end{aligned}
\tag{6.35}
$$

By chain rule from the nonlinear augmented time-varying system in (6.13), we have:

$$
\begin{aligned}
&\frac{dV(\tilde{X}_i(t), \overline{e}_i(t), t))}{dt} \\
&= \frac{\partial V(\tilde{X}_i(t), \overline{e}_i(t), t)}{\partial t} + \left(\frac{\partial V(\tilde{X}_i(t), \overline{e}_i(t), t)}{\partial [\tilde{X}_i^T(t)\ \overline{e}_i^T(t)]^T}\right)^T \\
&\times \left(\begin{bmatrix} \widetilde{F}_i(\widetilde{X}_i(t)) - L_i(\hat{\bar{X}}_i(t)) \widetilde{C}_i(\widetilde{X}_i(t)) \\ F_{e,i}(\overline{e}_i(t), t) \end{bmatrix} + \begin{bmatrix} \tilde{G}_i(\tilde{X}_i(t)) \\ G_{e,i}(\overline{e}_i(t), t) \end{bmatrix} U_i(t) \right. \\
&+ \begin{bmatrix} \bar{D}_i(\bar{X}_i(t)) & -L_i(\hat{\bar{X}}_i(t)) \\ D_{e,i}(\overline{e}_i(t), t) & 0 \end{bmatrix} \tilde{\bar{v}}_i^T(t) \\
&+ \left. \begin{bmatrix} \bar{F}_{-i}(\overline{X}_i(t)) \\ F_{e,-i}(\overline{e}_i(t), t) \end{bmatrix} \tilde{\bar{X}}_{-i}(t - \tau_{-i}(t)) \right), \ for\ i = 1, \ldots, N
\end{aligned}
\tag{6.36}
$$

Substituting (6.36) and $\tilde{\bar{G}}_i(\tilde{X}_i(t),\bar{e}_i(t),t)=\begin{bmatrix}\tilde{G}_i(\tilde{X}_i(t))\\ G_{e,i}(\bar{e}_i(t),t)\end{bmatrix}$ into (6.35), we have:

$$\begin{aligned}
J_i = & \min_{\substack{K_i(\hat{\bar{X}}_i(t),\bar{e}_i(t)),\\ L_i(\hat{\bar{x}}_i(t))}} \max_{\substack{\tilde{\bar{v}}_i(t),\\ \tilde{\bar{X}}_{-i}(t-\tau_{-i}(t))\\ \in \pounds_2[0,t_f]}} E\{-V(\tilde{X}_i(t_f),\bar{e}_i(t_f),t_f) \\
& +V(\tilde{X}_i(0),\bar{e}_i(0),0)+\int_0^{t_f}(\frac{\partial V(\tilde{X}_i(t),\bar{e}_i(t),t)}{\partial t} \\
& +\tilde{\bar{X}}_i^T(t)\bar{Q}_i\tilde{\bar{X}}_i(t)+U_i(t)R_iU_i(t)+(\frac{\partial V(\tilde{X}_i(t),\bar{e}_i(t),t)}{\partial[\tilde{X}_i^T(t)\ \bar{e}_i^T(t)]^T})^T \\
& \times\begin{bmatrix}\widetilde{F}_i(\widetilde{X}_i(t))-L_i(\widehat{X}_i(t))(\widetilde{C}_i(\widetilde{X}_i(t)))\\ F_{e,i}(\bar{e}_i(t),t)\end{bmatrix} \\
& +(\frac{\partial V(\tilde{X}_i(t),\bar{e}_i(t),t)}{\partial[\tilde{X}_i^T(t)\ \bar{e}_i^T(t)]^T})^T\tilde{\bar{G}}_i(\tilde{X}_i(t),\bar{e}_i(t),t)U_i(t) \\
& +(\frac{\partial V(\tilde{X}_i(t),\bar{e}_i(t),t)}{\partial[\tilde{X}_i^T(t)\ \bar{e}_i^T(t)]^T})^T\begin{bmatrix}\bar{D}_i(\bar{X}_i(t)) & -L_i(\hat{\bar{X}}_i(t))\\ D_{e,i}(\bar{e}_i(t),t) & 0\end{bmatrix} \\
& \times\tilde{\bar{v}}_i(t)+(\frac{\partial V(\tilde{X}_i(t),\bar{e}_i(t),t)}{\partial[\tilde{\bar{X}}_i^T(t)\ \bar{e}_i^T(t)]^T})^T\begin{bmatrix}\bar{F}_{-i}(\overline{X_i}(t))\\ F_{e,-i}(\bar{e}_i(t),t)\end{bmatrix}\times\tilde{\bar{X}}_{-i}(t-\tau_{-i}(t)-\rho_i^2\tilde{\bar{v}}_i^T(t)\tilde{\bar{v}}_i(t) \\
& -\rho_i^2\tilde{\bar{X}}_{-i}(t-\tau_{-i}(t)\tilde{\bar{X}}_{-i}(t-\tau_{-i}(t))dt\} \\
& \mathit{for}\ i=1,\ldots,N
\end{aligned} \tag{6.37}$$

By the completing square method, we get:

$$\begin{aligned}
J_i = & \min_{\substack{K_i(\hat{\bar{X}}_i(t),\bar{e}_i(t)),\\ L_i(\hat{\bar{x}}_i(t))}} \max_{\substack{\tilde{\bar{v}}_i(t),\\ \tilde{\bar{X}}_{-i}(t-\tau_{-i}(t))\\ \in \pounds_2[0,t_f]}} E\{-V(\tilde{X}_i(t_f),\bar{e}_i(t_f),t_f) \\
& +V(\tilde{X}_i(0),\bar{e}_i(0),0)+\int_0^{t_f}(\frac{\partial V(\tilde{X}_i(t),\bar{e}_i(t),t)}{\partial t}+\tilde{\bar{X}}_i^T(t)\bar{Q}_i\tilde{\bar{X}}_i(t) \\
& -\frac{1}{4}(\frac{\partial V(\tilde{X}_i(t),\bar{e}_i(t),t)}{\partial[\tilde{\bar{X}}_i^T(t)\ \bar{e}_i^T(t)]^T})^T\times\tilde{\bar{G}}_i(\tilde{X}_i(t),\bar{e}_i(t),t)R_i^{-1}\tilde{\bar{G}}_i^T(\tilde{X}_i(t),\bar{e}_i(t),t) \\
& \times(\frac{\partial V(\tilde{X}_i(t),\bar{e}_i(t),t)}{\partial[\tilde{\bar{X}}_i^T(t)\ \bar{e}_i^T(t)]^T})+(\frac{1}{2}\tilde{\bar{G}}_i^T(\tilde{X}_i(t),\bar{e}_i(t),t)\times(\frac{\partial V(\tilde{X}_i(t),\bar{e}_i(t),t)}{\partial[\tilde{\bar{X}}_i^T(t)\ \bar{e}_i^T(t)]^T}) \\
& +R_iU_i(t))^T R_i^{-1}\times(\frac{1}{2}\tilde{\bar{G}}_i^T(\tilde{X}_i(t),\bar{e}_i(t),t)(\frac{\partial V(\tilde{X}_i(t),\bar{e}_i(t),t)}{\partial[\tilde{X}_i^T(t)\ \bar{e}_i^T(t)]^T})+R_iU_i(t))
\end{aligned}$$

$$
\begin{aligned}
&+(\tfrac{\partial V(\tilde{X}_i(t),\bar{e}_i(t),t)}{\partial[\tilde{X}_i^T(t)\ \bar{e}_i^T(t)]^T})^T \times \begin{bmatrix} \widetilde{F}_i(\widetilde{X}_i(t)) - L_i(\widehat{\overline{X}_i}(t))(\widetilde{C}_i(\widetilde{X}_i(t))) \\ F_{e,i}(\bar{e}_i(t),t) \end{bmatrix} + \tfrac{1}{4\rho_i^2} \\
&\times(\tfrac{\partial V(\tilde{X}_i(t),\bar{e}_i(t),t)}{\partial[\tilde{X}_i^T(t)\ \bar{e}_i^T(t)]^T})^T \begin{bmatrix} \bar{D}_i(\bar{X}_i(t)) & -L_i(\hat{\bar{X}}_i(t)) \\ D_{e,i}(\bar{e}_i(t),t) & 0 \end{bmatrix} \\
&\times \begin{bmatrix} \bar{D}_i(\bar{X}_i(t)) & -L_i(\hat{\bar{X}}_i(t)) \\ D_{e,i}(\bar{e}_i(t),t) & 0 \end{bmatrix}^T (\tfrac{\partial V(\tilde{X}_i(t),\bar{e}_i(t),t)}{\partial[\tilde{X}_i^T(t)\ \bar{e}_i^T(t)]^T}) \\
&-\tfrac{1}{4\rho_i^2}(\tfrac{\partial V(\tilde{X}_i(t),\bar{e}_i(t),t)}{\partial[\tilde{X}_i^T(t)\ \bar{e}_i^T(t)]^T})^T \begin{bmatrix} \bar{F}_{-i}(\overline{X_i}(t)) \\ F_{e,-i}(\bar{e}_i(t),t) \end{bmatrix} \times \begin{bmatrix} \bar{F}_{-i}(\overline{X_i}(t)) \\ F_{e,-i}(\bar{e}_i(t),t) \end{bmatrix}^T (\tfrac{\partial V(\tilde{X}_i(t),\bar{e}_i(t),t)}{\partial[\tilde{X}_i^T(t)\ \bar{e}_i^T(t)]^T}) \\
&-(\tfrac{1}{2\rho_i^2} \times \begin{bmatrix} \bar{D}_i(\bar{X}_i(t)) & -L_i(\hat{\bar{X}}_i(t)) \\ D_{e,i}(\bar{e}_i(t),t) & 0 \end{bmatrix}^T (\tfrac{\partial V(\tilde{X}_i(t),\bar{e}_i(t),t)}{\partial[\tilde{X}_i^T(t)\ \bar{e}_i^T(t)]^T}) \\
&-\rho_i\tilde{\bar{v}}_i(t))^T \times (\tfrac{1}{2\rho_i^2} \begin{bmatrix} \bar{D}_i(\bar{X}_i(t)) & -L_i(\hat{\bar{X}}_i(t)) \\ D_{e,i}(\bar{e}_i(t),t) & 0 \end{bmatrix}^T \\
&\times(\tfrac{\partial V(\tilde{X}_i(t),\bar{e}_i(t),t)}{\partial[\tilde{X}_i^T(t)\ \bar{e}_i^T(t)]^T}) - \rho_i\tilde{\bar{v}}_i(t)) - (\tfrac{1}{2\rho_i^2} \begin{bmatrix} \bar{F}_{-i}(\overline{X_i}(t)) \\ F_{e,-i}(\bar{e}_i(t),t) \end{bmatrix}^T (\tfrac{\partial V(\tilde{X}_i(t),\bar{e}_i(t),t)}{\partial[\tilde{X}_i^T(t)\ \bar{e}_i^T(t)]^T}) \\
&\quad -\rho_i\tilde{\bar{X}}_{-i}(t-\tau_{-i}))^T (\tfrac{1}{2\rho_i^2} \begin{bmatrix} \bar{F}_{-i}(\overline{X_i}(t)) \\ F_{e,-i}(\bar{e}_i(t),t) \end{bmatrix}^T \\
&\quad \times(\tfrac{\partial V(\tilde{X}_i(t),\bar{e}_i(t),t)}{\partial[\tilde{X}_i^T(t)\ \bar{e}_i^T(t)]^T}) - \rho_i\tilde{\bar{X}}_{-i}(t-\tau_{-i})))dt\} \ for\ i=1,\dots,N
\end{aligned}
\tag{6.38}
$$

In (6.38) the optimal $U_i^*(t) = K_i^*(\hat{\bar{X}}_i(t),\bar{e}_i(t))$ in (6.17), the worst-case $\tilde{\bar{v}}_i^*(t)$ in (6.16) and $\tilde{\bar{X}}_{-i}^*(t-\tau_{-i}(t))$ in (6.18) make the square terms involved in (6.38) become zero to achieve $\min\limits_{U_i(t)=K_i(\hat{\bar{X}}_i(t),\bar{e}_i(t))} \max\limits_{\tilde{\bar{v}}_i(t),\tilde{\bar{X}}_{-i}(t-\tau_{-i}(t))}$. Then we can get:

$$
\begin{aligned}
J_i &= \min_{L_i(\hat{\bar{x}}_i(t))} E\{-V(\tilde{X}_i(t_f),\bar{e}_i(t_f),t_f) + V(\tilde{X}_i(0),\bar{e}_i(0),0) \\
&\quad + \int_0^{t_f} (\tfrac{\partial V(\tilde{X}_i(t),\bar{e}_i(t),t)}{\partial t} + \tilde{\bar{X}}_i^T(t)\bar{Q}_i\tilde{\bar{X}}_i(t) \\
&\quad -\tfrac{1}{4}(\tfrac{\partial V(\tilde{X}_i(t),\bar{e}_i(t),t)}{\partial[\tilde{X}_i^T(t)\ \bar{e}_i^T(t)]^T})^T \tilde{\bar{G}}_i(\tilde{X}_i(t),\bar{e}_i(t),t)R_i^{-1} \\
&\quad \times\tilde{\bar{G}}_i^T(\tilde{X}_i(t),\bar{e}_i(t),t)(\tfrac{\partial V(\tilde{X}_i(t),\bar{e}_i(t),t)}{\partial[\tilde{X}_i^T(t)\ \bar{e}_i^T(t)]^T})
\end{aligned}
\tag{6.39}
$$

$$+\left(\frac{\partial V(\tilde{X}_i(t),\bar{e}_i(t),t)}{\partial[\tilde{X}_i^T(t)\ \bar{e}_i^T(t)]^T}\right)^T\begin{bmatrix}\widetilde{F}_i(\widetilde{X}_i(t))-L_i(\widehat{\overline{X}}_i(t))\widetilde{C}_i(\widetilde{X}_i(t))\\ F_{e,i}(\bar{e}_i(t),t)\end{bmatrix}$$

$$+\frac{1}{4\rho_i^2}\left(\frac{\partial V(\tilde{X}_i(t),\bar{e}_i(t),t)}{\partial[\tilde{X}_i^T(t)\ \bar{e}_i^T(t)]^T}\right)^T\begin{bmatrix}\bar{D}_i(\bar{X}_i(t)) & -L_i(\hat{\bar{X}}_i(t))\\ D_{e,i}(\bar{e}_i(t),t) & 0\end{bmatrix}$$

$$\times\begin{bmatrix}\bar{D}_i(\bar{X}_i(t)) & -L_i(\hat{\bar{X}}_i(t))\\ D_{e,i}(\bar{e}_i(t),t) & 0\end{bmatrix}^T\left(\frac{\partial V(\tilde{X}_i(t),\bar{e}_i(t),t)}{\partial[[\tilde{X}_i^T(t)\ \bar{e}_i^T(t)]^T}\right)$$

$$+\frac{1}{4\rho_i^2}\left(\frac{\partial V(\tilde{X}_i(t),\bar{e}_i(t),t)}{\partial[[\tilde{X}_i^T(t)\ \bar{e}_i^T(t)]^T}\right)^T\begin{bmatrix}\bar{F}_{-i}(\overline{X}_i(t))\\ F_{e,-i}(\bar{e}_i(t),t)\end{bmatrix}$$

$$\times\begin{bmatrix}\bar{F}_{-i}(\overline{X}_i(t))\\ F_{e,-i}(\bar{e}_i(t),t)\end{bmatrix}^T\left(\frac{\partial V(\tilde{X}_i(t),\bar{e}_i(t),t)}{\partial[\tilde{X}_i^T(t)\ \bar{e}_i^T(t)]^T}\right)dt\}$$

$$for\ i=1,\ldots,N$$

By the fact:

$$\frac{\partial V(\tilde{X}_i(t),\bar{e}_i(t),t)}{\partial[\tilde{X}_i^T(t)\ \bar{e}_i^T(t)]^T}=\begin{bmatrix}\frac{\partial V(\tilde{X}_i(t),\bar{e}_i(t),t)}{\partial\tilde{X}_i(t)}\\ \frac{\partial V(\tilde{X}_i(t),\bar{e}_i(t),t)}{\partial\bar{e}_i(t)}\end{bmatrix} \tag{6.40}$$

we get:

$$\left(\frac{\partial V(\tilde{X}_i(t),\bar{e}_i(t),t)}{\partial[\tilde{X}_i^T(t),\bar{e}_i^T(t)]^T}\right)^T\begin{bmatrix}\widetilde{F}_i(\widetilde{X}_i(t))-L_i(\widehat{\overline{X}}_i(t))\widetilde{C}_i(\widetilde{X}_i(t))\\ F_{e,i}(\bar{e}_i(t),t)\end{bmatrix}$$
$$=\left(\frac{\partial V(\tilde{X}_i(t),\bar{e}_i(t),t)}{\partial\tilde{X}_i(t)}\right)^T(\widetilde{F}_i(\widetilde{X}_i(t))-L_i(\widehat{\overline{X}}_i(t))\widetilde{C}_i(\widetilde{X}_i(t))) \tag{6.41}$$
$$+\left(\frac{\partial V(\tilde{X}_i(t),\bar{e}_i(t),t)}{\partial\bar{e}_i(t)}\right)^T F_{e,i}(\bar{e}_i(t),t)$$
$$for\ i=1,\ldots,N$$

and

$$\left(\frac{\partial V(\tilde{X}_i(t),\bar{e}_i(t),t)}{\partial[\tilde{X}_i^T(t)\ \bar{e}_i^T(t)]^T}\right)^T\begin{bmatrix}\bar{D}_i(\bar{X}_i(t)) & -L_i(\hat{\bar{X}}_i(t))\\ D_{e,i}(\bar{e}_i(t),t) & 0\end{bmatrix}$$
$$\times\begin{bmatrix}\bar{D}_i(\bar{X}_i(t)) & -L_i(\hat{\bar{X}}_i(t))\\ D_{e,i}(\bar{e}_i(t),t) & 0\end{bmatrix}^T\left(\frac{\partial V(\tilde{X}_i(t),\bar{e}_i(t),t)}{\partial[\tilde{X}_i^T(t)\ \bar{e}_i^T(t)]^T}\right) \tag{6.42}$$
$$=\begin{bmatrix}\left(\frac{\partial V(\tilde{X}_i(t),\bar{e}_i(t),t)}{\partial\tilde{X}_i(t)}\right)^T & \left(\frac{\partial V(\tilde{X}_i(t),\bar{e}_i(t),t)}{\partial\bar{e}_i(t)}\right)^T\end{bmatrix}$$

$$\times\begin{bmatrix}\bar{D}_i(\bar{X}_i(t))\bar{D}_i^T(\bar{X}_i(t))+L_i\%(\widehat{\bar{X}_i}(t))L_i^T(\widehat{\bar{X}_i}(t)) & \bar{D}_i(\bar{X}_i(t))D_{e,i}^T(\bar{e}_i(t),t)\\ D_{e,i}(\bar{e}_i(t),t)\bar{D}_i^T(\bar{X}_i(t)) & D_{e,i}(\bar{e}_i\%(t),t)D_{e,i}^T(\bar{e}_i(t),t)\end{bmatrix}$$

$$\times\begin{bmatrix}\frac{\partial V(\tilde{X}_i(t),\bar{e}_i(t),t)}{\partial\tilde{X}_i(t)}\\ \frac{\partial V(\tilde{X}_i(t),\bar{e}_i(t),t)}{\partial\bar{e}_i(t)}\end{bmatrix}=(\frac{\partial V(\tilde{X}_i(t),\bar{e}_i(t),t)}{\partial\tilde{X}_i(t)})^T$$

$$\times(\bar{D}_i(\bar{X}_i(t))\bar{D}_i^T(\bar{X}_i(t))+L_i(\widehat{\bar{X}_i}(t))L_i^T(\widehat{\bar{X}_i}(t)))$$

$$\times(\frac{\partial V(\tilde{X}_i(t),\bar{e}_i(t),t)}{\partial\tilde{X}_i(t)})+(\frac{\partial V(\tilde{X}_i(t),\bar{e}_i(t),t)}{\partial\tilde{X}_i(t)})^T$$

$$\times\bar{D}_i(X_i(t))D_{e,i}(\bar{e}_i(t),t)^T(\frac{\partial V(\tilde{X}_i(t),\bar{e}_i(t),t)}{\partial\bar{e}_i(t)})$$

$$+(\frac{\partial V(\tilde{X}_i(t),\bar{e}_i(t),t)}{\partial\tilde{X}_i(t)})D_{e,i}(\bar{e}_i(t),t)\bar{D}_i^T(\bar{X}_i(t))$$

$$\times(\frac{\partial V(\tilde{X}_i(t),\bar{e}_i(t),t)}{\partial\bar{e}_i(t)})+(\frac{\partial V(\tilde{X}_i(t),\bar{e}_i(t),t)}{\partial\tilde{X}_i(t)})^T$$

$$\times D_{e,i}(\bar{e}_i(t),t)D_{e,i}(\bar{e}_i(t),t)^T(\frac{\partial V(\tilde{X}_i(t),\bar{e}_i(t),t)}{\partial\bar{e}_i(t)})$$

$$for\ i=1,\ldots,N$$

and

$$(\frac{\partial V(\tilde{X}_i(t),\bar{e}_i(t),t)}{\partial[\tilde{X}_i^T(t)\ \bar{e}_i^T(t)]^T})^T\begin{bmatrix}\bar{F}_{-i}(\overline{X_i}(t))\\ F_{e,-i}(\bar{e}_i(t),t)\end{bmatrix}\times\begin{bmatrix}\bar{F}_{-i}(\overline{X_i}(t))\\ F_{e,-i}(\bar{e}_i(t),t)\end{bmatrix}^T(\frac{\partial V(\tilde{X}_i(t),\bar{e}_i(t),t)}{\partial[\tilde{X}_i^T(t)\ \bar{e}_i^T(t)]^T})$$

$$=\begin{bmatrix}(\frac{\partial V(\tilde{X}_i(t),\bar{e}_i(t),t)}{\partial\tilde{X}_i(t)})^T & (\frac{\partial V(\tilde{X}_i(t),\bar{e}_i(t),t)}{\partial\bar{e}_i(t)})^T\end{bmatrix}$$

$$\times\begin{bmatrix}\bar{F}_{-i}(\overline{X_i}(t))\bar{F}_{-i}^T(\overline{X_i}(t)) & \bar{F}_{-i}(\overline{X_i}(t))F_{e,-i}^T(\bar{e}_i(t),t)\\ F_{e,-i}(\bar{e}_i(t),t)\bar{F}_{-i}^T(\overline{X_i}(t)) & F_{e,-i}(\bar{e}_i(t),t)F_{e,-i}^T(\bar{e}_i(t),t)\end{bmatrix}$$

$$\times\begin{bmatrix}\frac{\partial V(\tilde{X}_i(t),\bar{e}_i(t),t)}{\partial\tilde{X}_i(t)}\\ \frac{\partial V(\tilde{X}_i(t),\bar{e}_i(t),t)}{\partial\bar{e}_i(t)}\end{bmatrix}=(\frac{\partial V(\tilde{X}_i(t),\bar{e}_i(t),t)}{\partial\tilde{X}_i(t)})^T\times\bar{F}_{-i}(\overline{X_i}(t))\bar{F}_{-i}^T(\overline{X_i}(t))(\frac{\partial V(\tilde{X}_i(t),\bar{e}_i(t),t)}{\partial\tilde{X}_i(t)}) \tag{6.43}$$

$$+(\frac{\partial V(\tilde{X}_i(t),\bar{e}_i(t),t)}{\partial\tilde{X}_i(t)})^T\bar{F}_{-i}(\overline{X_i}(t))\times F_{e,-i}^T(\bar{e}_i(t),t)(\frac{\partial V(\tilde{X}_i(t),\bar{e}_i(t),t)}{\partial\bar{e}_i(t)})$$

$$+(\frac{\partial V(\tilde{X}_i(t),\bar{e}_i(t),t)}{\partial\tilde{X}_i(t)})^T F_{e,-i}(\bar{e}_i(t),t)\bar{F}_{-i}^T(\overline{X_i}(t))\times(\frac{\partial V(\tilde{X}_i(t),\bar{e}_i(t),t)}{\partial\bar{e}_i(t)})+(\frac{\partial V(\tilde{X}_i(t),\bar{e}_i(t),t)}{\partial\tilde{X}_i(t)})^T$$

$$\times F_{e,-i}(\bar{e}_i(t),t)F_{e,-i}^T(\bar{e}_i(t),t)(\frac{\partial V(\tilde{X}_i(t),\bar{e}_i(t),t)}{\partial\bar{e}_i(t)})$$

$$for\ i=1,\ldots,N$$

Substituting (6.40), (6.41), (6.42), and (6.43) into (6.39), we can get:

$$
\begin{aligned}
J_i = & \min_{L_i(\hat{\tilde{x}}_i(t))} E\{-V(\tilde{X}_i(t_f),\bar{e}_i(t_f),t_f)+V(\tilde{X}_i(0),\bar{e}_i(0),0) \\
& +\int_0^{t_f}(\frac{\partial V(\tilde{X}_i(t),\bar{e}_i(t),t)}{\partial t}+\bar{\tilde{X}}_i^T(t)\bar{Q}_i\bar{\tilde{X}}_i(t)-\tfrac{1}{4}(\frac{\partial V(\tilde{X}_i(t),\bar{e}_i(t),t)}{\partial[\tilde{X}_i^T(t)\ \bar{e}_i^T(t)]^T})^T\bar{\tilde{G}}_i(\tilde{X}_i(t),\bar{e}_i(t),t)R_i^{-1} \\
& \times\bar{\tilde{G}}_i^T(\tilde{X}_i(t),\bar{e}_i(t),t)(\frac{\partial V(\tilde{X}_i(t),\bar{e}_i(t),t)}{\partial[\tilde{X}_i^T(t)\ \bar{e}_i^T(t)]^T}) \\
& +(\frac{\partial V(\tilde{X}_i(t),\bar{e}_i(t),t)}{\partial\tilde{X}_i(t)})^T(\widetilde{F}_i(\widetilde{X}_i(t))-L_i(\widehat{\widetilde{X}}_i(t))\times\widetilde{C}_i(\widetilde{X}_i(t)) \\
& +(\frac{\partial V(\tilde{X}_i(t),\bar{e}_i(t),t)}{\partial\bar{e}_i(t)})^T F_{e,i}(\bar{e}_i(t),t)+\frac{1}{4\rho_i^2}(\frac{\partial V(\tilde{X}_i(t),\bar{e}_i(t),t)}{\partial\tilde{X}_i(t)})^T(\bar{D}_i\bar{X}_i(t)\bar{D}_i^T(\bar{X}_i(t)) \\
& +L_i(\widehat{\widetilde{X}}_i(t))L_i^T\widehat{\widetilde{X}}_i(t))(\frac{\partial V(\tilde{X}_i(t),\bar{e}_i(t),t)}{\partial\tilde{X}_i(t)}) \\
& +\frac{1}{4\rho_i^2}(\frac{\partial V(\tilde{X}_i(t),\bar{e}_i(t),t)}{\partial\tilde{X}_i(t)})^T\bar{D}_i(\bar{X}_i(t))D_{e,i}^T(\bar{e}_i(t),t) \\
& \times(\frac{\partial V(\tilde{X}_i(t),\bar{e}_i(t),t)}{\partial\bar{e}_i(t)})+\frac{1}{4\rho_i^2}(\frac{\partial V(\tilde{X}_i(t),\bar{e}_i(t),t)}{\partial\tilde{X}_i(t)})^T\times D_{e,i}(\bar{e}_i(t),t)\bar{D}_i^T\bar{X}_i(t)(\frac{\partial V(\tilde{X}_i(t),\bar{e}_i(t),t)}{\partial\bar{e}_i(t)}) \\
& +\frac{1}{4\rho_i^2}(\frac{\partial V(\tilde{X}_i(t),\bar{e}_i(t),t)}{\partial\tilde{X}_i(t)})^T D_{e,i}(\bar{e}_i(t),t)D_{e,i}^T(\bar{e}_i(t),t) \\
& \times(\frac{\partial V(\tilde{X}_i(t),\bar{e}_i(t),t)}{\partial\bar{e}_i(t)})+\frac{1}{4\rho_i^2}(\frac{\partial V(\tilde{X}_i(t),\bar{e}_i(t),t)}{\partial\tilde{X}_i(t)})^T\times\bar{F}_{-i}(\overline{X_i}(t))\bar{F}_{-i}^T(\overline{X_i}(t))(\frac{\partial V(\tilde{X}_i(t),\bar{e}_i(t),t)}{\partial\tilde{X}_i(t)}) \\
& +\frac{1}{4\rho_i^2}(\frac{\partial V(\tilde{X}_i(t),\bar{e}_i(t),t)}{\partial\tilde{X}_i(t)})^T\bar{F}_{-i}(\overline{X_i}(t))F_{e,-i}^T(\bar{e}_i(t),t) \\
& \times(\frac{\partial V(\tilde{X}_i(t),\bar{e}_i(t),t)}{\partial\bar{e}_i(t)})+\frac{1}{4\rho_i^2}(\frac{\partial V(\tilde{X}_i(t),\bar{e}_i(t),t)}{\partial\tilde{X}_i(t)})^T \\
& \times F_{e,-i}(\bar{e}_i(t),t)\bar{F}_{-i}^T(\overline{X_i}(t))(\frac{\partial V(\tilde{X}_i(t),\bar{e}_i(t),t)}{\partial\bar{e}_i(t)}) \\
& +\frac{1}{4\rho_i^2}(\frac{\partial V(\tilde{X}_i(t),\bar{e}_i(t),t)}{\partial\tilde{X}_i(t)})^T F_{e,-i}(\bar{e}_i(t),t)\times F_{e,-i}^T(\bar{e}_i(t),t)(\frac{\partial V(\tilde{X}_i(t),\bar{e}_i(t),t)}{\partial\bar{e}_i(t)})dt\} \\
= & \min_{L_i(\hat{\tilde{x}}_i(t))} E\{-V(\tilde{X}_i(t_f),\bar{e}_i(t_f),t_f)+V(\tilde{X}_i(0),\bar{e}_i(0),0) \\
& +\int_0^{t_f}(\frac{\partial V(\tilde{X}_i(t),\bar{e}_i(t),t)}{\partial t}+\bar{\tilde{X}}_i^T(t)\bar{Q}_i\bar{\tilde{X}}_i(t) \\
& -\tfrac{1}{4}(\frac{\partial V(\tilde{X}_i(t),\bar{e}_i(t),t)^T}{\partial[\tilde{X}_i^T(t)\ \bar{e}_i^T(t)})^T\bar{\tilde{G}}_i(\tilde{X}_i(t),\bar{e}_i(t),t)R_i^{-1} \\
& \times\bar{\tilde{G}}_i^T(\tilde{X}_i(t),\bar{e}_i(t),t)(\frac{\partial V(\tilde{X}_i(t),\bar{e}_i(t),t)^T}{\partial[\tilde{X}_i^T(t)\ \bar{e}_i^T(t)}) \\
& +(\frac{\partial V(\tilde{X}_i(t),\bar{e}_i(t),t)}{\partial\tilde{X}_i(t)})^T\widetilde{F}_i(\widetilde{X}_i(t))+(\frac{\partial V(\tilde{X}_i(t),\bar{e}_i(t),t)}{\partial\bar{e}_i(t)})^T F_{e,i}(\bar{e}_i(t),t) \\
& +\frac{1}{4\rho_i^2}(\frac{\partial V(\tilde{X}_i(t),\bar{e}_i(t),t)}{\partial\tilde{X}_i(t)})^T\bar{D}_i(\bar{X}_i(t))\bar{D}_i^T(\bar{X}_i(t))\times(\frac{\partial V(\tilde{X}_i(t),\bar{e}_i(t),t)}{\partial\tilde{X}_i(t)})
\end{aligned}
$$

$$+\frac{1}{4\rho_i^2}(\frac{\partial V(\bar{X}_i(t),\bar{e}_i(t),t)}{\partial \bar{X}_i(t)})^T \bar{D}_i(\bar{X}_i(t))D_{e,i}^T(\bar{e}_i(t),t)$$
$$\times(\frac{\partial V(\bar{X}_i(t),\bar{e}_i(t),t)}{\partial \bar{e}_i(t)})+\frac{1}{4\rho_i^2}(\frac{\partial V(\bar{X}_i(t),\bar{e}_i(t),t)}{\partial \bar{X}_i(t)})^T$$
$$\times D_{e,i}(\bar{e}_i(t),t)\bar{D}_i^T(\bar{X}_i(t))(\frac{\partial V(\bar{X}_i(t),\bar{e}_i(t),t)}{\partial \bar{e}_i(t)})$$
$$+\frac{1}{4\rho_i^2}(\frac{\partial V(\bar{X}_i(t),\bar{e}_i(t),t)}{\partial \bar{X}_i(t)})^T D_{e,i}(\bar{e}_i(t),t)D_{e,i}^T(\bar{e}_i(t),t)$$
$$\times(\frac{\partial V(\bar{X}_i(t),\bar{e}_i(t),t)}{\partial \bar{e}_i(t)})+\frac{1}{4\rho_i^2}(\frac{\partial V(\bar{X}_i(t),\bar{e}_i(t),t)}{\partial \bar{X}_i(t)})^T$$
$$\times\bar{F}_{-i}(\overline{X_i}(t))\bar{F}_{-i}^T(\overline{X_i}(t))(\frac{\partial V(\bar{X}_i(t),\bar{e}_i(t),t)}{\partial \bar{X}_i(t)}) \tag{6.44}$$
$$+\frac{1}{4\rho_i^2}(\frac{\partial V(\bar{X}_i(t),\bar{e}_i(t),t)}{\partial \bar{X}_i(t)})^T \bar{F}_{-i}(\overline{X_i}(t))F_{e,-i}^T(\bar{e}_i(t),t)$$
$$\times(\frac{\partial V(\bar{X}_i(t),\bar{e}_i(t),t)}{\partial \bar{e}_i(t)})+\frac{1}{4\rho_i^2}(\frac{\partial V(\tilde{X}_i(t),\bar{e}_i(t),t)}{\partial \tilde{X}_i(t)})^T$$
$$\times F_{e,-i}(\bar{e}_i(t),t)\bar{F}_{-i}^T(\overline{X_i}(t))(\frac{\partial V(\tilde{X}_i(t),\bar{e}_i(t),t)}{\partial \bar{e}_i(t)})$$
$$+\frac{1}{4\rho_i^2}(\frac{\partial V(\tilde{X}_i(t),\bar{e}_i(t),t)}{\partial \tilde{X}_i(t)})^T F_{e,-i}(\bar{e}_i(t),t)\times F_{e,-i}^T(\bar{e}_i(t),t)(\frac{\partial V(\tilde{X}_i(t),\bar{e}_i(t),t)}{\partial \bar{e}_i(t)})$$
$$+\frac{1}{4\rho_i^2}(L_i^T(\widehat{\overline{X}_i}(t))\times(\frac{\partial V(\tilde{X}_i(t),\bar{e}_i(t),t)}{\partial \tilde{X}_i(t)})-\frac{1}{2}\widetilde{C}_i(\widetilde{X}_i(t))^T)(L_i^T(\widehat{\overline{X}_i}(t))$$
$$\times(\frac{\partial V(\tilde{X}_i(t),\bar{e}_i(t),t)}{\partial \tilde{X}_i(t)})-\frac{1}{2}\widetilde{C}_i(\widetilde{X}_i(t)))-\frac{1}{16\rho_i^2}\widetilde{C_i^T}(\widetilde{X}_i(t))\widetilde{C}_i(\widetilde{X}_i(t)))dt\}$$
$$for\ i=1,\ldots,N$$

We can obtain the optimal observer gain $L_i^*(\widehat{\overline{X}_i}(t))$ as (6.19) from (6.44). Then, in the optimal observer gain case of (6.44), it can be rewritten as follows:

$$J_i = E\{-V(\tilde{X}_i(t_f),\bar{e}_i(t_f),t_f)+V(\tilde{X}_i(0),\bar{e}_i(0),0)$$
$$+\int_0^{t_f}(\frac{\partial V(\tilde{X}_i(t),\bar{e}_i(t),t)}{\partial t}+\bar{\tilde{X}}_i^T(t)\bar{Q}_i\bar{\tilde{X}}_i(t)$$
$$-\frac{1}{4}(\frac{\partial V(\tilde{X}_i(t),\bar{e}_i(t),t)}{\partial[\tilde{X}_i^T(t)\ \bar{e}_i^T(t)]^T})^T\bar{\tilde{G}}_i(\tilde{X}_i(t),\bar{e}_i(t),t)R_i^{-1}$$
$$\times\bar{\tilde{G}}_i^T(\tilde{X}_i(t),\bar{e}_i(t),t)(\frac{\partial V(\tilde{X}_i(t),\bar{e}_i(t),t)}{\partial[\tilde{X}_i^T(t)\ \bar{e}_i^T(t)]^T})+(\frac{\partial V(\tilde{X}_i(t),\bar{e}_i(t),t)}{\partial[\tilde{X}_i^T(t)\ \bar{e}_i^T(t)]^T})^T\bar{\tilde{F}}_i(\tilde{X}_i(t),\bar{e}_i(t),t) \tag{6.45}$$
$$+\frac{1}{4\rho_i^2}(\frac{\partial V(\tilde{X}_i(t),\bar{e}_i(t),t)}{\partial[\tilde{X}_i^T(t)\ \bar{e}_i^T(t)]^T})^T\bar{\tilde{D}}_i(\tilde{X}_i(t),\bar{e}_i(t),t)\times(\frac{\partial V(\tilde{X}_i(t),\bar{e}_i(t),t)}{\partial[\tilde{X}_i^T(t)\ \bar{e}_i^T(t)]^T})+\frac{1}{4\rho_i^2}(\frac{\partial V(\tilde{X}_i(t),\bar{e}_i(t),t)}{\partial[\tilde{X}_i^T(t)\ \bar{e}_i^T(t)]^T})^T$$
$$\times\bar{F}_{-i}(\widetilde{\overline{X}_i}(t))(\frac{\partial V(\tilde{X}_i(t),\bar{e}_i(t),t)}{\partial[\tilde{X}_i^T(t)\ \bar{e}_i^T(t)]^T})-\rho_i^2\widetilde{C_i^T}(\widetilde{X}_i(t))\widetilde{C}_i(\widetilde{X}_i(t)))dt\}$$
$$for\ i=1,\ldots,N$$

By HJIE$_i$ in (6.20) and (6.45), we get:

$$\begin{aligned} J_i &= E\{-V(\tilde{X}_i(t_f),\bar{e}_i(t_f),t_f)+V(\tilde{X}_i(0),\bar{e}_i(0),0)\} \\ &for\ i=1,\ldots,N \end{aligned} \tag{6.46}$$

where $V(\tilde{X}_i(t),\bar{e}_i(t),t_f)\geq 0$, then:

$$\begin{aligned} J_i &\leq E\{V(\tilde{X}_i(0),\bar{e}_i(0),0)\} \\ &for\ i=1,\ldots,N \end{aligned} \tag{6.47}$$

(b) If $\tilde{\bar{v}}_i(t)\in\pounds_2[0,\infty)$ and $\tilde{\bar{X}}_{-i}(t-\tau_{-i}(t))\in\pounds_2[0,\infty)$ and $V(\tilde{X}_i(0),\bar{e}_i(0),0)<\infty$, from the definition of J_i in (6.15), then (6.47) becomes:

$$\begin{aligned} &E\{\int_0^{t_f}(\tilde{\bar{X}}_i^T(t)\bar{Q}_i\tilde{\bar{X}}_i(t))+U_i^T(t)R_i(t)U_i(t)\} \\ &\leq E\{V(\tilde{X}_i(0),\bar{e}_i(0),0)+\rho_i^2\int_0^{t_f}(\tilde{\bar{v}}_i^T(t)\tilde{\bar{v}}_i(t) \\ &\quad+\tilde{\bar{X}}_{-i}^T(t-\tau_{-i}(t)))\tilde{\bar{X}}_{-i}^T(t-\tau_{-i}(t)))dt\}, \\ &for\ i=1,\ldots,N \end{aligned} \tag{6.48}$$

Since $\rho_i^2E\{\int_0^{t_f}(\tilde{\bar{v}}_i^T(t)\tilde{\bar{v}}_i(t)+\tilde{\bar{X}}_{-i}^T(t-\tau_{-i}(t)))\tilde{\bar{X}}_{-i}^T(t-\tau_{-i}(t)))dt\}$ is finite, from (6.48), it is seen that $E\{U_i^T(t)U_i(t)\}$, $E\{\tilde{X}_i^T(t))\tilde{X}_i(t)\}$, and $E\{\bar{e}_i^T(t)\bar{e}_i(t)\}$ approach to zero as $t_f\to\infty$ in the mean square error sense. This completes the proof.

Appendix 6B
Proof of Theorem 6.2

First, we suppose:

$$\begin{aligned}&\left(\frac{\partial V(\tilde{X}_i(t),\overline{e}_i(t),t)}{\partial[\tilde{X}_i^T(t)\ \overline{e}_i^T(t)\ t]^T}\right)_\varepsilon = \left(\frac{\partial V(\tilde{X}_i(t),\overline{e}_i(t),t)}{\partial[\tilde{X}_i^T(t)\ \overline{e}_i^T(t)\ t]^T}\right) + \Xi_i(\tilde{X}_i(t),\overline{e}_i(t),t)\\ &for\ i=1,\ldots,N\end{aligned} \tag{6.49}$$

where $\Xi_i(\tilde{X}_i(t),\overline{e}_i(t),t)$ is the error function between $\left(\frac{\partial V(\tilde{X}_i(t),\overline{e}_i(t),t)}{\partial[\tilde{X}_i^T(t)\ \overline{e}_i^T(t)\ t]^T}\right)_\varepsilon$ and $\left(\frac{\partial V(\tilde{X}_i(t),\overline{e}_i(t),t)}{\partial[\tilde{X}_i^T(t)\ \overline{e}_i^T(t)\ t]^T}\right)$.

By the fact that $\text{HJIE}_{\varepsilon,i} = \varepsilon_i(\theta_p^i(t))$ and $\text{HJIE}_i = 0$, $\varepsilon_i(\theta_p^i(t))$ can be rewritten as follows:

$$\begin{aligned}
\varepsilon_i(\theta_p^i(t)) &= HJIE_{i,\varepsilon} - HJIE_i\\
&= \left(\left(\frac{\partial V(\tilde{X}_i(t),\overline{e}_i(t),t)}{\partial[\tilde{X}_i^T(t)\ \overline{e}_i^T(t)\ t]^T}\right)_\varepsilon^T - \left(\frac{\partial V(\tilde{X}_i(t),\overline{e}_i(t),t)}{\partial[\tilde{X}_i^T(t)\ \overline{e}_i^T(t)\ t]^T}\right)^T\right)\\
&\times\begin{bmatrix}\tilde{\overline{F}}_i(\tilde{X}_i(t),\overline{e}_i(t),t)\\ 1\end{bmatrix} - \frac{1}{4}\left(\frac{\partial V(\tilde{X}_i(t),\overline{e}_i(t),t)}{\partial[\tilde{X}_i^T(t)\ \overline{e}_i^T(t)\ t]^T}\right)_\varepsilon^T\\
&\times\begin{bmatrix}\tilde{\overline{G}}_i(\tilde{X}_i(t),\overline{e}_i(t),t)\\ 0\end{bmatrix} R_i^{-1}\begin{bmatrix}\tilde{\overline{G}}_i(\tilde{X}_i(t),\overline{e}_i(t),t)\\ 0\end{bmatrix}^T\\
&\times\left(\frac{\partial V(\tilde{X}_i(t),\overline{e}_i(t),t)}{\partial[\tilde{X}_i^T(t)\ \overline{e}_i^T(t)\ t]^T}\right)_\varepsilon + \frac{1}{4}\left(\frac{\partial V(\tilde{X}_i(t),\overline{e}_i(t),t)}{\partial[\tilde{X}_i^T(t)\ \overline{e}_i^T(t)\ t]^T}\right)^T\\
&\times\begin{bmatrix}\tilde{\overline{G}}_i(\tilde{X}_i(t),\overline{e}_i(t),t)\\ 0\end{bmatrix} R_i^{-1}\begin{bmatrix}\tilde{\overline{G}}_i(\tilde{X}_i(t),\overline{e}_i(t),t)\\ 0\end{bmatrix}^T\\
&\times\left(\frac{\partial V(\tilde{X}_i(t),\overline{e}_i(t),t)}{\partial[\tilde{X}_i^T(t)\ \overline{e}_i^T(t)\ t]^T}\right) + \frac{1}{4\rho_i^2}\left(\frac{\partial V(\tilde{X}_i(t),\overline{e}_i(t),t)}{\partial[\tilde{X}_i^T(t)\ \overline{e}_i^T(t)\ t]^T}\right)_\varepsilon^T\\
&\times\begin{bmatrix}\tilde{\overline{D}}_i(\tilde{X}_i(t),\overline{e}_i(t),t) & 0\\ 0 & 0\end{bmatrix}\left(\frac{\partial V(\tilde{X}_i(t),\overline{e}_i(t),t)}{\partial[\tilde{X}_i^T(t)\ \overline{e}_i^T(t)\ t]^T}\right)_\varepsilon\\
&- \frac{1}{4\rho_i^2}\left(\frac{\partial V(\tilde{X}_i(t),\overline{e}_i(t),t)}{\partial[\tilde{X}_i^T(t)\ \overline{e}_i^T(t)\ t]^T}\right)^T\begin{bmatrix}\tilde{\overline{D}}_i(\tilde{X}_i(t),\overline{e}_i(t),t) & 0\\ 0 & 0\end{bmatrix}
\end{aligned} \tag{6.50}$$

$$\times(\frac{\partial V(\tilde{X}_i(t),\bar{e}_i(t),t)}{\partial[\tilde{X}_i^T(t)\ \bar{e}_i^T(t)\ t]^T})+\frac{1}{4\rho_i^2}(\frac{\partial V(\tilde{X}_i(t),\bar{e}_i(t),t)}{\partial[\tilde{X}_i^T(t)\ \bar{e}_i^T(t)\ t]^T})_\varepsilon^T$$

$$\times\begin{bmatrix}\tilde{\bar{F}}_{-i}(\widetilde{\widetilde{X}}_i(t),\bar{e}_i(t),t) & 0\\ 0 & 0\end{bmatrix}(\frac{\partial V(\tilde{X}_i(t),\bar{e}_i(t),t)}{\partial[\tilde{X}_i^T(t)\ \bar{e}_i^T(t)\ t]^T})_\varepsilon$$

$$-\frac{1}{4\rho_i^2}(\frac{\partial V(\tilde{X}_i(t),\bar{e}_i(t),t)}{\partial[\tilde{X}_i^T(t)\ \bar{e}_i^T(t)\ t]^T})^T\begin{bmatrix}\tilde{\bar{F}}_{-i}(\widetilde{\widetilde{X}}_i(t),\bar{e}_i(t),t) & 0\\ 0 & 0\end{bmatrix}\times(\frac{\partial V(\tilde{X}_i(t),\bar{e}_i(t),t)}{\partial[\tilde{X}_i^T(t)\ \bar{e}_i^T(t)\ t]^T})$$

$$for\ i=1,\ldots,N$$

By (6.49), we get:

$$\varepsilon_i(\theta_p^i(t))=\Xi_i^T(\tilde{X}_i(t),\bar{e}_i(t),t)\begin{bmatrix}\widetilde{\widetilde{F}}_i(\tilde{X}_i(t),\bar{e}_i(t),t)\\ 1\end{bmatrix}$$

$$-\frac{1}{4}\Xi_i^T(\tilde{X}_i(t),\bar{e}_i(t),t)\begin{bmatrix}\tilde{\bar{G}}_i(\tilde{X}_i(t),\bar{e}_i(t),t)\\ 0\end{bmatrix}R_i^{-1}$$

$$\times\begin{bmatrix}\tilde{\bar{G}}_i(\tilde{X}_i(t),\bar{e}_i(t),t)\\ 0\end{bmatrix}^T(\frac{\partial V(\tilde{X}_i(t),\bar{e}_i(t),t)}{\partial[\tilde{X}_i^T(t)\ \bar{e}_i^T(t)\ t]^T})$$

$$-\frac{1}{4}(\frac{\partial V(\tilde{X}_i(t),\bar{e}_i(t),t)}{\partial[\tilde{X}_i^T(t)\ \bar{e}_i^T(t)\ t]^T})^T\begin{bmatrix}\tilde{\bar{G}}_i(\tilde{X}_i(t),\bar{e}_i(t),t)\\ 0\end{bmatrix}R_i^{-1}$$

$$\times\begin{bmatrix}\tilde{\bar{G}}_i(\tilde{X}_i(t),\bar{e}_i(t),t)\\ 0\end{bmatrix}^T\Xi_i(\tilde{X}_i(t),\bar{e}_i(t),t)$$

$$-\frac{1}{4}\Xi_i^T(\tilde{X}_i(t),\bar{e}_i(t),t)\begin{bmatrix}\tilde{\bar{G}}_i(\tilde{X}_i(t),\bar{e}_i(t),t)\\ 0\end{bmatrix}R_i^{-1}$$

$$\times\begin{bmatrix}\tilde{\bar{G}}_i(\tilde{X}_i(t),\bar{e}_i(t),t)\\ 0\end{bmatrix}^T\Xi_i(\tilde{X}_i(t),\bar{e}_i(t),t)$$

$$+\frac{1}{4\rho_i^2}\Xi_i^T(\tilde{X}_i(t),\bar{e}_i(t),t)\begin{bmatrix}\tilde{\bar{D}}_i(\tilde{X}_i(t),\bar{e}_i(t),t) & 0\\ 0 & 0\end{bmatrix}$$

$$\times(\frac{\partial V(\tilde{X}_i(t),\bar{e}_i(t),t)}{\partial[\tilde{X}_i^T(t)\ \bar{e}_i^T(t)\ t]^T})+\frac{1}{4\rho_i^2}(\frac{\partial V(\tilde{X}_i(t),\bar{e}_i(t),t)}{\partial[\tilde{X}_i^T(t)\ \bar{e}_i^T(t)\ t]^T})^T$$

$$\times\begin{bmatrix}\tilde{\bar{D}}_i(\tilde{X}_i(t),\bar{e}_i(t),t) & 0\\ 0 & 0\end{bmatrix}\Xi_i(\tilde{X}_i(t),\bar{e}_i(t),t)$$

$$+\frac{1}{4\rho_i^2}\Xi_i^T(\tilde{X}_i(t),\bar{e}_i(t),t)\begin{bmatrix}\tilde{\bar{D}}_i(\tilde{X}_i(t),\bar{e}_i(t),t) & 0\\ 0 & 0\end{bmatrix}$$

$$\times \Xi_i(\tilde{X}_i(t),\bar{e}_i(t),t) + \frac{1}{4\rho_i^2}\Xi_i^T(\tilde{X}_i(t),\bar{e}_i(t),t) \times \begin{bmatrix} \tilde{\bar{F}}_{-i}(\tilde{\bar{X}}_i(t),\bar{e}_i(t),t) & 0 \\ 0 & 0 \end{bmatrix} \left(\frac{\partial V(\tilde{X}_i(t),\bar{e}_i(t),t)}{\partial[\tilde{X}_i^T(t)\ \bar{e}_i^T(t)\ t]^T}\right)$$

$$+\frac{1}{4\rho_i^2}\left(\frac{\partial V(\tilde{X}_i(t),\bar{e}_i(t),t)}{\partial[\tilde{X}_i^T(t)\ \bar{e}_i^T(t)\ t]^T}\right)^T \begin{bmatrix} \tilde{\bar{F}}_{-i}(\tilde{\bar{X}}_i(t),\bar{e}_i(t),t) & 0 \\ 0 & 0 \end{bmatrix} \times \Xi_i(\tilde{X}_i(t),\bar{e}_i(t),t) + \frac{1}{4\rho_i^2}\Xi_i^T(\tilde{X}_i(t),\bar{e}_i(t),t) \times \begin{bmatrix} \tilde{\bar{F}}_{-i}(\tilde{\bar{X}}_i(t),\bar{e}_i(t),t) & 0 \\ 0 & 0 \end{bmatrix} \Xi_i(\tilde{X}_i(t),\bar{e}_i(t),t) \quad \text{for } i=1,\ldots,N \tag{6.51}$$

By the symmetric property, we have following equations:

$$\Xi_i^T(\tilde{X}_i(t),\bar{e}_i(t),t)\begin{bmatrix} \tilde{\bar{G}}_i(\tilde{X}_i(t),\bar{e}_i(t),t) \\ 0 \end{bmatrix} R_i^{-1} \times \begin{bmatrix} \tilde{\bar{G}}_i(\tilde{X}_i(t),\bar{e}_i(t),t) \\ 0 \end{bmatrix}^T \left(\frac{\partial V(\tilde{X}_i(t),\bar{e}_i(t),t)}{\partial[\tilde{X}_i^T(t)\ \bar{e}_i^T(t)\ t]^T}\right) = \left(\frac{\partial V(\tilde{X}_i(t),\bar{e}_i(t),t)}{\partial[\tilde{X}_i^T(t)\ \bar{e}_i^T(t)\ t]^T}\right)^T \begin{bmatrix} \tilde{\bar{G}}_i(\tilde{X}_i(t),\bar{e}_i(t),t) \\ 0 \end{bmatrix} R_i^{-1} \times \begin{bmatrix} \tilde{\bar{G}}_i(\tilde{X}_i(t),\bar{e}_i(t),t) \\ 0 \end{bmatrix}^T \Xi_i(\tilde{X}_i(t),\bar{e}_i(t),t), \quad \text{for } i=1,\ldots,N \tag{6.52}$$

$$\Xi_i^T(\tilde{X}_i(t),\bar{e}_i(t),t)\begin{bmatrix} \tilde{\bar{D}}_i(\tilde{X}_i(t),\bar{e}_i(t),t) & 0 \\ 0 & 0 \end{bmatrix} \times \left(\frac{\partial V(\tilde{X}_i(t),\bar{e}_i(t),t)}{\partial[\tilde{X}_i^T(t)\ \bar{e}_i^T(t)\ t]^T}\right) = \left(\frac{\partial V(\tilde{X}_i(t),\bar{e}_i(t),t)}{\partial[\tilde{X}_i^T(t)\ \bar{e}_i^T(t)\ t]^T}\right)^T \begin{bmatrix} \tilde{\bar{D}}_i(\tilde{X}_i(t),\bar{e}_i(t),t) & 0 \\ 0 & 0 \end{bmatrix} \Xi_i(\tilde{X}_i(t),\bar{e}_i(t),t), \quad \text{for } i=1,\ldots,N \tag{6.53}$$

and

$$\Xi_i^T(\tilde{X}_i(t),\bar{e}_i(t),t)\begin{bmatrix}\tilde{\bar{F}}_{-i}(\tilde{\bar{X}}_i(t),\bar{e}_i(t),t) & 0\\ 0 & 0\end{bmatrix}\times(\frac{\partial V(\tilde{X}_i(t),\bar{e}_i(t),t)}{\partial[\tilde{X}_i^T(t)\ \bar{e}_i^T(t)\ t]^T})$$

$$=(\frac{\partial V(\tilde{X}_i(t),\bar{e}_i(t),t)}{\partial[\tilde{X}_i^T(t)\ \bar{e}_i^T(t)\ t]^T})^T\begin{bmatrix}\tilde{\bar{F}}_{-i}(\tilde{\bar{X}}_i(t),\bar{e}_i(t),t) & 0\\ 0 & 0\end{bmatrix}\Xi_i(\tilde{X}_i(t),\bar{e}_i(t),t), \tag{6.54}$$

$$for\ i=1,\ldots,N$$

By (6.52), (6.53), and (6.54), then (6.51) becomes:

$$\begin{aligned}\varepsilon_i(\theta_p^i(t)) = \Xi_i^T(\tilde{X}_i(t),\bar{e}_i(t),t)\{&\begin{bmatrix}\tilde{\bar{F}}_i(\tilde{X}_i(t),\bar{e}_i(t),t)\\ 1\end{bmatrix}\\ &-\tfrac{1}{2}\begin{bmatrix}\tilde{\bar{G}}_i(\tilde{X}_i(t),\bar{e}_i(t),t)\\ 0\end{bmatrix}R_i^{-1}\begin{bmatrix}\tilde{\bar{G}}_i(\tilde{X}_i(t),\bar{e}_i(t),t)\\ 0\end{bmatrix}^T\\ &\times(\frac{\partial V(\tilde{X}_i(t),\bar{e}_i(t),t)}{\partial[\tilde{X}_i^T(t)\ \bar{e}_i^T(t)\ t]^T})-\tfrac{1}{4}\begin{bmatrix}\tilde{\bar{G}}_i(\tilde{X}_i(t),\bar{e}_i(t),t)\\ 0\end{bmatrix}R_i^{-1}\\ &\times\begin{bmatrix}\tilde{\bar{G}}_i(\tilde{X}_i(t),\bar{e}_i(t),t)\\ 0\end{bmatrix}^T\Xi_i(\tilde{X}_i(t),\bar{e}_i(t),t)+\frac{1}{2\rho_i^2}\\ &\times\begin{bmatrix}\tilde{\bar{D}}_i(\tilde{X}_i(t),\bar{e}_i(t),t) & 0\\ 0 & 0\end{bmatrix}(\frac{\partial V(\tilde{X}_i(t),\bar{e}_i(t),t)}{\partial[\tilde{X}_i^T(t)\ \bar{e}_i^T(t)\ t]^T})\\ &+\frac{1}{4\rho_i^2}\begin{bmatrix}\tilde{\bar{D}}_i(\tilde{X}_i(t),\bar{e}_i(t),t) & 0\\ 0 & 0\end{bmatrix}\Xi_i(\tilde{X}_i(t),\bar{e}_i(t),t)\\ &+\frac{1}{2\rho_i^2}\begin{bmatrix}\tilde{\bar{F}}_i(\tilde{X}_i(t),\bar{e}_i(t),t)\\ 1\end{bmatrix}(\frac{\partial V(\tilde{X}_i(t),\bar{e}_i(t),t)}{\partial[\tilde{X}_i^T(t)\ \bar{e}_i^T(t)\ t]^T})\\ &+\frac{1}{4\rho_i^2}\begin{bmatrix}\tilde{\bar{F}}_i(\tilde{X}_i(t),\bar{e}_i(t),t)\\ 1\end{bmatrix}\Xi_i(\tilde{X}_i(t),\bar{e}_i(t),t)\},\\ &for\ i=1,\ldots,N\end{aligned} \tag{6.55}$$

If $\varepsilon_i(\theta_p^i(t))\rightarrow 0$ in (6.23), then (6.55) $\rightarrow 0$, too. Clearly, the term:

$$\begin{bmatrix}\tilde{\bar{F}}_i(\tilde{X}_i(t),\bar{e}_i(t),t)\\ 1\end{bmatrix}-\tfrac{1}{2}\begin{bmatrix}\tilde{\bar{G}}_i(\tilde{X}_i(t),\bar{e}_i(t),t)\\ 0\end{bmatrix}R_i^{-1}$$

$$\times\begin{bmatrix}\tilde{\bar{G}}_i(\tilde{X}_i(t),\bar{e}_i(t),t)\\ 0\end{bmatrix}^T(\frac{\partial V(\tilde{X}_i(t),\bar{e}_i(t),t)}{\partial[\tilde{X}_i^T(t)\ \bar{e}_i^T(t)\ t]^T})$$

$$
\begin{aligned}
&\begin{bmatrix} \tilde{\bar{\bar{F}}}_i(\tilde{X}_i(t),\bar{e}_i(t),t) \\ 1 \end{bmatrix} - \frac{1}{2}\begin{bmatrix} \tilde{\bar{G}}_i(\tilde{X}_i(t),\bar{e}_i(t),t) \\ 0 \end{bmatrix} R_i^{-1} \\
&\times \begin{bmatrix} \tilde{\bar{G}}_i(\tilde{X}_i(t),\bar{e}_i(t),t) \\ 0 \end{bmatrix}^T \left(\frac{\partial V(\tilde{X}_i(t),\bar{e}_i(t),t)}{\partial[\tilde{X}_i^T(t)\ \bar{e}_i^T(t)\ t]^T}\right) \\
&-\frac{1}{4}\begin{bmatrix} \tilde{\bar{G}}_i(\tilde{X}_i(t),\bar{e}_i(t),t) \\ 0 \end{bmatrix} R_i^{-1} \begin{bmatrix} \tilde{\bar{G}}_i(\tilde{X}_i(t),\bar{e}_i(t),t) \\ 0 \end{bmatrix}^T \times \Xi_i(\tilde{X}_i(t),\bar{e}_i(t),t) \\
&+\frac{1}{2\rho_i^2}\begin{bmatrix} \tilde{\bar{D}}_i(\tilde{X}_i(t),\bar{e}_i(t),t) & 0 \\ 0 & 0 \end{bmatrix} \times \left(\frac{\partial V(\tilde{X}_i(t),\bar{e}_i(t),t)}{\partial[\tilde{X}_i^T(t)\ \bar{e}_i^T(t)\ t]^T}\right) \\
&+\frac{1}{4\rho_i^2}\begin{bmatrix} \tilde{\bar{D}}_i(\tilde{X}_i(t),\bar{e}_i(t),t) & 0 \\ 0 & 0 \end{bmatrix} \times \Xi_i(\tilde{X}_i(t),\bar{e}_i(t),t) \\
&+\frac{1}{2\rho_i^2}\begin{bmatrix} \tilde{\bar{\bar{F}}}_i(\tilde{X}_i(t),\bar{e}_i(t),t) \\ 1 \end{bmatrix} \times \left(\frac{\partial V(\tilde{X}_i(t),\bar{e}_i(t),t)}{\partial[\tilde{X}_i^T(t)\ \bar{e}_i^T(t)\ t]^T}\right) \\
&+\frac{1}{4\rho_i^2}\begin{bmatrix} \tilde{\bar{\bar{F}}}_i(\tilde{X}_i(t),\bar{e}_i(t),t) \\ 1 \end{bmatrix} \times \Xi_i(\tilde{X}_i(t),\bar{e}_i(t),t)
\end{aligned}
$$

in $\{.\}$ of (6.55) is not equal to zero for all $\tilde{X}_i(t)$ and $\bar{e}_i(t)$. If (6.55) $\to 0$, $\Xi_i(\tilde{X}_i(t), \bar{e}_i(t),t)$ must approach to zero. In this situation, from (6.49), it deduces that $\left(\frac{\partial V(\tilde{X}_i(t),\bar{e}_i(t),t)}{\partial[\tilde{X}_i^T(t)\ \bar{e}_i^T(t)\ t]^T}\right)_\varepsilon \to \left(\frac{\partial V(\tilde{X}_i(t),\bar{e}_i(t),t)}{\partial[\tilde{X}_i^T(t)\ \bar{e}_i^T(t)\ t]^T}\right)$. According to Theorem 6.1, the HJIE-reinforcement DNN-based robust decentralized H_∞ attack-tolerant observer-based tracking control scheme in Figure 6. 3 can approach to the theoretical robust decentralized H_∞ attack-tolerant observer-based team formation tracking control design. The proof is complete.

REFERENCES

[1] M. Zhu, X. Du, X. Zhang, H. Luo, and G. Wang, "Muti-quadrotor UAV rapid-assessment task-assignment problem in a post-earthquake scenario," *IEEE Access*, vol. 7, pp. 74542–74557, June 2019.

[2] N. H. Motlagh, M. Bagaa, and T. Taleb, "Quadrotor UAV-based IoT platform: A crowd surveillance use case," *IEEE Communications Magazine*, vol. 55, no. 2, pp. 128–134, Feb. 2017.

[3] K. Dorling, J. Heinrichs, G. G. Messier, and S. Magierowski, "Vehicle routing problems for drone delivery," *IEEE Transactions on Systems, Man, and Cybernetics: Systems*, vol. 47, no. 1, pp. 70–85, Jan. 2017.

[4] F. Liao, R. Teo, J. L. Wang, X. Dong, F. Lin, and K. Peng, "Distributed formation and reconfiguration control of VTOL quadrotor UAVs," *IEEE Transactions on Control Systems Technology*, vol. 25, no. 1, pp. 270–277, Jan. 2017.

[5] P. Ogren, M. Egerstedt, and X. Hu, "A control Lyapunov function approach to multiagent coordination," *IEEE Transactions on Robotics and Automation*, vol. 18, no. 5, pp. 847–851, Oct. 2002.

[6] B.-S. Chen, C.-P. Wang, and M.-Y. Lee, "Stochastic robust team tracking control of multi-quadrotor UAV networked system under Wiener and Poisson random fluctuations," *IEEE Transactions on Cybernetics*, vol. 51, no. 12, pp. 5786–5799, Dec. 2021.

[7] J. R. T. Lawton, R. W. Beard, and B. J. Young, "A decentralized approach to formation maneuvers, " *IEEE Transactions on Robotics and Automation*, vol. 19, no. 6, pp. 933–941, Dec. 2003.

[8] M.-Y. Lee, B.-S. Chen, Y. Chang, and C.-L. Hwang, "Stochastic robust team formation tracking design of multi-VTOL-quadrotor UAV networked control system in smart city under time-varying delay and random fluctuation," *IEEE Access*, vol. 8, pp. 131310–131326, 2020.

[9] M.-Y. Lee, B.-S. Chen, C.-Y. Tsai, and C.-L. Hwang, "Stochastic H_∞ robust decentralized tracking control of large-scale team formation quadrotor UAV network system with time-varying delay and packet dropout under interconnected couplings and wiener fluctuations," *IEEE Access*, vol. 9, pp. 41976–41997, 2021. http://doi.org/10.1109/ACCESS. 2021.3065127.

[10] L. Chettri and R. Bera, "A comprehensive survey on internet of things (IoT) toward 5g wireless systems," *IEEE Internet of Things Journal*, vol. 7, no. 1, pp. 16–32, Jan. 2020.

[11] W. C. Cheng, X. Zhang, and H. L. Zhang, "Statistical-QoS driven energy efficiency optimization over green 5G mobile wireless networks," *IEEE Journal on Selected Areas in Communications*, vol. 34, no. 12, pp. 3092–3107, Dec. 2016.

[12] Y. Shi and B. Yu, "Output feedback stabilization of networked control systems with random delays modeled by Markov chains," *IEEE Transactions on Automatic Control*, vol. 54, no. 7, pp. 1668–1674, July 2009.

[13] G. P. Liu, Y. Xia, D. Rees, and W. Hu, "Design and stability criteria of networked predictive control systems with random network delay in the feedback channel," *IEEE Transactions on Systems, Man, and Cybernetics, Part C (Applications and Reviews)*, vol. 37, no. 2, pp. 173–184, Mar. 2007.

[14] H. Gao and T. Chen, "Network-basedH_∞ output tracking control," *IEEE Transactions on Automatic Control*, vol. 53, no. 3, pp. 655–667, Apr. 2008.

[15] B. Xue, S. Li, and Q. Zhu, "Moving horizon state estimation for networked control systems with multiple packet dropouts," *IEEE Transactions on Automatic Control*, vol. 57, no. 9, pp. 2360–2366, Sep. 2012.

[16] R. Lu, Y. Xu, and R. Zhang, "A new design of model predictive tracking control for networked control system under random packet loss and uncertainties," *IEEE Transactions on Industrial Electronics*, vol. 63, no. 11, pp. 6999–7007, Nov. 2016.

[17] B. Genge, P. Haller, and I. Kiss, "Cyber-security-aware network design of industrial control systems," *IEEE Systems Journal*, vol. 11, no. 3, pp. 1373–1384, Sep. 2017.

[18] H. Zhang, P. Cheng, L. Shi, and J. Chen, "Optimal DoS attack scheduling in wireless networked control system," *IEEE Transactions on Control Systems Technology*, vol. 24, no. 3, pp. 843–852, May 2016.

[19] N. Liu, J. Zhang, H. Zhang, and W. Liu, "Security assessment for communication networks of power control systems using attack graph and MCDM," *IEEE Transactions on Power Delivery*, vol. 25, no. 3, pp. 1492–1500, July 2010.

[20] A. Shui, W. Chen, P. Zhang, S. Hu, and X. Huang, "Review of fault diagnosis in control systems," in *Professional Chinese Control and Decision Conference*, Guilin, China, 2009, pp. 5324–5329.

[21] S. Harshavarthini, R. Sakthivel, and C. K. Ahn, "Finite-time reliable attitude tracking control design for nonlinear quadrotor model with actuator faults," *Nonlinear Dynamics*, vol. 96, no. 4, pp. 2681–2692, June 2019.
[22] H. Yang and S. Yin, "Descriptor observers design for Markov jump systems with simultaneous sensor and actuator faults," *IEEE Transactions on Automatic Control*, vol. 64, no. 8, pp. 3370–3377, Aug. 2019.
[23] Q. Jia, W. Chen, Y. Zhang, and H. Li, "Fault reconstruction and fault tolerant control via learning observers in Takagi–Sugeno fuzzy descriptor systems with time delays," *IEEE Transactions on Industrial Electronics*, vol. 62, no. 6, pp. 3885–3895, June 2015.
[24] Z. Gao and S. X. Ding, "Fault estimation and fault-tolerant control for descriptor systems via proportional, multiple-integral and derivative observer design," *IET Control Theory & Applications*, vol. 1, no. 5, pp. 1208–1218, Sep. 2007.
[25] T. Basar and P. Bernhard, *H_∞ Optimal Control and Related Minimax Design Problems: A Dynamic Game Approach*, Boston, MA: Birkhäuser, 1995.
[26] J. W. Helton and M. R. James, *Extending H_∞ Control to Nonlinear Systems*, Philadelphia, PA: SIAM, 1999.
[27] J. Yong and X. Y. Zhou, *Stochastic Control: Hamiltonian Systems and HJB Equation*, New York: Springer, 1999.
[28] W. Zhang, L. Xie, and B. S. Chen, *Stochastic H2/H_∞ Control: A Nash Game Approach*, CRC Press, Boca Raton, United States, 2017.
[29] B.-S. Chen, M.-Y. Lee, and X.-H. Chen, "Security-enhanced filter design for stochastic systems under malicious attack via smoothed signal model and multiobjective estimation method," *IEEE Transactions on Signal Processing*, vol. 68, pp. 4971–4986, 2020.
[30] H. Gao, J. Lam, L. Xie, and C. Wang, "New approach to mixed H2/H_∞ filtering for polytopic discrete-time systems," *IEEE Transactions on Signal Processing*, vol. 53, no. 8, pp. 3183–3192, Aug. 2005.
[31] X. L. Lin, C. F. Wu, and B. S. Chen, "Robust H_∞ adaptive fuzzy tracking control for MIMO nonlinear stochastic Poisson jump diffusion systems," *IEEE Transactions on Cybernetics*, vol. 49, no. 8, pp. 3116–3130, Aug. 2019.
[32] B.-S. Chen, M.-Y. Lee, and T.-H. Lin, "DNN-based H_∞ control scheme of nonlinear time-varying dynamic systems with external disturbance and its application to UAV tracking design," *IEEE Access*, vol. 9, pp. 69635–69653, 2021.
[33] B.-S. Chen and P.-H. Wu, "Robust H_∞ observer-based reference tracking control design of nonlinear stochastic systems: HJIE embedded deep learning approach," *IEEE Access*, vol. 10, pp. 39889–39911, 2022.
[34] D. P. Kingma and J. Ba, "Adam: A method for stochastic optimization," *arXiv preprint arXir: 1412.6980*, ICLR, San Diego, pp. 1–15, Dec. 2014.
[35] S. X. Du, A. Dekka, and B. Wu, *Modular Multilevel Converters: Analysis, Control, and Applications*, Hoboken, NJ: Wiley-IEEE Press, Jan. 2018.
[36] Y. N. Dauphin, H. D. Vries, J. Chung, and Y. Bengio, "RMSProp and equilibrated adaptive learning rates for non-convex optimization," *Machine Learning, arXiv preprint arXiv:1502.04390*, Feb. 2015.
[37] A. H. Khan, X. Cao, S. Li, V. N. Katsikis, and L. Liao, "BAS-ADAM: An ADAM based approach to improve the performance of beetle antennae search optimizer," *IEEE/CAA Journal of Automatica Sinica*, vol. 7, no. 2, pp. 461–471, Mar. 2020.
[38] C. S. Tseng and B. S. Chen, "H_∞ decentralized fuzzy model reference tracking control design for nonlinear interconnected systems," *IEEE Transactions on Fuzzy Systems*, vol. 9, no. 6, pp. 795–809, Dec. 2001.

7 Integrating Local Motion Planning and Robust Decentralized Fault-Tolerant Tracking Control for Search and Rescue Task of Hybrid UAVs and Biped Robots Team System

7.1 INTRODUCTION

In recent years, the unmanned vehicle (UV) has attracted attention due to the advances in communication technology, sensing devices, and computing power. It not only reduces labor costs and brings convenience to life, but also more importantly can replace some dangerous jobs for humans. Due to these benefits, it has been widely used in many scenarios, such as S&R, battlefield, logistics and transportation, and surveillance [1]. Compared with a single UV, multiple UVs can perform more complex tasks and are more robust due to a large number of agents [2]. However, the cost is that the design of such a multi-agent system (MAS) becomes more intricate as there are more problems to be resolved, such as formation, collision avoidance between agents, task allocation, and cooperation between agents [3]. In addition to the number increment, a heterogeneous multi-agent system (HMAS) combining various types of UV is also valued [4]. Compared with homogeneous MAS, it can adapt to a wider variety of application scenarios because each agent has different aptitudes.

To construct an unmanned HMAS, three required key capabilities are perception, decision-making, and control. Perception is to obtain information through the sensor (e.g., localization or computer vision), decision-making is to make decisions through the sensor information, and control is to execute the decision through the actuator. In this chapter, we focus on decision-making and control problem only.

Three main problems of decision-making in an unmanned HMAS are task allocation, path planning and collision avoidance. Task allocation is to optimally assign

DOI: 10.1201/9781003545965-7

tasks to each agent under some constraints such as agent capabilities, fuel cost, and time cost [5]. Path planning is to optimally plan paths for each agent while subject to constraints such as agent kinodynamic properties, distance, and obstacles' collision [6]. Collision avoidance is to avoid collision with obstacles. Although collision avoidance is often concerned in path planning, the collision avoidance system is also independently studied because of the requirements for the safety and reliability of the actual system [7].

Although there are many types of UVs to make up an unmanned HMAS, unmanned aerial vehicle (UAV) and unmanned ground vehicle (UGV) have been the subject of major recent research because of their availability and applicability. Additionally, the complementarity between them also makes such a system more potential [8]. In other words, UAV is widely used in reconnaissance due to the high mobility. However, the carrying capacity of UAV is very low compared to UGV since there is no ground support. In contrast, UGV has higher carrying capacity but is easily restricted by ground obstacles and cannot move at high speed. For these reasons, a hybrid UAV-UGV team system will be more appealing. To discuss more concretely, we consider a hybrid UAV-UGV team system for S&R usage. For the need for search mobility, we choose quadrotor aircraft as UAV. In order to deal with the complex terrain of the S&R environment, we choose biped robot as UGV. Even though other types of UGVs like wheeled robots and vehicles are easier to handle than biped robot, the high degree of freedom and the compatibility of the human environment still makes it a good candidate of UGV in a S&R system.

Most of the literature focus on only one specific problem in such an unmanned multi-agent S&R system, such as task allocation problem, path planning problem or control problem. Additionally, few literatures illustrate the relationship between these problems. This leads this chapter to propose a system architecture of performing S&R tasks for each agent in URTS to illustrate these problems and the relationship between them. The flowchart of the system architecture is also given in Figure 7.1. We divide it into five main hierarchical: (i) task allocation, (ii) global path planning, (iii) behavior decision, (iv) local motion planning, and (v) reference tracking control. It is because the URTS needs to be able to assign different tasks to agents to perform first. After a task is assigned, if the task is to reach a goal location, a path to reach it needs to be planned. To make agent move on the path, a behavior corresponding to the environment is required to be determined. Then, a local motion corresponding to the behavior of the agent needs to be planned. Finally, a controller must be designed to track the trajectory of the motion. In order to further limit the scope of the study, this chapter will focus on the latter two processes. But to illustrate how the whole system works, the first three steps are also briefly stated.

The local motion planning is the bridge between path planning and tracking control since the path found by path planning algorithm and the path enforced to follow by a controller are not necessarily the same. The reason is that path planning algorithm usually treats the agent as a point, while the actual agent in the physical world is a mechanical system for the tracking control design. A mechanical system means that there exist kinodynamic constraints. This makes certain paths impossible to follow for an actual agent, such as paths that are not smooth, have too large curvature, or require too large velocity and acceleration. Although some literatures directly tackle the kinodynamic constraints on path planning problem [9], this chapter splits path planning into three steps: (i) global path planning, (ii) behavior decision, and (iii) local motion planning to deal with complex S&R terrain. Through this decomposition, we can focus on the local motion planning of specific behaviors. The local motion planning of flying behavior for UAV and walking behavior for robot is studied in this chapter, especially the latter. The local motion planning of biped robot walking, that is, stable walking pattern generation, is a popular research topic due to its challenge [10].

The reference tracking design is to control an agent to follow a desired reference trajectory. There are many reference control strategies for MAS. According to the way of the design of controller, it can be divided into centralized control and decentralized control in control field [11]. Centralized control means there exists a powerful central controller in MAS to gather the state information of MAS and send the control command back to each agent to reach a global goal. Due to the powerful nature of the central controller, control commands can be determined well and quickly. But when it fails, the whole system will be completely paralyzed. In contrast, decentralized control means that each agent has its own controller to collect and control the agent's own state information. Under this architecture, although the global goal cannot be achieved, the possibility of paralyzing the entire system due to the failure of the controller can be avoided.

Besides, the formation control is also a topic in MAS [12]. Its purpose is to keep a MAS in a formation while moving for some allocated tasks. Although formation control provides a simple framework for the reference control of a large number of agents, considering the complexity of the disaster relief environment, formation will make the application of URTS inflexible. It is because we expect that agents in URTS need to organize multiple teams of different scales and types to deal with multiple tasks of different scales and types in a disaster relief environment. In this situation, it is more reasonable to treat each agent in URTS as an independent individual to be controlled to follow a specific trajectory for its allocated task to form a team formation.

In order to cope with the fault in the actual system, the fault-tolerant control (FTC) has also been widely studied. According to the way of handling the fault, it can be divided into the passive FTC and the active FTC [13]. The passive FTC treats the fault as an unknown system perturbation and designs a control law to tolerate it. In contrast, the active FTC will first estimate and identify the fault and then compensate it through the feedback controller. Despite the extra complexity in controller design, the active FTC will outperform the passive FTC due to the extra estimation steps. Based on the foregoing, a robust H_∞ decentralized observer-based feedforward reference tracking FTC scheme is proposed to deal with the control problem in hybrid URTS.

The contributions of this study are described as follows:

(i) A system architecture of performing S&R tasks for each agent in hybrid URTS is proposed so that the local motion planning and reference tracking control issues involved in each agent of hybrid URTS and their systematic relationships can be defined and resolved.

(ii) A transformation from the path planned by a roadmap-based path planning algorithm to the reference trajectory required for S&R task of team formation tracking control design is proposed to enable some common roadmap-based path planning algorithms to guide the team formation tracking control of agents in hybrid URTS to reach the goal location without obstacle collision.

(iii) A general agent dynamic model and a novel feedforward linearization control scheme are proposed so that the robust H_∞ decentralized observer-based feedforward reference tracking FTC problems of the heterogeneous agents in URTS, that is, UAVs and biped robots, can be solved simultaneously to achieve S&R task under external disturbance and actuator and sensor fault.

The remainder of the chapter is organized as follows. In Section 7.2, a system architecture of performing S&R tasks for each agent in hybrid URTS is proposed and the function and relationship among its components, that is, Task Allocation, Global Path Planning, Behavior Decision, Local Motion Planning and Reference Tracking Control are described. In Section 7.3, the dynamics models of agents in URTS are given to plan the motion of UAV flying and biped robot walking behavior and the control strategy of agents. In Section 7.4, a robust H_∞ decentralized observer-based feedforward reference tracking FTC is proposed for the agents in URTS with the help of a general agent dynamic model. In Section 7.5, a simulation example is given to illustrate the operation of the system architecture of performing S&R tasks for each agent in hybrid URTS and to verify the effectiveness of the proposed tracking control method. Section 7.6 concludes the chapter.

Notation 7.1: $diag(A_1, A_2, \ldots, A_n)$: a block diagonal matrix with main diagonal blocks $A_1, A_2, \ldots, A_n$. A^T: transpose of A. $A > 0$: a positive definite matrix. (a_n): a sequence. (a_{k_n}): a subsequence of a sequence (a_n). $[a_{j,k}]$: a matrix with the entries $a_{j,k}$ in the jth row and kth column. $|S|$: size of a set S. $\otimes$: Kronecker product. I_n: n-dimension identity matrix. $x(t) \in L_2[0, t_f]$ if $\int_0^{t_f} x^T(t)x(t)dt < \infty$. $Sym(A)$: sum of a matrix A and its transpose, that is, $Sym(A) = A + A^T$.

7.2 PRELIMINARIES OF URTS FOR S&R USAGE

The URTS will start with a given S&R area, and end with the S&R task completed. The URTS is composed of N_T teams and a ground station. Each team contains N_A agents with 1 UAV and $N_A - 1$ robots. Hence, the jth agent in the ith team is denoted as $\alpha_{i,j}$, where $i = 1,2,\ldots,N_T$ and $j = 1,2,\ldots,N_A$. The UAVs are chosen as the first agents in each team, that is, $\alpha_{i,1}, i = 1,2,\ldots,N_T$. Each agent has environmental sensing capability and load capability, while the ground station is responsible for computing and decision-making. Besides, there are communication channels between agents and ground station through wireless network.

To complete search tasks, each team is designed to be responsible for a small area of the overall S&R area, and each agent will be assigned an appropriate path to cover the unsearched area. To complete rescue tasks, whenever a target (e.g., victims or disaster area) is found by the machine vision of nearby agent, the ground station will assign some agents to the location of target.

If a task is to reach a location of certain goal, we need to find a collision-free path to reach it. Hence, each agent will also sense distance-related information about its surrounding and send it back to the ground station. The ground station will combine this information with the goal location determined by the task and then plan a path to avoid obstacles and other agents nearby through a path planning algorithm.

We need to determine specific behavior to follow the path found by path planning algorithm according to the terrain situation especially for agents with complex mechanical systems like robots if the algorithm treat the agent as a point. Since such a system has a high degree of freedom, there are many ways to follow the same path (e.g., a robot can walk or run to follow the same path). Furthermore, agents in URTS are not always following the path. Sometimes they need the behavior such as stop to look around, get supplies and put supplies. To meet these needs, a behavioral decision is necessary.

In order for the agent to make a behavior, we must plan a corresponding motion by local motion planning. The motion is a prescribed reference path for an actual mechanical system to follow. Finally, a tracking controller is designed for each agent to follow this desired reference path.

The agents overall have the same system architecture in URTS except for some subprocess differences. Followed by the concept in Reference [14], a system architecture of performing S&R tasks for each agent in hybrid URTS is proposed as shown in Figure 7.1. The detailed functions of remaining five blocks, that is, Task Allocation, Global Path Planning, Behavior Decision, Local Motion Planning, and Reference Tracking Control, will be explained in the following sections.

7.2.1 Task Allocation

In the URTS, it can be expected that each agent $\alpha_{i,j}$ will be assigned to several specific tasks $T_k, k \in \mathbb{Z}$, such as searching a specific area, or delivering supplies to disaster area, etc. However, the number of agents and tasks is more than one, each agent has different capabilities (e.g., moving speed or load capacity) and state (e.g., the relative distance between the agent itself and the target or the amount of supplies carried), and each task has different characteristics (e.g., urgency, position, or amount of supplies needed). Therefore, the results of task allocation can be "good or bad", which leads us to finding the optimal allocation. This problem is referred to a task allocation problem or multi-robot task allocation problem. A problem formulation and a mathematical model of problem can be found in Reference [15]. Although many different formulations and models have been employed to solve the task allocation problem, the common goal is to find a set of agent-task pairs $(\alpha_{i,j}, T_k)$ to achieve a specific cost function. In this chapter, we assume that the tasks have been properly assigned so that every agent knows a goal configuration q_{goal} it needs to go at every moment. q_{goal} then passes to next block, i.e., the global path planning block, as shown in Figure 7.1.

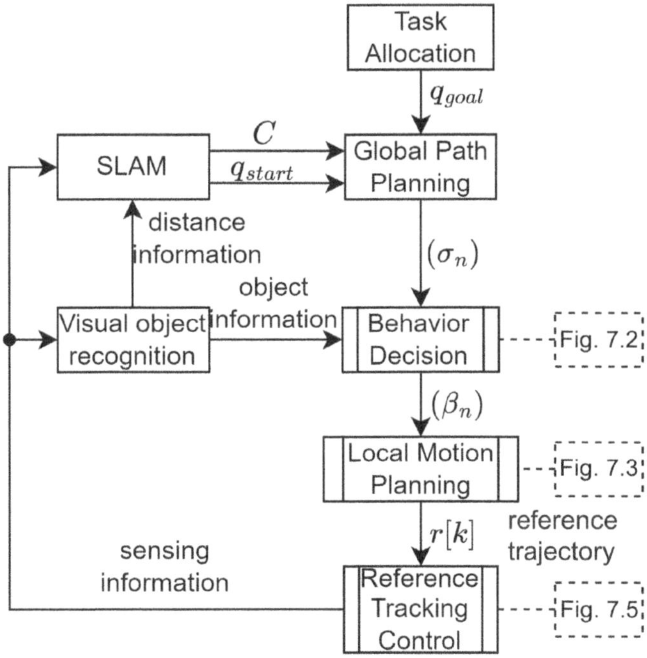

FIGURE 7.1 The system architecture of performing S&R tasks for each agent in hybrid URTS. The two blocks on left-hand side are used to convert the low-level sensor information into high-level information. The Simultaneous Localization and Mapping (SLAM) block converts sensor information and distance information into the current configuration q_{start} of agents and an occupancy map C of URTS. The visual object recognition block provides distance information and object information through the analysis of sensor information. The object information provides agent machine vision that enables it to determine an appropriate behavior (e.g., a robot can see an obstacle and decide to climb through it). The five blocks on right hand side are the flowchart of an agent performing a S&R task. From the top to the bottom, it is the decision of the goal configuration q_{qoal} of agents, the planning of the path (σ_n) of agents, the decision of the behavior (β_n) of agents corresponding to the path, the planning of the reference trajectory $r[k]$ of agents corresponding to the behavior, and the low-level reference tracking control of agents. A block with two vertical lines inside in flowchart represents a predefined process which has more detailed subprocesses. Behavior decision block is described in detail in Figure 7.2. Local motion planning block to generate the desired reference is described in detail in Figure 7.3. Reference Tracking Control block is described in detail in Figure 7.5.

Remark 7.1 The task allocation block is like a commander since it is used to assign task for agents. Thus, if the real S&R system has human experts as commanders, he can replace its job or make decisions together with it to maximize the rescue value.

Remark 7.2 Although each agent has its own team, agents can also work across teams. For example, if the result given by the task allocation algorithm contains the agent-task pairs $(\alpha_{1,5}, T_1)$ and $(\alpha_{2,2}, T_1)$, then the agent $\alpha_{1,5}$ in team 1 and the agent $\alpha_{2,2}$ in team 2 will execute task T_1 together.

7.2.2 Global Path Planning

After a goal configuration q_{goal} is assigned for each agent, next step is to find a collision-free path from current configuration q_{start} to arrive it. There must exist multiple feasible paths to go. Similar to task allocation, we usually want to find an optimal path. There are several path planning algorithms to handle this problem. Due to the developmental and universal nature of roadmap-based path planning algorithm, this chapter considers it as the path planning method of URTS. This method attempts to discretize the search space into interconnected roads and find the path on it.

According to the way of pathfinding, it can be divided into multi-query planner and single-query planner [16]. Multi-query planner will first construct a roadmap and then use a graph search method on it to query the best path, such as Probability Road Map (PRM), Visibility Graph, and Voronoi Diagrams [17]. Single-query planner will complete the pathfinding by constructing and querying simultaneously, such as Rapidly-exploring Random Tree (RRT), Expansice Space Tree (EST), and Ariadne's Clew [16]. However, the environment is dynamic rather static for URTS, so some extra structures need to impose on the aforementioned planner. Some common dynamic planners can also be found in Reference [16], such as PRM with D* search algorithm, dynamic RRT, and extended RRT.

Remark 7.3 To avoid agents colliding with each other, the concept of multi-agent path planning is proposed [18]. However, URTS operates in a large environment, so the probability of collision is small, and the agents have the ability to communicate. Therefore, an alternative solution is to use single-agent path planning together with the mechanism of waiting for the other agent to pass first in the event of a collision.

Furthermore, the constraints imposed by the mechanical structure are needed to consider within pathfinding process mentioned by other literatures, but it will be left to local motion planning block to deal with. The reason is that URTS works on a complex environment and therefore requires a variety of behaviors to respond, and different behaviors have different constraints (e.g., curvature constraints between running and walking behavior are expected to be different for a biped robot). In this case, a unified planner will become overly complex and impracticable. Therefore, we divide path planning into three subprocesses, that is, global path planning, behavior decision and local motion planning. The global path planning block becomes a global planner to regard an agent as a point without kinodynamic constraints. This enables the common roadmap-based path planning algorithms mentioned above to be used for global path planning in UTRS. The local motion planning block becomes a local planner to consider the motion planning of a specific behavior.

By treating a roadmap-based path planning algorithm as a black box, the output is a sequence (or waypoints), and the three inputs are current configuration q_{start}, goal configuration q_{goal}, and configuration space (or occupancy map) $\mathcal{C}$. q_{start} and $\mathcal{C}$ are obtained by SLAM, and q_{goal} is obtained by the previous block, Task Allocation block, as shown in Figure 7.1. $\mathcal{C}$ is a space containing all possible configurations of agents which are composed of free space $\mathcal{C}_{free}$ and obstacle space $\mathcal{C}_{obs}$, where

$\mathcal{C} = \mathcal{C}_{free} \cup \mathcal{C}_{obs}$ and $\mathcal{C}_{free} \cap \mathcal{C}_{obs} = \varnothing$. For a simpler explanation of how the URTS works, the following assumptions are made.

Assumption 7.1 The locating ability of the URTS is perfect so every agent can know its current configuration q_{start}.

Assumption 7.2 A Task Allocation algorithm is already designed so that every agent can know its goal configuration q_{goal}.

Assumption 7.3 The URTS is supposed to have a perfect real-time mapping ability, so a real-time configuration space $\mathcal{C}$ can be obtained.

Assumption 7.4 UAVs do not consider obstacle collision, so the path of UAVs can be directly assigned rather than found by planner. Robots do not consider obstacle collision in the direction perpendicular to the ground.

From above assumptions, a path of agent can be expressed as a sequence:

$$(\sigma_n), n \in \mathbb{Z} \cap [1, k_f], \sigma_n \in \mathcal{C} \tag{7.1}$$

where k_f is the time step when reaching goal. (σ_n) then passes to next block, that is, the behavior decision block, as shown in Figure 7.1.

Remark 7.4 Since the path planning is dynamic, (σ_n) is composed of multiple segments actually. Let (σ_{k_n}) be the subsequence of (σ_n), where k_n is the time step when a replanning decision occurs. Then, the segments of path from the result of the replanning in time step k_n can be expressed as sequences $(\sigma_m), m \in \mathbb{Z} \cap [k_n, k_{n+1})$. For agents, the replanning decision can be due to a goal changing that is made by human or Task Allocation block. For robot, it can be due to a collision detected by a dynamic roadmap-based planner.

7.2.3 Behavior Decision

The global path planning block tells agents where to go but not how since we regard the agent as a point. Therefore, agents need to determine an appropriate behavior to follow the path (σ_n) according to the machine vision provided by the object information. Taking robot as an example, it may walk, run, climb, or jump to follow the path (σ_n) according to real scenario. These behaviors with changing position are classified as "Moving" behavior in this chapter. Besides, the agents in the hybrid URTS do not always move. Sometimes they have to suspend to take an action (e.g., getting and putting supplies, rotating in place to collect more environment information) or deal with some unexpected situations (e.g., no path found, the robot falls). These behaviors without changing position are classified as "Action" behavior in this chapter. More behaviors can be added so that the agent can have more ways to act with environment but there must be a corresponding behavior every moment, otherwise the agent will lose control. The sequence of these behaviors can be expressed as:

$$(\beta_n), n \in \mathbb{Z} \cap [1, k_f], \beta_n \in \mathcal{B} \tag{7.2}$$

where $\mathcal{B}$ is the set of behaviors. It means that the path (σ_n) is divided into many segments and each segment corresponds to a specific behavior. By the object information in Figure 7.1, an appropriate behavior can be decided. However, it will be a rather complicated project, so this chapter will not discuss in detail. The flowchart of behavior decision and the behavior set $\mathcal{B}$ of agents in URTS can be roughly described in Figure 7.2. (β_n) then passes to next block, that is, the local motion planning block, as shown in Figure 7.1.

7.2.4 Local Motion Planning

After a specific behavior $\beta_n \in \mathcal{B}$ is determined, the next step is to plan a motion to achieve the specific behavior. Local motion planning block is like the global path planning block but with smaller scale and higher resolution and precision. Collision checking is needed since we consider agent as a point in the global path planning block but it is a real mechanical body here. Furthermore, the kinodynamic constraint is handled in this block. Although motion planning and path planning are separated into two blocks, the technologies involved are similar and often with the same notion in other literature. Therefore, the output of this block, that is, reference path $r[k]$, is also a path (it is also a sequence but this chapter uses the notation of discrete signal $r[k]$ to represent it). The flowchart of local motion planning block is shown in Figure 7.3.

To limit the scope of this study, we only focus on the motion planning of flying behavior of UAV and walking behavior of robot. A more detailed description will be given in the next section. $r[k]$ then passes to next block, that is, reference tracking control block, as shown in Figure 7.1.

7.2.5 Reference Tracking Control

To analyze the reference tracking control problem in the continuous time domain, the reference path $r[k]$ will be first converted to a continuous signal $r'(t)$ by D/A convertor with a timescale, that is, sampling period. By analyzing the dynamic model of each agent, a desired reference trajectory $r(t)$ is planned. $r(t)$ describes the position and orientation that need to be reached over time by a machine system governed by a dynamic equation. Note that the path and trajectory are distinguished in some literature. Different from path (e.g., (σ_n) or $r[k]$), a trajectory $r(t)$ has considered the time in physic world. We also distinguish them in this way in this chapter.

If each agent in hybrid URTS can track each own reference trajectory $r(t)$, then they can move in the physical world as we expect. To this end, a reference tracking control method needs to be designed. It will be discussed in detail in Section 7.4. In addition, the sensing information collected by the sensors in this block is not only used for reference tracking control but also fed back to the high-level block, as shown in Figure 7.1.

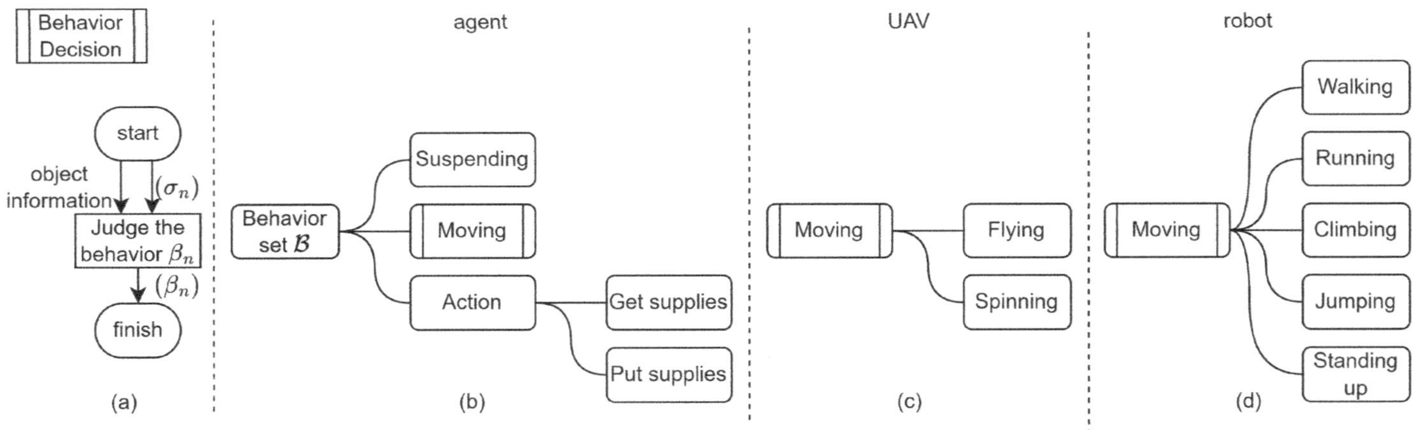

FIGURE 7.2 (a) The flowchart of behavior decision block in Figure 7.1 of each agent in URTS. The behavior to be taken at every moment by each agent will be decided in this block. (b) The behavior set of each agent in URTS. The leaves of the tree structure are the possible behaviors an agent can take. For UAVs, the moving behavior set is described in part (c). For robots, the moving behavior set is described in part (d). (c) The moving behavior set of each UAV. (d) The moving behavior set of each robot.

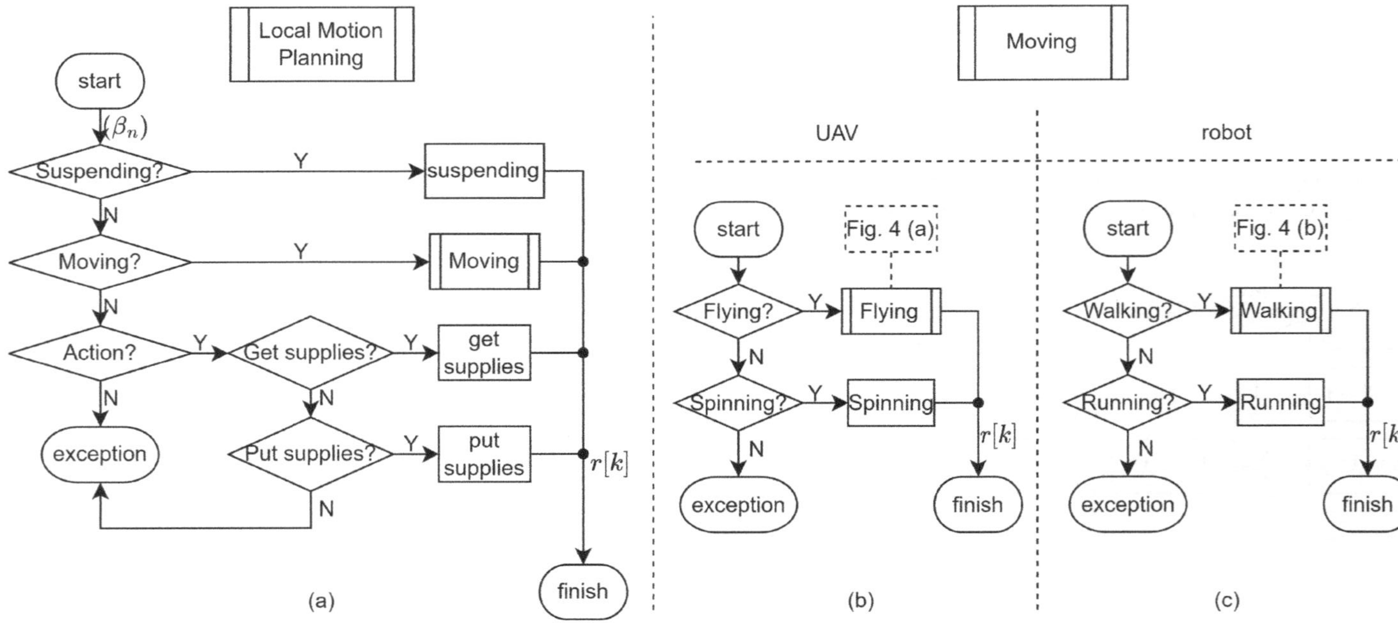

FIGURE 7.3 (a) The flowchart of local motion planning block in Figure 7.1 of each agent in URTS. Some basic behaviors of UAV and robot mentioned in Figure 7.2 are given to futher illustrate the flow of this block. The corresponding motion planning of an agent will be executed according to the behavior determined by behavior decision block. The exception terminator in figure represents an exceptional condition that performs unconsidered behaviors. The reference path $r[k]$ is a sequence (or discrete signal) planned by a specific behavior block. For UAVs, moving block is described in part (b). For robots, moving block is described in part (c). (b) Moving block of each UAV. Flying block in it is described detailly in Figure 7.4. (c) Moving block of each robot. Walking block in it is described in detail in Figure 7.4.

7.3 SYSTEM DESCRIPTION OF UAVS AND BIPED ROBOTS IN URTS

In order to plan a reference trajectory $r(t)$ for the local motion of UAV and robot in URTS, their dynamic models must be given first. After the system description of UAV and robot in URTS, the local motion planning of flying and walking behavior, as shown in Figure 7.4, will be discussed subsequently and separately in the next two subsections.

Before the discussion of the motion planning of these two behaviors of UAV and robot in URTS, the following assumption is maded.

Assumption 7.3.1 The space between obstacles is large enough to eliminate the need for collision checking again, and the average speed of agents is slow enough to ignore the nonholonomic constraints (the velocity and acceleration constraints).

However, for these two behaviors, there exists an inevitable holonomic constraint on the curvature of local motion. Although an accurate reference trajectory without breaking the curvature constraint can be planned, it is not easy to solve this problem. Additionally, it is not necessary for these two behaviors in URTS since they are used to move from one location to another, while the effect of error during moving caused by breaking the curvature constraint is relatively insignificant. As an alternative, this problem can be handled by curve fitting which can be regarded as a post-process of the path (σ_n). The result of post-process, that is, smoothed path (σ_n'), will not cause collision by *Assumption* 7.3.1. The post-process will appear in the beginning of motion planning process of these two behaviors, as shown in Figure 7.4.

7.3.1 Local Motion Planning of Flying of UAV

A dynamic model about how an UAV in URTS moves in the physical world is given first. By Newton-Euler equation, the dynamic model of each UAV in URTS can be formulated as [19]:

$$\begin{bmatrix} f_u \\ \tau_u \end{bmatrix} = \begin{bmatrix} mI & 0 \\ 0 & J \end{bmatrix}\begin{bmatrix} \ddot{X} \\ \ddot{\Theta} \end{bmatrix} + \begin{bmatrix} 0 \\ \dot{\Theta}\times(J\dot{\Theta}) \end{bmatrix} + \begin{bmatrix} f_g \\ 0 \end{bmatrix} + \begin{bmatrix} K_F & 0 \\ 0 & K_\tau \end{bmatrix}\begin{bmatrix} \dot{X} \\ \dot{\Theta} \end{bmatrix} \tag{7.3}$$

where $J = \begin{bmatrix} J_x & 0 & 0 \\ 0 & J_y & 0 \\ 0 & 0 & J_z \end{bmatrix}$, $K_F = \begin{bmatrix} K_x & 0 & 0 \\ 0 & K_y & 0 \\ 0 & 0 & K_z \end{bmatrix}$, $K_\tau = \begin{bmatrix} K_{\tau_x} & 0 & 0 \\ 0 & K_{\tau_y} & 0 \\ 0 & 0 & K_{\tau_z} \end{bmatrix}$,

$f_u = \begin{bmatrix} f_x \\ f_y \\ f_z \end{bmatrix} = R(\Theta)\begin{bmatrix} 0 \\ 0 \\ F \end{bmatrix}$, $\tau_u = \begin{bmatrix} \tau_x \\ \tau_y \\ \tau_z \end{bmatrix}$, $X = \begin{bmatrix} x \\ y \\ z \end{bmatrix}$, $\Theta = \begin{bmatrix} \phi \\ \theta \\ \psi \end{bmatrix}$, $f_g = \begin{bmatrix} 0 \\ 0 \\ -mg \end{bmatrix}$,

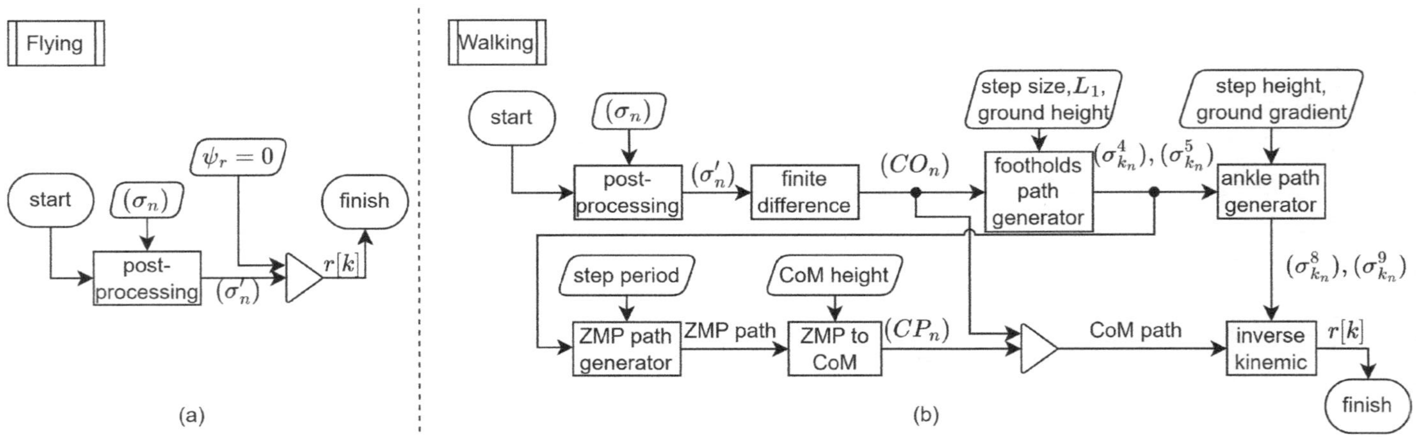

FIGURE 7.4 The flowchart of flying block and walking block for local motion planning in Figure 7.3. In both blocks, a smoothed path (σ'_n) is obtained by post-processing first. Then the respective motion of behavior is planned. (a) For the flying block in Figure 7.3 of each UAV, by combining the smoothed path (σ'_n) and setting $\phi_d = 0$ (let UAV fly without spinning), we can plan the reference path $r[k]$. (b) For the walking block in Figure 7.3 of each robot, following the process in Section IV-B, we can plan the reference path $r[k]$.

$R(\Theta)=R_z(\psi)R_y(\theta)R_x(\phi)$, $\quad R_z(\psi)=\begin{bmatrix} form1\cos\psi & -\sin\psi & 0 \\ \sin\psi & \cos\psi & 0 \\ 0 & 0 & 1 \end{bmatrix}$,

$R_y(\theta)=\begin{bmatrix} \cos\theta & 0 & \sin\theta \\ 0 & 1 & 0 \\ -\sin\theta & 0 & \cos\theta \end{bmatrix}$, $R_x(\phi)=\begin{bmatrix} 1 & 0 & 0 \\ 0 & \cos\phi & -\sin\phi \\ 0 & \sin\phi & \cos\phi \end{bmatrix}$. g is the gravity acceleration, m and J are the mass and inertia matrix of UAV, respectively, τ_u and F are the total torque and force acting on UAV, respectively, Θ is the Euler angles in body frame as shown in Figure 2.8 in Reference [19], X is the position of center of mass (CoM) in inertial frame, K_τ and K_F are the aerodynamic damping coefficients, and $R(\Theta)$ is the intrinsic rotation matrix from body frame to inertial frame. This model treats the UAV as a mass point and can control the total force F and the total torque τ_u. For UAV, the reference trajectory $r(t)=[x_r(t),y_r(t),z_r(t),\phi_r(t),\theta_r(t),\psi_r(t)]^T \in \mathbb{R}^6$ is in the task space, where the subscript r denotes the reference.

Remark 7.5 Since the UAV (quadrotor) has four rotors, we only have four control input. To simplify the model, the UAV dynamic model in (7.3) considers the actuator control input $u'(t)=[F,\tau_u^T]^T=[F,\tau_x,\tau_y,\tau_z]^T$ as the equivalent control input for the four rotors. Besides, since $u'(t)\in\mathbb{R}^4$ has two less degrees of freedom than $r(t)\in\mathbb{R}^6$, UAV is an underactuated system. This makes the two degrees of freedom in the reference trajectory of UAV not be assigned arbitrarily but be inversely calculated through other degrees of freedom that can be assigned arbitrarily by the dynamic equation in (7.3).

Now, suppose the UAV flying behavior occurs between time step k_1 and k_2, that is, $\beta_n = flying$, $n\in\mathbb{Z}\cap[k_1,k_2]$. The corresponding path (σ_n), $n\in\mathbb{Z}\cap[k_1,k_2]$ will be smoothed first by linear interpolation and then by cubic spline interpolation, which gives the smoothed path $(\sigma_{n'})$, $n\in\mathbb{Z}\cap[k_1,k_2+d(k_2-k_1)]$, $\sigma_{n'}\in\mathbb{R}^3$, where $d\in\mathbb{Z}$ is the interpolation density. Then, $(\sigma_{n'})$ will be the position reference path $[x_r[k],y_r[k],z_r[k]]^T$. Subsequently, we consider the orientation reference path, row angle $\phi_r[k]$, pitch angle $\theta_r[k]$ and yaw angle $\psi_r[k]$. $\phi_r[k]$ is set to zero since there is no need for spinning when flying. $\theta_r[k]$ and $\psi_r[k]$ cannot be planned beforehand since UAV is an underactuated system, which will be discussed in the next section. Finally, the reference path $r[k]$ of UAV can be planned by combining them together, that is, $r[k]=[x_r[k],y_r[k],z_r[k],\phi_r[k]]^T\in\mathbb{R}^4$. The flowchart of flying motion planning is shown in Figure 7.4a.

7.3.2 Local Motion Planning of Walking of Robot

By Lagrange equation, the dynamic model of a biped robot in URTS can be formulated as:

$$\tau_R = M_R(q)\ddot{q}+C(q,\dot{q})\dot{q}+G(q) \tag{7.4}$$

where τ_R is the total torque on revolute joints, $q, \dot{q}, \ddot{q} \in \mathbb{R}^{12}$ are angular position, angular velocity, and angular acceleration vector of revolute joints, $M_R(q) \in \mathbb{R}^{12\times 12}$ is the inertia matrix, $C(q, \dot{q}) \in \mathbb{R}^{12}$ is the Coriolis and centripetal force vector, and $G(q) \in \mathbb{R}^{12}$ is the gravitational force vector. The detailed kinematic and dynamic parameters can be found in the online source [20]. For biped robot, the reference trajectory $r(t) = q_r(t) \in \mathbb{R}^{12}$ is planned in the joint space. Furthermore, the walking of biped robot suffers from the falling problem, i.e, how to plan a stable walking pattern to prevent robot from falling. These make the motion planning of walking behavior of robot more difficult. In this chapter, a Three-Dimensional Linear Inverted Pendulum Model (3D-LIPM) [21] is used for the motion planning of walking behavior of each biped robot in URTS.

With 3D-LIPM, we can significantly reduce the amount of computation. Let us define the body frame of biped robot as $\{\widehat{X_b}, \widehat{Y_b}, \widehat{Z_b}\}$ as in Reference [20]. Taking the forward direction of biped robot as $\widehat{X_b}$ direction, the left direction as $\widehat{Y_b}$ direction, and the torso direction as $\widehat{Z_b}$ direction in body frame, "Falling" means the moments on the robot in $\widehat{X_b}$ and $\widehat{Y_b}$ directions are not zero. More accurately, the biped robot will not fall if the zero moment point (ZMP) lies in the support polygon, that is, the convex hull of face of supported foots. The ZMP in $\widehat{X_b}$ direction can be described as [22] (the ZMP in the $\widehat{Y_b}$ direction is in the same form):

$$x_{zmp} = \frac{\sum_{i=1}^{12} (m_i(\ddot{z}_i + g)x_i - m_i \ddot{x}_i z_i - I_{iy}\ddot{\Omega}_{iy})}{\sum_{i=1}^{12} m_i(\ddot{z}_i + g)} \tag{7.5}$$

where m_i is the CoM, x_i, z_i are the linear position components, I_{iy} is the inertial component, and $\ddot{\Omega}_{iy}$ is the angular acceleration component of link i. However, it is difficult to directly calculate the analytical solution of $q_r(t)$ through (7.5), since there exists a complex coordinate transformation between $q_r(t)$ and $x_i, z_i, \ddot{\Omega}_{iy}$. At the same time, it is necessary to ensure that x_{zmp} falls in the support polygon which also has a relationship with $q_r(t)$. To simplify this complex problem, an approximate solution can be derived through 3D-LIPM. We plan CoM reference first and then obtain $q_r(t)$ by using inverse kinematic (IK) with given step size, step height, step period, and CoM height. Many researchers have used this method to avoid complex calculations for ZMP of the actual robot dynamic model. Although there exists a model error between the actual dynamic model and 3D-LIPM, the planning process will be more simple.

Following the same motion planning step in UAV, the smoothed path $(\sigma_{n'})$ for biped robot can be obtained at first. For the convenience of explanation, suppose walking has occured between time step 1 and N, that is, $n \in \mathbb{Z} \cap [1, N]$. Note that $(\sigma_{n'})$ is not actual CoM reference in robot case since CoM of robot needs to be "swinging" for balance. Despite that, $\sigma_{n'}$ tells the biped robot the position to go to so that the $\widehat{X_b}$

direction can be obtained by doing finite difference on $(\sigma_{n'})$ due to the expectation that the biped robot will move forward (rather than sideways or backward). To keep torso upright, the $\widehat{Z_b}$ direction is equal to the z-axis in the inertial frame $\widehat{Z_g}$. Given $\widehat{X_b}$ and $\widehat{Z_b}$, $\widehat{Y_b}$ can be obtained obviously through cross product. The sequence of body frame, that is, CoM orientation path (CO_n) then be planned through the above steps.

Remark 7.6 A frame (or homogeneous transformation) in $\mathbb{R}^3$ can be determined by giving the "position" and "orientation" with respect to a reference frame. That is, given the frames of two joints in links with known kinematics, the frames of joints between them can be found by IK. Hence, we need to find the position path and orientation path, which compose the desired path of the frame.

Let us denote the x and y components of $\sigma_{n'}$ in $(\sigma_{n'})$ as the sequence $(\sigma_n^1), \sigma_n^1 \in \mathbb{R}^2$. The left and right "envelopes", (σ_n^2) and (σ_n^3), of (σ_n^1) with a fixed distance L_1 can be planned by (σ_n^1) and (CO_n) through the geometric relation among $(\sigma_n^i), i = 1,2,3$, where L_1 is the feet width (or shoulder width). Then the x and y components of the left and right foothold paths, $(\sigma_{k_n}^2)$ and $(\sigma_{k_n}^3)$, respectively, can be planned by a given step size, which are the subsequence of (σ_n^2) and (σ_n^3), respectively. Finally, the left and right foothold paths $(\sigma_{k_n}^i), \sigma_{k_n}^i \in \mathbb{R}^3, i = 4,5$ are planned by adding the z component which is given by ground height.

After foothold paths are planned, ankle position path can also be planned by the given step height which is customized by the designer or based on the height of the obstacle to be crossed. Taking the left foothold path as an example, x and y components of the highest position of ankle during stride are set as the middle point of two footholds $\sigma_{k_m}^2$ and $\sigma_{k_{m+1}}^2$ where $m \in \mathbb{Z} \cap [1, M-1]$ with $k_1 = 1$ and $k_M = N$, and the z component is given by the step height. By using the cubic spline interpolation, we have the left and right ankle position paths, (σ_n^6) and (σ_n^7). To keep the soles of the feet on the ground, the ankle orientation path can be planned by the gradient of ground. Finally, the left and right ankle paths, (σ_n^8) and (σ_n^9), are planned by combining the position and orientation path together. So far, the remaining work is to find out the CoM path and then to combine with the ankle path to calculate the joint path through IK.

To plan the CoM position path (CP_n), ZMP path needs to be planned first. ZMP path can be planned through foothold paths $(\sigma_{k_n}^4)$ and $(\sigma_{k_n}^5)$ since ZMP needs to lie in the support face and the foothold path points out when the feet are on the ground. Suppose the CoM height z_c of biped robot is kept constant when walking, then the biped robot model can be regarded as an 3D-LIPM [21]:

$$\begin{aligned} \ddot{x}_c &= \frac{g}{z_c}(x_c - p_x) \\ \ddot{y}_c &= \frac{g}{z_c}(y_c - p_y) \end{aligned} \tag{7.6}$$

where (x_c, y_c, z_c) is the position of CoM of the inverted pendulum, g is the gravity acceleration, and (p_x, p_y) is the position of ZMP on the x-y plane. Since z_c, g, and

(p_x, p_y) are given, (x_c, y_c) can be solved. From the dynamic equations in the x and y directions in (7.6), it can be found that they are decoupled and thus can be calculated separately. Therefore, only the solution in the x direction is given below (the y direction as the same). To solve it, a method is proposed to convert it to a servo problem [23]:

$$\begin{aligned} \dot{\overline{x}}_c &= A\overline{x}_c + Bu_s \\ y_s &= C\overline{x}_c \end{aligned} \tag{7.7}$$

where $\overline{x}_c = \begin{bmatrix} x_c \\ \dot{x}_c \\ \ddot{x}_c \end{bmatrix}$, $A = \begin{bmatrix} 0 & 1 & 0 \\ 0 & 0 & 1 \\ 0 & 0 & 0 \end{bmatrix}$, $B = \begin{bmatrix} 0 \\ 0 \\ 1 \end{bmatrix}$, and $C = \begin{bmatrix} 1 & 0 & -z_c / g \end{bmatrix}$. Our goal is to find a control input u_s in order that the output y_s can track the ZMP reference trajectory p_x so that the state x_c, that is, the solution of ODE in (7.6) can be obtained, that is, the CoM position path can be planned. Unlike conventional methods, the problem is solved by the optimal control. The system is discretized first and the discrete LQ optimal tracker is employed to achieve the output tracking. The formulation of the discrete LQ optimal tracker can be found in Table 4.4–1 in Reference [24]. The CoM position path (CP_n) can then be obtained by combining x_c and y_c with z_c.

By combining the CoM orientation path (CO_n) with the position path (CP_n), the CoM path can be obtained. Finally, the joint path, that is, reference path $r[k] \in \mathbb{R}^{12}$ of robot can be found by solving IK. The flowchart of walking motion planning is shown in Figure 7.4b.

7.4 REFERENCE TRACKING CONTROL OF EACH AGENT IN HYBRID URTS

Through the previous step, the planning of the reference path $r[k]$ of each agent in hybrid URTS has been completed. However, in order to calculate the unplannable reference paths, that is, $\phi_r[k]$ and $\theta_r[k]$, of UAV and analyze the reference trajectory tracking problem in the continuous domain, we have to transform $r[k]$ into the reference trajectory $r(t)$. Before converting the planned reference path $r[k]$ to desired reference trajectory $r(t)$, we first convert the UAV dynamic model in (7.3) and robot dynamic model in (7.4) into a form called agent dynamic model in a hybrid team to analyze their team formation tracking control problems together. Through some appropriate variable transformations, we have:

$$M(x(t))\ddot{x}(t) + H(x(t), \dot{x}(t)) = u(t) \tag{7.8}$$

where $u(t) \in \mathbb{R}^n$ is the control input vector, $x(t) \in \mathbb{R}^n$ is the state vector, $M(x(t)) \in \mathbb{R}^{n \times n}$ is the inertia matrix, and $H(x(t), \dot{x}(t)) \in \mathbb{R}^n$ is the non-inertial force vector. For each UAV in hybrid URTS, we have $u(t) = \begin{bmatrix} f_u \\ \tau_u \end{bmatrix}$, $x(t) = \begin{bmatrix} X \\ \Theta \end{bmatrix}$, $M(x(t)) = \begin{bmatrix} mI & 0 \\ 0 & J \end{bmatrix}$,

$H(x(t),\dot{x}(t)) = \begin{bmatrix} 0 \\ \dot{\Theta}\times(J\dot{\Theta}) \end{bmatrix} + \begin{bmatrix} f_g \\ 0 \end{bmatrix} + \begin{bmatrix} K_F & 0 \\ 0 & K_\tau \end{bmatrix}\begin{bmatrix} \dot{X} \\ \dot{\Theta} \end{bmatrix}$, and $n = 6$ by UAV dynamics in (7.3). For each robot in hybrid URTS, we have $u(t) = \tau_R$, $x(t) = q$, $M(x(t)) = M_R(q)$, $H(x(t),\dot{x}(t)) = C(q,\dot{q})\dot{q} + G(q)$, and $n = 12$ by robot dynamics in (7.4). The feedforward linearization control law for an agent in URTS is proposed as:

$$u(t) = M(r(t))(\ddot{r}(t) + u_{fb}(t)) + H(r(t),\dot{r}(t)) \tag{7.9}$$

where $r(t) \in \mathbb{R}^n$ is the desired reference trajectory, $M(r(t))$, $\ddot{r}(t)$, and $H(r(t),\dot{r}(t))$ are the feedforward control terms for canceling system nonlinearity, and $u_{fb}(t)$ is the feedback control law to be further designed for improving system tracking robustness.

Remark 7.7 In Section 3.2.2 of Reference [19], the dynamic system of UAV is divided into inner loop system and outer loop system to analyze the reference tracking problem of UAV. In order to improve the problem that the control gains in four state equations (z, ϕ, θ , and ψ) of inner loop system and the two states of outer loop system (x and y) in Reference [19] has no design specifications, we changed the controller design method so that the control gains of the six states of UAV can be designed together via the given specifications, that is, design the feedforward linearization control law $u(t) \in \mathbb{R}^6$ to achieve the observer-based tracking specification in (7.24). This can reduce the time cost of manually adjusting control gains. At the same time, the tracking control problem of UAV and robot can be analyzed simultaneously.

Remark 7.8 The difference between the feedforward linearization control law $u(t)$ in (7.9) of this chapter and the traditional feedback linearization control (or computed torque control) is that we use the "feedforward" linearization control, that is, use $M(r(t))$ and $H(r(t),\dot{r}(t))$ in (7.9) instead of $M(x(t))$ and $H(x(t),\dot{x}(t))$. This is because the state $x(t)$ is assumed to be unavailable in this chapter so $x(t)$ cannot be used in feedforward linearization control law $u(t)$. While the asymptotical tracking control is achieved, that is, $x(t) \to r(t)$, the feedforward linearization control will be achieved too.

To complete the planning of reference trajectory $r(t)$ of each agent, a D/A converter is used to transform the reference path $r[k]$ (output of local motion planning) into a continuous signal $r'(t)$, as shown in Figure 7.5. For UAV, we get $r'(t) = [x_r, y_r, z_r, \psi_r]^T \in \mathbb{R}^4$. Besides, for an UAV in hybrid URTS, it can be seen that the control input $u(t) = [f_u^T, \tau_u^T]^T = [f_x, f_y, f_z, \tau_x, \tau_y, \tau_z]^T \in \mathbb{R}^6$ we design in (7.9) is different from the actuator control input $u'(t) = [F, \tau_x, \tau_y, \tau_z]^T \in \mathbb{R}^4$ for UAV since UAV is an underactuated system. The two degrees of freedom we reserved in Section 7.3.1, that is, ϕ_r and θ_r, are just to solve this problem. By substituting $\Theta = \begin{bmatrix} \phi_r \\ \theta_r \\ \psi_r \end{bmatrix}$ into $\begin{bmatrix} f_x \\ f_y \\ f_z \end{bmatrix} = R(\Theta)\begin{bmatrix} 0 \\ 0 \\ F \end{bmatrix}$ from UAV dynamics in (7.3), the three unknown variables F, ϕ_r, and θ_r can be found from these three equations using inverse dynamic because the component forces f_x, f_y, and f_z, and the yaw angle ψ_r are given. Combining ϕ_r and θ_r with $r'(t)$, we obtain $r(t) = [x_r, y_r, z_r, \phi_r, \theta_r, \psi_r]^T$. At the same time, we find $u'(t)$. For biped robot, we directly have $r(t) = r'(t) \in \mathbb{R}^{12}$ and $u'(t) = u(t) \in \mathbb{R}^{12}$ since biped robot is a fully actuated system. So far, the transformation from reference path $r[k]$ to reference trajectory $r(t)$ for each agent in hybrid URTS is done. We define the process of converting $r'(t)$ and $u(t)$ into $r(t)$ and $u'(t)$ mentioned above as the reference generation block in Figure 7.5.

Remark 7.9 Since this chapter analyzes each agent $\alpha_{i,j}$ individually, the subscripts i and j of the corresponding variables are omitted for the convenience of notation in the following. For example, $x_{i,j}(t)$ is omitted as $x(t)$ in (7.8), $r_{i,j}(t)$ is omitted as $r(t)$ in (7.9), and so on.

Remark 7.10 Let $r_{i,j}(t)$ be the reference trajectory $r(t)$ of the agent $\alpha_{i,j}$, $r_{i,j}[k]$ be the reference path $r[k]$ of the agent $\alpha_{i,j}$, $(\beta_n)_{i,j}$ be the behavior sequence (β_n) of the agent $\alpha_{i,j}$, $(\sigma_n)_{i,j}$ be the collision-free path (σ_n) of the agent $\alpha_{i,j}$, $q_{goal,i,j}$ be the goal configuration q_{goal} of the agent $\alpha_{i,j}$, and so on. As long as $\alpha_{i,j}$ can track $r_{i,j}(t)$, the hybrid URTS can work as we expect since the previous blocks, that is, task allocation, global path planning, behavior decision, and local motion planning, have completed their respective responsibilities and planned their corresponding values, $q_{goal,i,j}$, $(\sigma_n)_{i,j}$, $(\beta_n)_{i,j}$, and $r_{i,j}[k]$. That is, $r_{i,j}(t)$ is the reference trajectory that can accomplish the specific task, follow the specific path and perform the specific behavior.

To make the model more realistic, the following disturbances encountered in actual scenarios are considered:

(i) For each agent, there exists coupling effect due to co-channel interference in communication between agents [25].
(ii) For each agent, there exists cyber-attack on communication network between agents and ground station.
(iii) For each agent, there exists sensor noise.
(iv) For each UAV, there exists vortex coupling.
(v) For each UAV, there exists wind disturbance [26].
(vi) For each robot, there exists ground reaction force [27].

Let $x_{i,j}(t)$ denote the state vector of agents $\alpha_{i,j}$ where $i = 1,2,\ldots,N_T$, $j = 1,2,\ldots,N_A$. The coupling effect $c_{i,j}(t)$ on each agent from other agents in hybrid URTS can be represented as:

$$c_{i,1}(t) = \sum_{k=1,k\neq i}^{N_T} D_{i,1,k}(x_{i,1}(t))x_{k,1}(t) \tag{7.10}$$

for UAV $\alpha_{i,1}$ and

$$c_{i,j}(t) = \sum_{k=1,k\neq j}^{N_A} D_{i,j,k}(x_{i,j}(t))x_{i,k}(t) \tag{7.11}$$

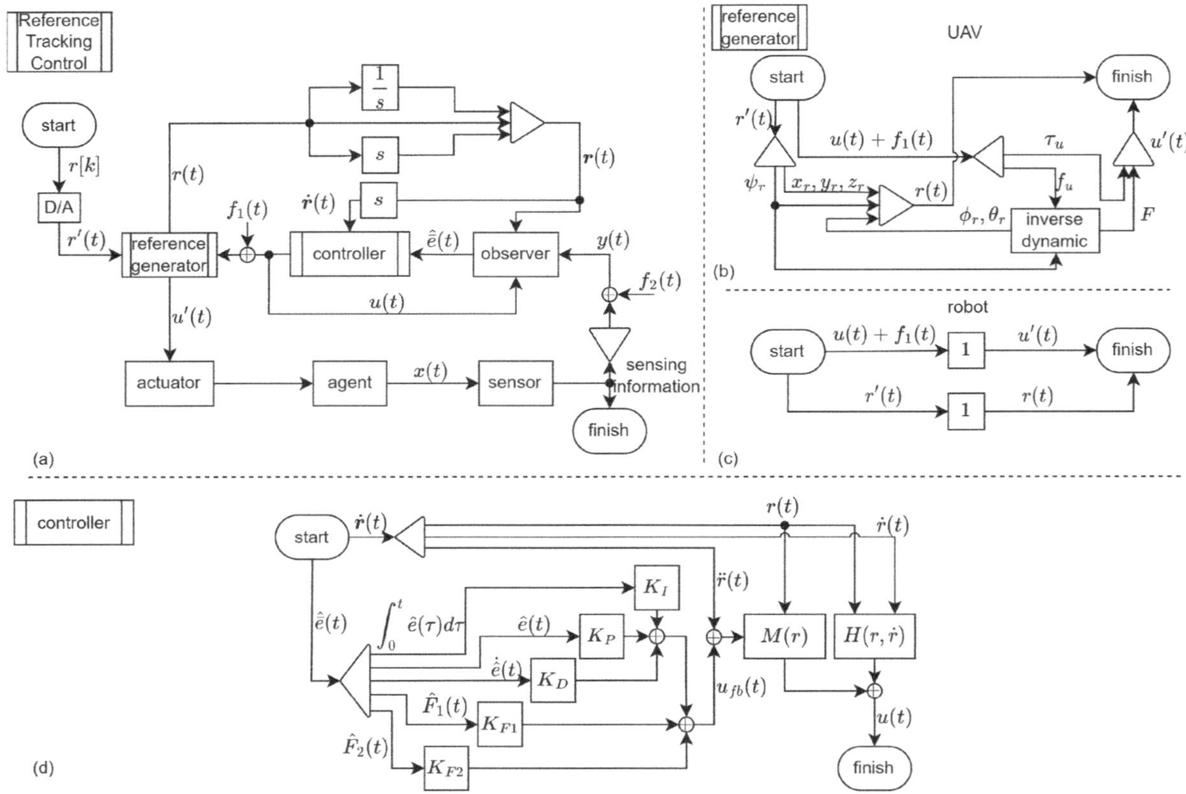

FIGURE 7.5 (a) The flowchart of Reference Tracking Control block in Figure 7.1 of each agent in hybrid URTS. The controller block is described in part (d). The controller block (the proposed general H_∞ decentralized observer-based feedforward reference tracking FTC scheme) is designed for a fully actuated agent dynamic model in (7.8), while the UAV is an underactuated system. It makes the designed feedforward linearization control law $u(t) \in \mathbb{R}^6$ and the actuator control input $u'(t) \in \mathbb{R}^4$ different for UAV. The reference generator block is introduced to deal with this problem. For UAVs, the reference generator block is described in part (b). For robots, the reference generator block is described in part (c). (b) The reference generator block for each UAV. $u'(t)$ and $r(t)$ can be calculated from $u(t)$ and $r'(t)$ by inverse dynamic through the UAV dynamic model in (7.3). (C) The reference generator block for each robot. $u'(t) = u(t)$ and $r(t) = r'(t)$ since the robot dynamic model in (7.4) is fully actuated. (d) The proposed general H_∞ decentralized observer-based feedforward reference tracking FTC scheme for each agent in hybrid URTS.

for biped robot $\alpha_{i,j}$ where $i = 1,2,\ldots,N_T$ and $j = 2,3,\ldots,N_A$ [25]. For the convenience of notation, we use $c(t)$ to represent $c_{i,j}(t)$. Since the ground station is responsible for the calculation, the calculated control command in (7.9) will be transmitted to the agent through the network channel in hybrid URTS. Therefore, the coupling effect due to co-channel interference and the cyber-attack signal will deteriorate the control command. In addition, the wind disturbance and the ground reaction force will apply extra force on an agent in (7.8). Therefore, through an appropriate conversion, the above disturbances can be equivalent to two disturbance forces $c(t)+d_1(t)\in\mathbb{R}^n$ where $d_1(t)$ is the non-coupling external disturbance. The nominal system in (7.8) of an agent in hybrid URTS then can be rewritten as the following real system:

$$M(x(t))\ddot{x}(t)+H(x(t),\dot{x}(t)) = u(t)+c(t)+d_1(t) \tag{7.12}$$

Now, substituting the feedforward linearization control law $u(t)$ in (7.9) into (7.12) and subtracting $M(x(t))\ddot{r}(t)$ from the left and right sides, we have:

$$\begin{aligned} &M(x(t))(\ddot{x}(t)-\ddot{r}(t))+H(x(t),\dot{x}(t)) \\ &= (M(r(t))-M(x(t)))\ddot{r}(t)+M(r(t))u_{fb}(t)+H(r(t),\dot{r}(t))+c(t)+d_1(t) \end{aligned} \tag{7.13}$$

By multiplying $M(x(t))^{-1}$ on both sides of (7.13) and with some arrangements, we have the tracking error differential equation as follows:

$$\ddot{e}(t) = u_{fb}(t)+f_1(t) \tag{7.14}$$

where $f_1(t) = M(x(t))^{-1}(-\Delta M(\ddot{r}(t)+u_{fb}(t))-\Delta H+c(t)+d_1(t))\in\mathbb{R}^n$ is considered as the actuator fault signal, $\Delta M \triangleq M(x(t))-M(r(t))$ and $\Delta H \triangleq H(x(t),\dot{x}(t)) - H(r(t),\dot{r}(t))$ are the error terms from feedforward compensation, and $e(t)=x(t)-r(t)$ is the tracking error. Let us denote $e(t) = \begin{bmatrix} \int_0^t e^T(\tau)d\tau & e^T(t) & \dot{e}^T(t) \end{bmatrix}^T \in\mathbb{R}^{3n}$, the tracking error differential equation in (7.14) can be rewritten as the following linear tracking error system:

$$\dot{e}(t) = Ae(t)+B(u_{fb}(t)+f_1(t)) \tag{7.15}$$

where $A = A_0\otimes I_n, B = B_0\otimes I_n$ with $A_0 = \begin{bmatrix} 0 & 1 & 0 \\ 0 & 0 & 1 \\ 0 & 0 & 0 \end{bmatrix}, B_0 = \begin{bmatrix} 0 \\ 0 \\ 1 \end{bmatrix}$.

Through the above analysis, the tracking control problem of the nonlinear system of each agent in hybrid URTS with external disturbance in (7.12) is transformed into the regulation problem of the linear tracking error system in (7.15) with actuator fault signal $f_1(t)$ by the feedforward linearization control law $u(t)$ in (7.9). The remaining step is to design an appropriate feedback control law $u_{fb}(t)$ in (7.15) to make the linear tracking error system robustly stable. In a real system, the feedback information is measured by sensor, i.e, the state $x(t)$ in (7.8) is unavailable. At the same time, the sensor noise on sensor also needs to be considered as mentioned before. Since the sensor information will be transmitted back to the ground station for calculating control command through the network channel in URTS, not only the sensor noise but also the cyber-attack signals are concerned. We consider the effect of them as a sensor fault in this chapter. Assuming $x(t)=\begin{bmatrix}\int_0^t x^T(\tau)d\tau & x^T(t) & \dot{x}^T(t)\end{bmatrix}^T$, the measurement output equation can be described as:

$$y(t)=Cx(t)+B_2 f_2(t) \tag{7.16}$$

where $y(t)\in\mathbb{R}^l$ is the output vector, $C\in\mathbb{R}^{l\times 3n}$ is the output matrix, and $B_2\in\mathbb{R}^{l\times o}$ is the input matrix of sensor fault signal $f_2(t)\in\mathbb{R}^o$.

Remark 7.11 We consider the integral of the agent's state $\int_0^t x^T(\tau)d\tau$ to be measurable because $\int_0^t x^T(\tau)d\tau$ can be calculated indirectly via agent's state $x(t)$ and an integrator. For UAV, $x(t)$ can be measured by GPS and inertial measurement unit. For biped robot, $x(t)$ can be measured by the sensor of motor on the joints.

Let us define $r(t)=\begin{bmatrix}\int_0^t r^T(\tau)d\tau & r^T(t) & \dot{r}^T(t)\end{bmatrix}^T$ to modify the output equation in (7.16) and combine it with (7.15), we have the following tracking error dynamic system of an agent in the hybrid URTS:

$$\begin{aligned}\dot{e}(t)&=Ae(t)+B(u_{fb}(t)+f_1(t))\\ y(t)&=Ce(t)+Cr(t)+B_2 f_2(t)\end{aligned} \tag{7.17}$$

To deal with the fault signals $f_i(t), i=1,2$, a smoothing signal model is introduced [28]:

$$\begin{aligned}\dot{F}_i(t)&=A_i F_i(t)+v_i(t)\\ f_i(t)&=C_i F_i(t)\end{aligned} \tag{7.18}$$

where $F_i(t)=\begin{bmatrix}f_i^T(t) & f_i^T(t-h) & \dots & f_i^T(t-wh)\end{bmatrix}^T\in\mathbb{R}^{(w_i+1)n_i}$, $A_i\in\mathbb{R}^{(w_i+1)n_i\times(w_i+1)n_i}$, $v_i(t)$ is the model error, $C_i=\begin{bmatrix}1 & 0 & \dots & 0\end{bmatrix}\otimes I_{n_i}$, and w_i is the window size of smoothing signal model with $n_1=n$ and $n_2=o$. Substituting (7.18) into (7.17), we get the following augmented tracking error system of an agent in hybrid URTS:

$$\begin{aligned}\dot{\bar{e}}(t)&=\bar{A}\bar{e}(t)+\bar{B}u_{fb}(t)+\bar{v}(t)\\ y(t)&=\bar{C}\bar{e}(t)+Cr(t)\end{aligned} \tag{7.19}$$

where $\bar{e}(t)=\begin{bmatrix}e(t)\\ F_1(t)\\ F_2(t)\end{bmatrix}$ is the augmented tracking error vector, $\bar{A}=\begin{bmatrix}A & BC_1 & 0\\ 0 & A_1 & 0\\ 0 & 0 & A_2\end{bmatrix}$, $\bar{B}=\begin{bmatrix}B\\ 0\\ 0\end{bmatrix}$, $\bar{C}=\begin{bmatrix}C & 0 & B_2C_2\end{bmatrix}$, and $\bar{v}(t)=\begin{bmatrix}0\\ v_1(t)\\ v_2(t)\end{bmatrix}$. Since the fault signals become a state variable of the augmented tracking error system of an agent in (7.19), their corruption on the tracking error dynamic system in (7.17) can be avoided. A Luenberger observer is proposed to estimate them and $e(t)$ simultaneously to achieve an active FTC by the following estimation system:

$$\begin{aligned}\dot{\hat{\overline{e}}}(t) &= \overline{A}\hat{\overline{e}}(t)+\overline{B}u_{fb}(t)-L(y(t)-\hat{y}(t))\\ \hat{y}(t) &= \overline{C}\hat{\overline{e}}(t)+Cr(t)\end{aligned} \tag{7.20}$$

Assumption 7.4.1 The augmented tracking error system (7.19) of an agent is observable, that is, $rank\begin{bmatrix} zI-\overline{A} \\ \overline{C}\end{bmatrix} = 3n+(w_1+1)n+(w_2+1)o, \forall z \in eig(\overline{A})$.

The feedback control law $u_{fb}(t)$ of each agent in hybrid URTS can then be designed as follows:

$$u_{fb}(t) = K\hat{\overline{e}}(t) \tag{7.21}$$

where K is the control gain.

Remark 7.12 Let $u_{fb}(t) = K\hat{\overline{e}}(t) = [K_I, K_P, K_D, K_{F_1}, K_{F_2}]\ [\int_0^t \hat{e}^T(\tau)d\tau,\ \hat{e}^T(t),\ \dot{\hat{e}}^T(t), \widehat{F_1}^T(t), \widehat{F_2}^T(t)]^T$, we can find that the control gain K is composed of the PID control gains K_I, K_P, and K_D for $\int_0^t \hat{e}(\tau)d\tau$, $\hat{e}(t)$ and $\dot{\hat{e}}(t)$, respectively, and the fault control gains K_{F_i} for $F_i(t)$ with $i=1,2$. This shows that the feedback control law $u_{fb}(t)$ has fault-tolerant capability and PID control characteristic.

Let us define the augmented estimation error $\tilde{e}(t)=\overline{e}(t)-\hat{\overline{e}}(t)$, the augmented estimation error system can be obtained by (7.19) and (7.20):

$$\dot{\tilde{e}}(t) = \overline{A}\tilde{e}(t)+L\overline{C}\tilde{e}(t) = (\overline{A}+L\overline{C})\tilde{e}(t) \tag{7.22}$$

Combining (7.19), (7.21), and (7.22), we have the following augmented tracking and estimation error system of each agent in the hybrid URTS:

$$\dot{\tilde{x}}(t) = \tilde{A}\tilde{x}(t)+\tilde{v}(t) \tag{7.23}$$

where $\tilde{x}(t) = \begin{bmatrix}\overline{e}(t)\\ \tilde{e}(t)\end{bmatrix}$, $\tilde{A} = \begin{bmatrix}\overline{A}+\overline{B}K & -\overline{B}K \\ 0 & \overline{A}+L\overline{C}\end{bmatrix}$, and $\tilde{v}(t) = \begin{bmatrix}\overline{v}(t)\\ \overline{v}(t)\end{bmatrix}$

In order to enable the designed control gain K in (7.21) and observer gain L in (7.20) to achieve a specific performance for the augmented system in (7.23) under the disturbance $\overline{v}(t)$, the robust H_∞ decentralized observer-based tracking control strategy below a prescribed disturbance attenuation level ρ^2 for each agent in hybrid URTS is given as follows:

$$\frac{\int_0^{t_f} (\bar{e}^T(t)Q_1\bar{e}(t)+\tilde{e}^T(t)Q_2\tilde{e}(t)+u_{fb}^T(t)Ru_{fb}(t))dt - V(\tilde{x}(0))}{\int_0^{t_f} \tilde{v}^T(t)\tilde{v}(t)dt} \le \rho^2 \quad (7.24)$$

where t_f is the final time, $Q_1 \ge 0$ is the weighting matrix of tracking error, $Q_2 \ge 0$ is the weighting matrix of estimation error, $R > 0$ is the weighting matrix of control effort, $V(\tilde{x}(0))$ is the initial condition effect on the augmented tracking and estimation error system in (7.23) which is to be extracted from the H_∞ decentralized team formation tracking control performance, and $\tilde{v}(t)$ is the total disturbance whose effect on $\bar{e}(t)$, $\tilde{e}(t)$ and $u_{fb}(t)$ is needed to be attenuated. If we can find the control gain K and observer gain L such that (7.24) holds, then the effect of total disturbance $\tilde{v}(t)$ on augmented tracking error $\bar{e}(t)$ and augmented estimation error $\tilde{e}(t)$ can be attenuated to a prescribed level ρ^2 from the viewpoint of energy. Before analyzing the robust H_∞ decentralized observer-based tracking control problem of each agent in (7.24), the following lemmas are given:

Lemma 7.1 [29]. For any matriices X and Y with appropriate dimensions, and matrix $R = R^T > 0$ the following inequality holds:

$$X^TY + Y^TX \le X^TR^{-1}X + Y^TRY \quad (7.25)$$

Lemma 7.2 (Schur Complement [29]). For the matrices $X = X^T$and $Y = Y^T$ and matrix R with appropriate dimensions, the following statement is true:

$$\begin{bmatrix} X & R \\ R^T & Y \end{bmatrix} > 0 \Leftrightarrow Y > 0, X - RY^{-1}R^T > 0 \quad (7.26)$$

Then, the following theorem is given.

Theorem 7.1 (i) If there exists matrices $P = P^T > 0, K, L$ such that the following Riccati-like matrix inequality holds:

$$Q + P\tilde{A} + \tilde{A}^TP + \tilde{K}^TR\tilde{K} + \frac{1}{\rho^2}PP \le 0 \quad (7.27)$$

where $\tilde{K} = [K - K]$, $Q = \begin{bmatrix} Q_1 & 0 \\ 0 & Q_2 \end{bmatrix}$, then the H_∞ decentralized observer-based team formation tracking control strategy in (7.24) of each agent in the hybrid URTS can be achieved. (ii) If $\tilde{v}(t)$ is of finite energy, that is, $\tilde{v}(t) \in L_2[0,\infty)$, then $\bar{e}(t)$, $\tilde{e}(t)$ and $u_{fb}(t)$ all quadratically asymptotically tend to 0, that is, $\bar{e}^T(t)\bar{e}(t) \to 0$, $\tilde{e}^T(t)\tilde{e}(t) \to 0$, and $u_{fb}^T(t)u_{fb}(t) \to 0$ as $t \to \infty$.

Remark 7.13 Since the Riccati-like inequality in (7.27) of each agent $\alpha_{i,j}$ in URTS has not involved the system information of other agents (e.g., the coupling term $c_{i,j}(t)$

in (7.10) and (7.11)), therefore, the robust decentralized team formation tracking control can be achieved.

Proof. (i) Choosing the Lyapunov function $V(\tilde{x}(t)) = \tilde{x}^T(t)P\tilde{x}(t)$ for the augmented system (7.23) with $P = P^T > 0$, we have:

$$\begin{aligned}
&\int_0^{t_f} (\tilde{x}^T(t)Q\tilde{x}(t) + u_{fb}^T(t)Ru_{fb}(t))dt \\
&= V(\tilde{x}(0)) - V(\tilde{x}(t_f)) + \int_0^{t_f} (\tilde{x}^T(t)Q\tilde{x}(t) + u_{fb}^T(t)Ru_{fb}(t) + \dot{V}(\tilde{x}(t)))dt \quad (7.28)\\
&\le V(\tilde{x}(0)) + \int_0^{t_f} (\tilde{x}^T(t)Q\tilde{x}(t) + u_{fb}^T(t)Ru_{fb}(t) + Sym(\dot{\tilde{x}}^T(t)P\tilde{x}(t)))dt
\end{aligned}$$

By (7.23) and Lemma 7.1, we have:

$$\begin{aligned}
Sym(\dot{\tilde{x}}^T(t)P\tilde{x}(t)) &= Sym((\tilde{A}\tilde{x}(t) + \tilde{v}(t))^T P\tilde{x}(t)) \\
&= \tilde{x}^T(t)(P\tilde{A} + \tilde{A}^T P + \frac{1}{\rho^2}PP)\tilde{x}(t) + \rho^2\tilde{v}^T(t)\tilde{v}(t)
\end{aligned} \quad (7.29)$$

Substituting (7.21), (7.29), and $\tilde{x}^T(t)Q\tilde{x}(t) = \bar{e}^T(t)Q_1\bar{e}(t) + \tilde{e}^T(t)Q_2\tilde{e}(t)$ into (7.28), we get:

$$\begin{aligned}
&\int_0^{t_f} (\bar{e}^T(t)Q_1\bar{e}(t) + \tilde{e}^T(t)Q_2\tilde{e}(t))dt + u_{fb}^T(t)Ru_{fb}(t))dt \\
&\le V(\tilde{x}(0)) + \int_0^{t_f} (\tilde{x}^T(t)(Q + P\tilde{A} + \tilde{A}^T P + \tilde{K}^T R\tilde{K} + \frac{1}{\rho^2}PP)\tilde{x}(t) \\
&+ \rho^2\tilde{v}^T(t)\tilde{v}(t))dt
\end{aligned}$$

Thus, if (7.27) holds, then (7.24) also holds.

(ii) From (7.24), $\tilde{v}(t) \in L_2[0,\infty)$ and $V(\tilde{x}(0))$ is finite, and we get $\int_0^{t_f} (\bar{e}^T(t)Q_1\bar{e}(t) + \tilde{e}^T(t)Q_2\tilde{e}(t) + u_{fb}^T(t)Ru_{fb}(t))dt \le \rho^2 \int_0^{t_f} \tilde{v}^T(t)\tilde{v}(t)dt + V(\tilde{x}(0)) < \infty$. Therefore, $\bar{e}^T(t)\bar{e}(t) \to 0$, $\tilde{e}^T(t)\tilde{e}(t) \to 0$ and $u_{fb}^T(t)u_{fb}(t) \to 0$ as $t \to \infty$.

Although the sufficient condition (7.27) for the existence of the H_∞ decentralized observer-based tracking control strategy in (7.24) have been found, it can not be solved easily since it is a bilinear matrix inequality (BMI) and there exists strong coupling between the designed variables K and L. To solve the issue, a two-step design procedure is exploited as follows.

Step 1. First, let the Lyapunov function of augmented system (7.23) be the sum of two Lyapunov function of subsystems (7.19) and (7.22), that is, $V(\tilde{x}(t)) = \tilde{x}^T(t)P\tilde{x}(t) = \bar{e}^T(t)P_1\bar{e}(t) + \tilde{e}^T(t)P_2\tilde{e}(t)$. Substituting $P = \begin{bmatrix} P_1 & 0 \\ 0 & P_2 \end{bmatrix}$ and $Q = \begin{bmatrix} Q_1 & 0 \\ 0 & Q_2 \end{bmatrix}$ into (7.27), we get:

$$\begin{bmatrix} Q_1 & 0 \\ 0 & Q_2 \end{bmatrix} + Sym\left(\begin{bmatrix} P_1 & 0 \\ 0 & P_2 \end{bmatrix}\begin{bmatrix} \bar{A}+\bar{B}K & -\bar{B}K \\ 0 & \bar{A}+L\bar{C} \end{bmatrix}\right)$$
$$+\begin{bmatrix} K^T RK & -K^T RK \\ -K^T RK & K^T RK \end{bmatrix} + \frac{1}{\rho^2}\begin{bmatrix} P_1P_1 & 0 \\ 0 & P_2P_2 \end{bmatrix} \tag{7.30}$$
$$= \begin{bmatrix} M_{11} & -P_1\bar{B}K - K^T RK \\ * & M_{22} \end{bmatrix} < 0$$

where $M_{11} = Q_1 + Sym(P_1(\bar{A}+\bar{B}K)) + K^T RK + \frac{1}{\rho^2}P_1P_1$, $M_{22} = Q_2 + Sym(P_2(\bar{A}+L\bar{C})) + K^T RK + \frac{1}{\rho^2}P_2P_2$. By the fact that $\begin{bmatrix} M_{11} & -P_1\bar{B}K - K^T RK \\ * & M_{22} \end{bmatrix} < 0 \Rightarrow M_{11} < 0, M_{22} < 0$, the inequality $M_{11} < 0$ is used to find P_1, K. Premultiplying and postmultiplying $M_{11} < 0$ by $W_1 = P_1^{-1}$ and applying Lemma 7.2, we obtain:

$$\begin{bmatrix} Sym(\bar{A}W_1 + \bar{B}Y_1) + \frac{1}{\rho^2} & W_1{}^{1/2}\sqrt{Q_1} & Y_1^T \\ * & -I & 0 \\ * & * & -R^{-1} \end{bmatrix} < 0 \tag{7.31}$$

where $Y_1 = KW_1$. By solving the LMI in (7.31), we can obtain W_1, Y_1.

Step 2. Substituting $P_1 = W_1^{-1}$ and $K = Y_1W_1^{-1}$ found in *Step* 1 into (7.30) and applying Lemma 7.2, we obtain:

$$\begin{bmatrix} M_{11} & -P_1\bar{B}K - K^T RK & P_2 \\ * & Q_2 + Sym(P_2\bar{A} + Y_2\bar{C}) + K^T RK & 0 \\ * & * & -\rho^2 I \end{bmatrix} < 0 \tag{7.32}$$

where $Y_2 = P_2L$. By solving the LMI (7.32), we can obtain P_2, Y_2.

If we want to find the optimal H_∞ decentralized observer-based tracking control strategy for the augmented tracking and estimation error system in (7.23) of each agent in hybrid URTS, we need to solve the following LMIs-constrained optimization problem:

$$\begin{aligned} &\rho^{*2} = \min_{P,K,L} \rho^2 \\ &s.t. (7.31), (7.32) \end{aligned} \tag{7.33}$$

The design procedure of the optimal decentralized H_∞ observer-based feedforward reference tracking FTC scheme for each agent in (7.12) is summarized as follows:

(i) Apply the feedforward control in (7.9) to obtain the linearized tracking error dynamic system in (7.17) for each agent in hybrid URTS.
(ii) Construct the smoothing signal models (7.18) for the actuator fault $f_1(t)$ and sensor fault $f_2(t)$. Embed these smoothing signal models into the linearized system (7.17) to get the augmented tracking error system of each agent in (7.19).
(iii) Construct the robust observer-based FTC law in (7.20) and (7.21) for each agent.
(iv) Solve the LMIs-constrained optimization problem (7.33) by the two-step design procedure to obtain the control gain K and observer gain $L = P_2^{-1}Y_2$ for observer-based controller in (7.20) and (7.21) of each agent in the hybrid URTS.

The overall flowchart of reference tracking control of each agent in hybrid URTS is shown in Figure 7.5. The reference generator is used to compute the desired reference trajectory $r(t)$ and actuator control input $u'(t)$ for each agent according to the continuous signal $r'(t)$ obtained by D/A and the feedforward linearization control law $u(t)$ we design in (7.9). Passing $r(t)$ through the integrator and differentiator, we get $r(t)$. $r(t)$ is then passed to observer to calculate the error $e(t)$. Its differential, $r(t)$, is then inputted to controller for feedforward control. The sensor measures not only the agent's own information (e.g., position or velocity) but also environmental information. The former, measurement output $y(t)$, is passed to observer in (7.20) to get the estimation $\hat{e}(t)$ for feedback control. The latter is passed back to the high-level block for positioning, mapping and object recognition.

Remark 7.14 By the proposed agent dynamics model in (7.8) and the introduction of reference generator block in Figure 7.5 b and c, a general H_∞ decentralized observer-based feedforward reference tracking FTC scheme for each agent $\alpha_{i,j}$ in hybrid URTS can be designed as shown in the controller block in Figure 7.5d. The decentralized architecture also ensures the scalability of URTS. More specifically, let us introduce subscripts i and j to the corresponding variables of each agent $\alpha_{i,j}, i = 1,2,...,N_T, j = 1,2,...,N_A$ (e.g., the state $x_{i,j}(t)$, the reference trajectory $r_{i,j}(t)$, the control gain $K_{i,j}$, etc.). It can be seen that the number of teams $N_T > 0$ and the number of agents in a team $N_A > 0$ are scalable.

Although the control gain in (7.21) and observer gain in (7.20) for each agent in URTS can already be found through the previous steps, the calculation speed of solving the matrix inequality (7.27) and the online calculation speed of controller and observer can be further improved by reducing the dimensionality of the system. Observing the matrices A, B, C, B_2 in the linearized system (7.17), it can be further split into n subsystems for each agent ($n = 6$ for UAV and $n = 12$ for robot) if the matrices C, B_2 in the output equation (7.16) have the same form to A, B and $l = l_0 n, o = o_0 n$, that is, $C = C_0 \otimes I_n, B_2 = B_{2,0} \otimes I_n$ where $C_0 \in \mathbb{R}^{l_0 \times 3}, B_{2,0} \in \mathbb{R}^{l_0 \times o_0}$. Let us decompose the error $e(t) = \sum_{i=1}^{n} e_i(t) \otimes \mathbf{e}_i$, the control $u_{fb}(t) = \sum_{i=1}^{n} u_{fb,i}(t) \otimes \mathbf{e}_i$, the actuator fault $f_1(t) = \sum_{i=1}^{n} f_{1,i}(t) \otimes \mathbf{e}_i$, the output $y(t) = \sum_{i=1}^{n} y_i(t) \otimes \mathbf{e}_i$, and the sensor fault $f_2(t) = \sum_{i=1}^{n} f_{2,i}(t) \otimes \mathbf{e}_i$ in (7.17) where $e_i(t) \in \mathbb{R}^3$, $u_{fb,i}(t) \in \mathbb{R}$, $f_{1,i}(t) \in \mathbb{R}, y_i(t) \in \mathbb{R}^{l_0}$, $f_{2,i}(t) \in \mathbb{R}^{o_0}$, and $\mathbf{e}_i$ is standard unit column vectors in $\mathbb{R}^n$, we get the n subsystems for each agent:

$$\begin{aligned} \dot{e}_i(t) &= A_0 e_i(t) + B_0 (u_{fb,i}(t) + f_{1,i}(t)) \\ y_i(t) &= C_0 e_i(t) + C_0 r_i(t) + B_{2,0} f_{2,i}(t) \end{aligned} \tag{7.34}$$

where $i = 1, 2, ..., n$.

Remark 7.15 If the linearized system (7.17) can be split into n subsystems, this means that the error $e_i(t)$ of each state variable $x(t)$ of each agent in (7.8), can be measured independently via n sensors to obtain the independent outputs $y_i(t)$. In actual UAV system or biped robot system, this is usually done.

By Theorem 7.1 again, the form of subsystems in (7.34) shows that we can find the control gain $K_i \in \mathbb{R}^{1 \times s}$ and observer gain $L_i \in \mathbb{R}^{s \times l_0}$ of the ith subsystem (7.34) that achieve the H_∞ decentralized observer-based tracking control performance with a prescribed attenuation level ρ_i, where $s = 3 + (w_1 + 1) + (w_2 + 1) o_0$. The original control gain K of the original agent system can be reconstructed by $K = \begin{bmatrix} k_1 & k_2 & \dots & k_n \end{bmatrix}^T, k_i = K_i^T \otimes \mathbf{e}_i$. The original observer gain L can be reconstructed in the same way.

In this case, the calculation speed of finding gains K, L of each agent can be improved since the dimensionality is decreased. Furthermore, the online calculation speed of controller and observer can also be improved since there are more zeros in the control gain K and observer gain L found by this method while maintaining the same estimation and tracking robustness. More clearly, the number of elements in matrix K, that is, the number of scalar gains, changes from $n \times sn$ to $n(1 \times s)$. For L, it changes from $sn \times l_0 n$ to $n(s \times l_0)$. The number of scalar gains to be designed is significantly reduced.

7.5 SIMULATION RESULTS

In this section, a specific S&R procedure for URTS is given to illustrate the proposed system architecture of performing S&R tasks for each agent in hybrid URTS and demonstrate the effectiveness of local motion planning and reference control strategy in hybrid URTS. First, an S&R area divided into N_T areas, $area_i$, where $i = 1, 2, ..., N_T$, is given as shown in Figure 7.6. To simplify the description, we will focus on the UAV

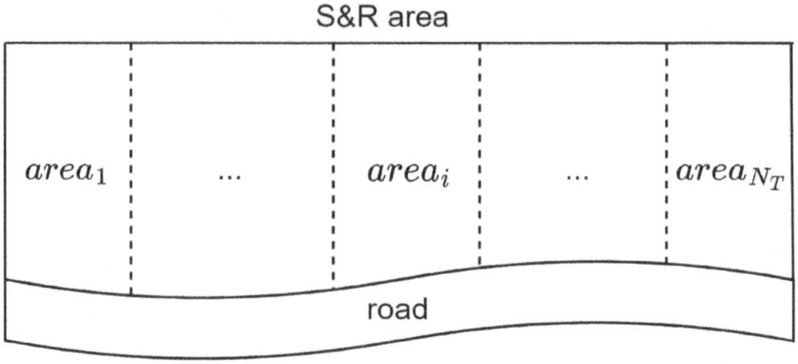

FIGURE 7.6 An example of an S&R area in URTS. This area is divided into N_T areas, and $team_i$ is responsible for $area_i$.

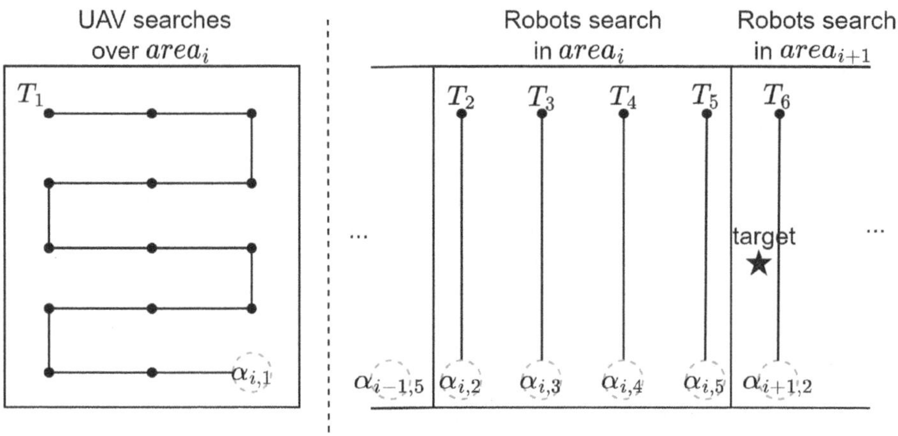

FIGURE 7.7 The allocation of search tasks in the ith team and $(i+1)$th team at the beginning. The search tasks $T_j, j=1,2,...,6$ are allocated by the task allocation block. The content of the search tasks is to reach some consecutive goals q_{goal} (black dots in figure) so that the straight path between goals q_{goal} can cover the unsearched area. For UAV, the sequence formed by q_{goal} is directly the path (σ_n) due to the no-collision assumption, *Assumption* 7.4. For robots, q_{goal} will be passed to the global path planning block together with current configuration q_{start} and occupancy map $\mathcal{C}$ to find collision-free paths (σ_n).

and robot in the ith team and the $(i+1)$th team. Suppose each team has five agents, that is, $N_A = 5$, then we can denote the ith team as a set, that is, $team_i = \{\alpha_{i,j} \mid j = 1,2,...,5\}$.

At the beginning, the task allocation block will assign the agents in $team_i$ with some search tasks in $area_i$ to build the occupancy map and find targets. The search task is assumed to be allocated by dividing the unsearched region, as shown in Figure 7.7. Representing the search tasks in Figure 7.7 as a set $task_1 = \{T_j \mid j = 1,2,...,5\} \cup \{T_6\}$, the proper agent-task pairs' $allocation_1 = \{(\alpha_{i,j}, T_j) \mid j = 1,2,...,5\} \cup \{(\alpha_{i+1,2}, T_6)\}$ can be obtained through the task allocation block. Suppose a goal is found after a while, as shown in Figure 7.7. At this point, we have a rescue task T_7. The new task list $task_2 = task_1 \cup \{T_7\}$ is obtained by updating the old one. If the ground station assigns $\alpha_{i,5}$ and $\alpha_{i+1,2}$ to perform T_7 through the task allocation algorithm, then we have the new allocation $allocation_2 = (allocation_1 - \{(\alpha_{i,5},T_5),(\alpha_{i+1,2},T_6)\}) \cup \{(\alpha_{i,5},T_7),(\alpha_{i+1,2},T_7)\}$. Until the S&R task is over, the task allocation block will continuously work in the similar way.

To illustrate the global path planning block and behavior decision block, we choose the pairs $(\alpha_{i,5},T_7)$ and $(\alpha_{i,1},T_1)$ as example. For the UAV $\alpha_{i,1}$, the path $(\sigma_n), n \in \mathbb{Z} \cap [1,k_f], k_f = 16$ is directly assigned as shown in Figure 7.8 without going through the global path planning block by *Assumption* 7.4. The behavior sequence (β_n) is set as $\beta_n = flying, n \in \mathbb{Z} \cap [1,k_f]$. For the robot $\alpha_{i,5}$, we have the goal configuration q_{goal} from the task T_7. With the current configuration q_{start} and configuration space $\mathcal{C}$ obtained by SLAM, the path $(\sigma_n), n \in \mathbb{Z} \cap [1,k_f], k_f = 27$ can be planned as shown in Figure 7.9. The behavior sequence (β_n) is set as $\beta_n = walking$ for $n \in \mathbb{Z} \cap [1,5]$, $\beta_n = climbing$ for $n \in \mathbb{Z} \cap [6,15]$ and $\beta_n = running$ for $n \in \mathbb{Z} \cap [16,27]$. We choose the walking behavior $\beta_n, n \in \mathbb{Z} \cap [1,5]$ to illustrate the local motion planning block of robots.

Through the previous planning steps and the help of reference generator block, the reference trajectory $r(t)$ of flying behavior for each UAV and walking behavior

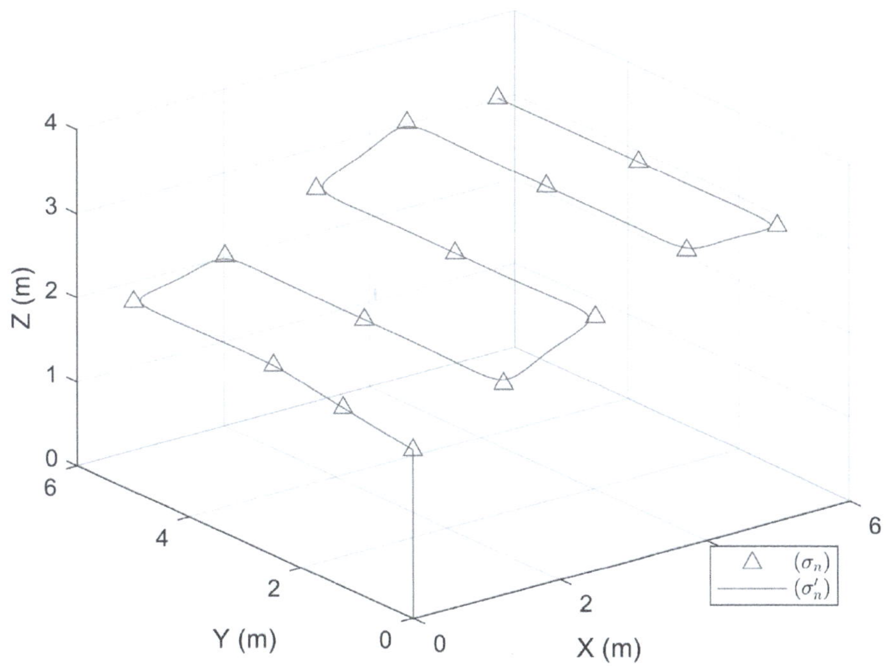

FIGURE 7.8 The result of local motion planning of flying behavior in $(\beta_n), n \in \mathbb{Z} \cap [1,16]$ of UAV $\alpha_{i,1}$ performing the search task T_1.

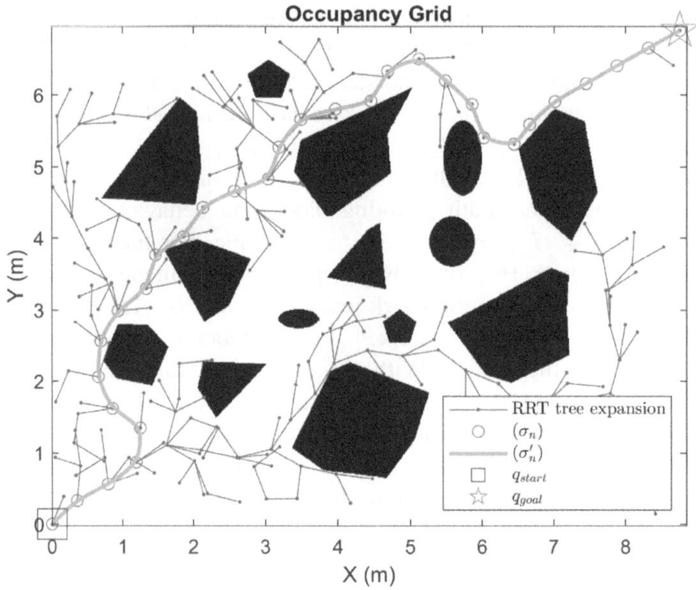

FIGURE 7.9 The global path planning result of the robot $\alpha_{i,5}$ performing the rescue task T_7 using RRT algorithm. The block polygons represent the obstacle space $\mathcal{C}_{obs}$.

After (β_n) is set, we can plan the reference path $r[k]$ by local motion planning block. Following the procedure in Figure 7.4, the results of local motion planning of UAV flying and robot walking are shown in Figures 7.8 and 7.10, respectively.

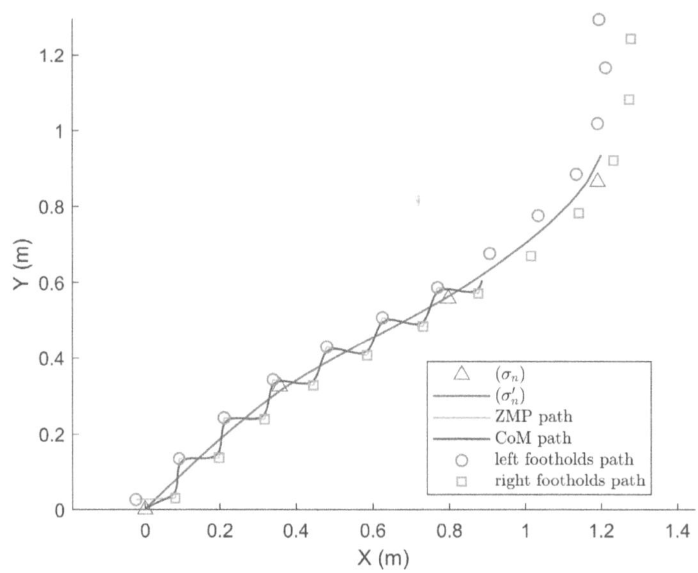

FIGURE 7.10 The top view of result of local motion planning of walking behavior in $(\beta_n), n \in \mathbb{Z} \cap [1,3]$ of robot $\alpha_{i,5}$ performing the rescue task T_7. The joint path, that is, reference path $r[k]$ will be obtained by solving IK with CoM, left foothold and right foothold paths.

for each biped robot has been planned in URTS. The remaining parameters are set as follows:

Agents

(1) System parameters: Initial value $\tilde{x}(0) = [e(0)^T, F_1(0)^T, F_2(0)^T, \tilde{e}(0)]^T = [[0.1X,...,0.1X]^T, 0,0,0]^T$ where $X \sim \mathcal{N}(0,1)$. $C = C_0 \otimes I_n$ where $C_0 = I_3$.

(2) Designed parameters of subsystems: $\rho^* = 30$. Actuator and sensor window size of smoothing signal model are $w_1 = 3$ and $w_2 = 4$, respectively.

UAVs

(1) System parameters: $B_2 = B_{2,0} \otimes I_n$ where $B_{2,0} = [0, 0.1, 1]^T$. $g = 9.81$, $m = 2$, $J_x = J_y = 1.25$, $Jz = 2.2$, $K_x = K_y = K_z = 0.01$, $K_\phi = K_\theta = K_\psi = 0.012$.

(2) Designed parameters of subsystems: $Q_1 = 10 * diag(diag(1,100,10),0,0)$, $Q_2 = diag(0.1diag([1,100,10]), diag(1,0.1,0.01), 20diag(1,0.1,0.01,0.001))$, $R = 0.02$.

(3) Couplings from other agents in (7.10): $c(t) = \sum_{k=1,k\neq i}^{N_T} D_{i,1,k}(x_{i,1}(t))x_{k,1}(t) \in \mathbb{R}^6$ where $D_{i,1,k}(x_{i,1}(t)) = diag(x_{i,1,1}(t),\ldots,x_{i,1,6}(t))$ with $x_{i,1}(t) = [x_{i,1,1}(t), \ldots, x_{i,1,6}(t)]^T$.

(4) External disturbance: $d_1(t) = [100\sin(3t),...,100\sin(3t)]^T \in \mathbb{R}^6$.

(5) Sensor fault: $f_2(t)$ is set as a smoothed square wave, as shown in Figure 7.12.

Robots:

(1) System parameters: $B_2 = B_{2,0} \otimes I_n$ where $B_{2,0} = [0,0,1]^T$. Remaining parameters can be found in Appendix E in Reference [20].

(2) Designed parameters of subsystems:

$Q_1 = 50 * diag(diag(1,100,10),0,0)$, $Q_2 = 5diag(diag([1,100,10])$, $diag(1,0.1,0.01)$, $diag(1,0.1,0.01,0.001))$, $R = 0.002$.

(3) Couplings from other agents in (7.11): $c(t) = \sum_{k=1,k\neq j'}^{N_A} D_{i,j',k}(x_{i,j'}(t))x_{i,k}(t) \in \mathbb{R}^{12}$ where $D_{i,j',k}(x_{i,j'}(t)) = diag(x_{i,1,1}(t),\ldots,x_{i,1,12}(t))$ with $x_{i,j'}(t) = [x_{i,j',1}(t),\ldots, x_{i,j',12}(t)]^T$.

(4) External disturbance:
$d_1(t) = [10\sin(3t),...,10\sin(3t)]^T \in \mathbb{R}^{12}$.

(5) Sensor fault: $f_2(t)$ is set as a smoothed square wave, as shown in Figure 7.16.

Remark 7.16 Since we set $C = C_0 \otimes I_n$ and $B_2 = B_{2,0} \otimes I_n$, the proposed method of decomposing each agent model into n subsystems is used to find K and L. For the

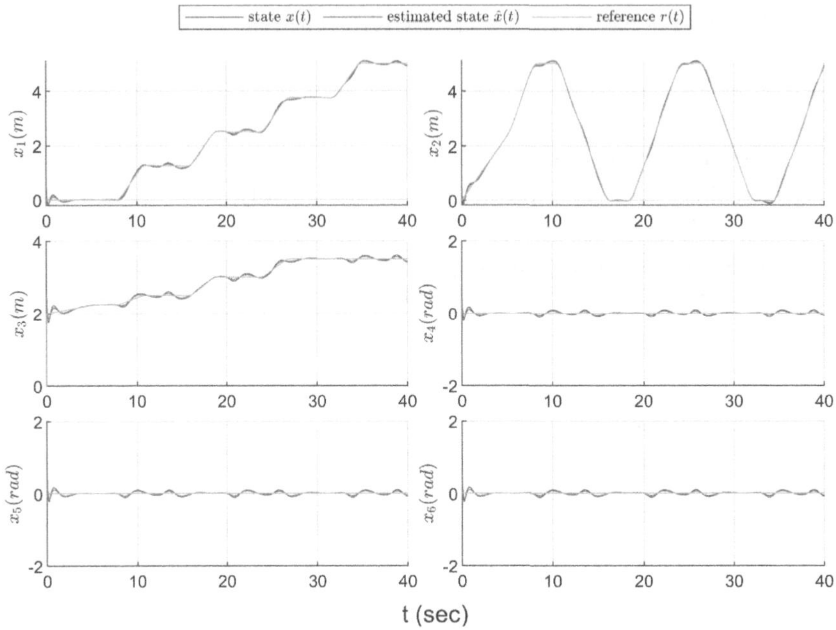

FIGURE 7.11 The trajectories of components of reference $r(t)$, state $x(t)$, and estimated state $\hat{x}(t)$ of the UAV $\alpha_{1,1}$.

convenience of solving the LMIs-constrained optimization problem in (7.33), we set the same design parameters Q_1, Q_2, R, and ρ^* for each subsystem as shown above.

The simulation results of tracking and estimation in reference tracking control block of the UAV $\alpha_{1,1}$ and the robot $\alpha_{1,5}$ in $team_1$ are given as follows:

UAV $\alpha_{1,1}$: The position trajectories of reference $r(t)$, state $x(t) = e(t) + r(t)$ and estimated state $\hat{x}(t) = \hat{e}(t) + r(t)$ are shown in Figure 7.11. The estimation of actuator fault $f_1(t)$ in (7.15) is shown in Figure 7.12. The estimation of sensor fault $f_2(t)$ in (7.16) is shown in Figure 7.13. The actuator control input $u'(t)$ is shown in Figure 7.14.

Robot $\alpha_{1,5}$: The position trajectories of reference $r(t)$, state $x(t) = e(t) + r(t)$ and estimated state $\hat{x}(t) = \hat{e}(t) + r(t)$ are shown in Figure 7.15. The estimation of actuator fault $f_1(t)$ in (7.15) is shown in Figure 7.16. The estimation of sensor fault $f_2(t)$ in (7.16) is shown in Figure 7.16. The actuator control input $u'(t)$ is shown in Figure 7.17.

In Figure 7.11, the tracking and estimation errors of UAV position and attitude reach the steady state all within 2 s. There is a brief jitter when the sensor fault signal changes drastically, but it returns quickly to the steady state. In Figure 7.15, the tracking and estimation errors of robot joint angles immediately reach and maintain steady state under the influence of fault signals. In Figures 7.12 and 7.16, the results show that the actuator fault $f_1(t)$, that is, feedforward control errors and disturbances, can be

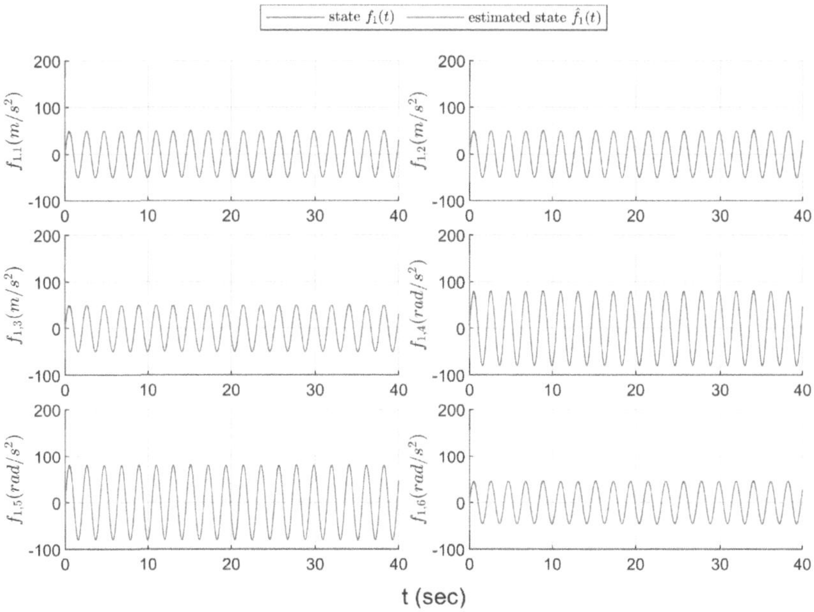

FIGURE 7.12 The estimation of components of actuator fault $f_1(t)$ of the UAV $\alpha_{1,1}$.

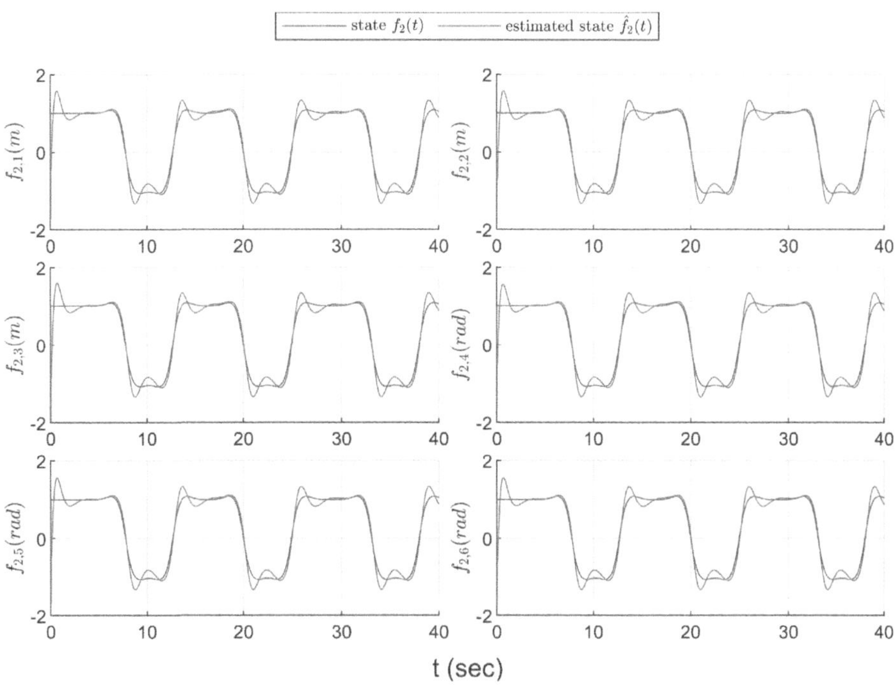

FIGURE 7.13 The estimation of components of actuator fault $f_1(t)$ of the UAV $\alpha_{1,1}$.

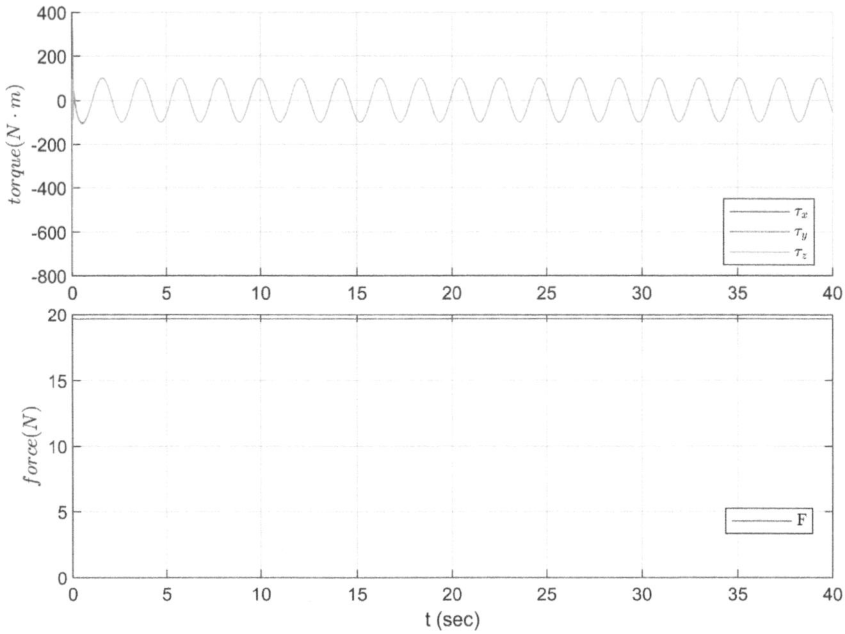

FIGURE 7.14 The actuator control input $u'(t)$ of the UAV $\alpha_{1,1}$.

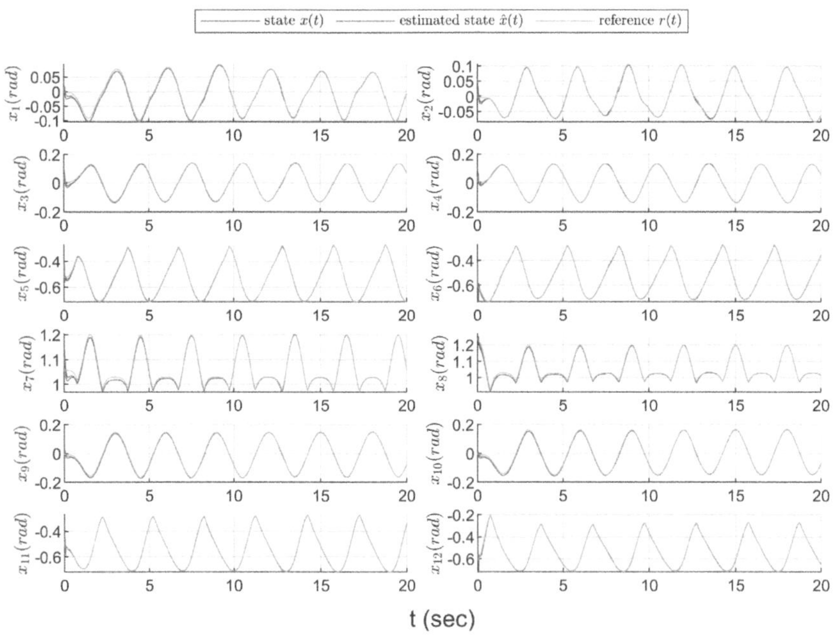

FIGURE 7.15 The trajectories of components of reference $r(t)$, state $x(t)$, and estimated state $\hat{x}(t)$ of the robot $\alpha_{1,5}$.

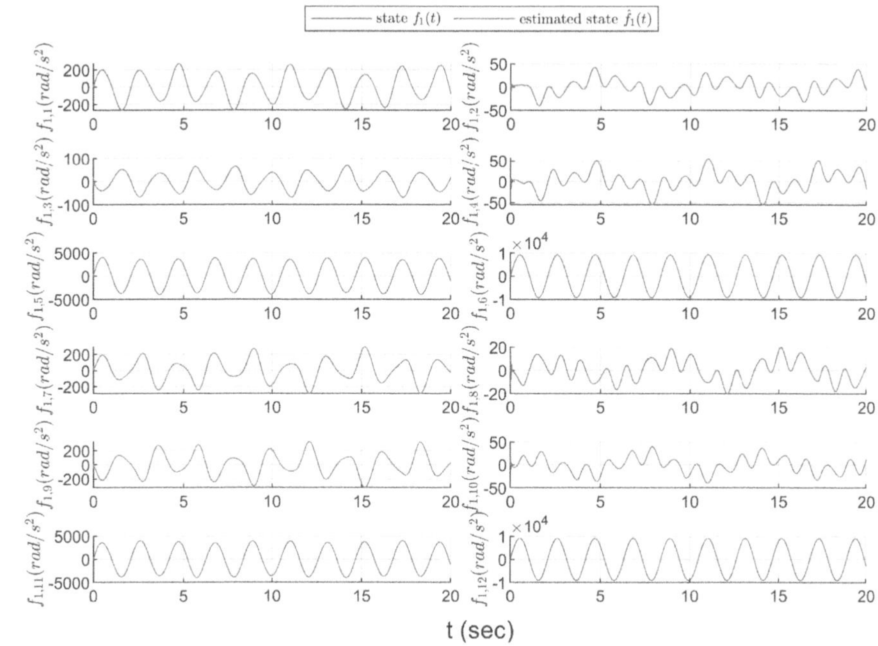

FIGURE 7.16 The estimation of components of actuator fault $f_1(t)$ of the robot $\alpha_{1,5}$.

effectively estimated. However, in Figures 7.13 and 7.16, the estimation of sensor fault $f_2(t)$ has an overshot phenomenon when there is a large change and returns to a steady state after about 2 s. This is because the sensor fault $f_2(t)$ in $y(t)$ will directly influence on the estimation of Luenberger observer in (7.20). In Figures 7.14 and 7.17, the actuator control input $u'(t)$ have high frequency and high amplitude at the initial instance due to the high gain characteristic of the robust control. After that, they maintain the sine wave shape to offset the estimated actuator fault value. Besides, it can be seen that the total force F of UAV remains constant against the gravity in Figure 7.14.

In order to show the effect of active FTC based on the proposed embedded smoothing model method, a traditional PID computed torque controller without FTC is used for comparison [30]. In order to only focus on the effect of FTC, the method in Reference [30] was revised to observer-based output feedback control law, that is, the control law become $u(t) = M(r(t))(\ddot{r}(t) + u_{fb}(t)) + H(r(t), \dot{r}(t))$ where $u_{fb}(t) = [K_I, K_P, K_D][\int_0^t \hat{e}^T(\tau)d\tau, \hat{e}^T(t), \dot{\hat{e}}^T(t)]^T$. The results are shown in Figure 7.19 for a UAV and Figure 7.20 for a robot. From Figure 7.19 and Figure 7.20, it can be clearly seen that the influence of the actuator fault $f_1(t)$ on the tracking error is revealed. The influence of the sensor fault $f_2(t)$ is relatively insignificant because its value is much smaller than $f_1(t)$ by our simulation setting. However, the effect of fluctuation $f_2(t)$ with a period of 10 s (corresponding to the period of the smoothed square wave) can still be seen from the UAV attitude state variables x_4, x_5, and x_6 in Figure 7.19 and the robot joint state variables x_5, x_6, x_{11}, and x_{12} in Figure 7.20.

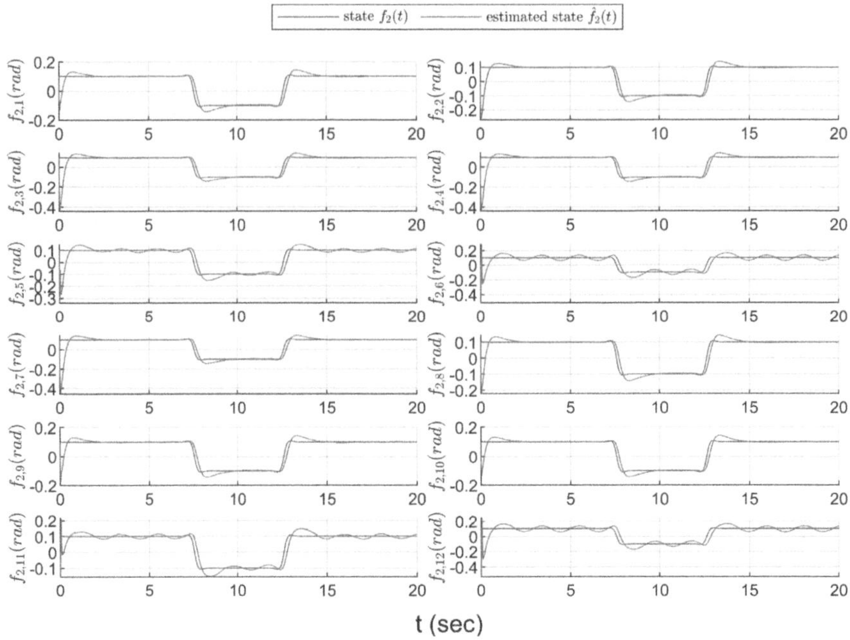

FIGURE 7.17 The estimation of components of sensor fault $f_2(t)$ of the robot $\alpha_{1,5}$.

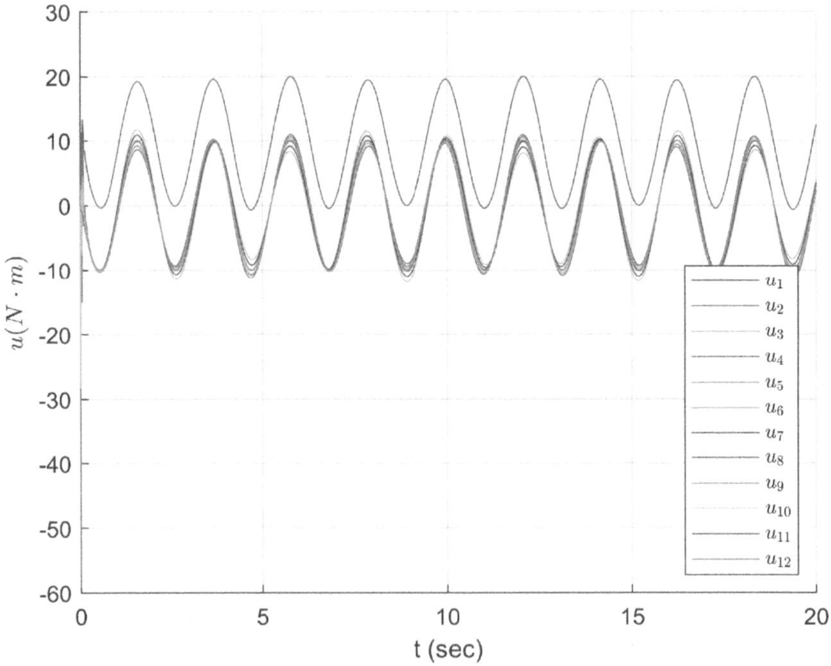

FIGURE 7.18 The actuator control input $u'(t)$ of the robot $\alpha_{1,5}$.

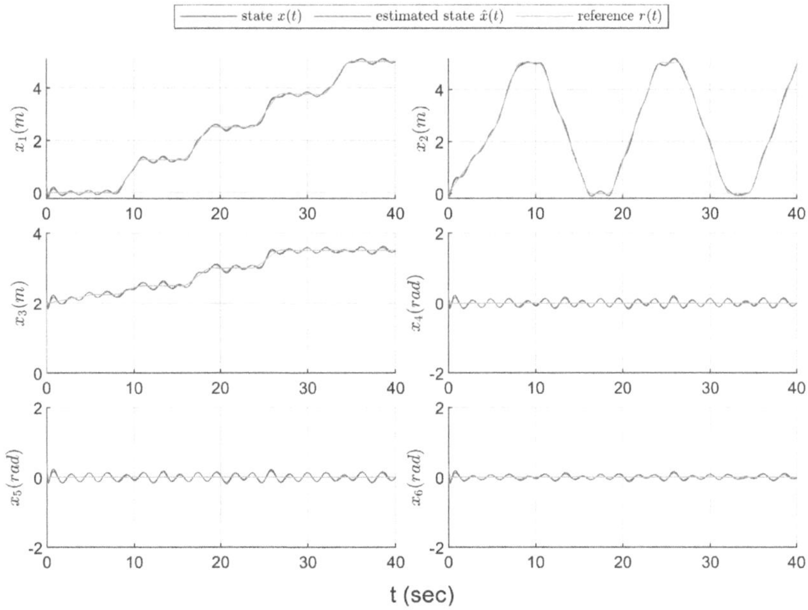

FIGURE 7.19 The trajectories of reference $r(t)$, state $x(t)$, and estimated state $\hat{x}(t)$ of the UAV $\alpha_{1,1}$ by the traditional PID computed torque controller without FTC [30].

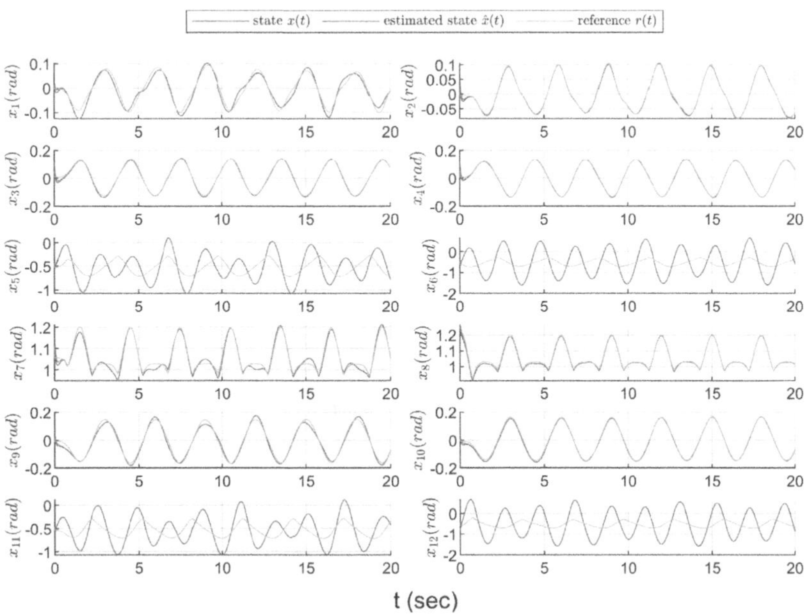

FIGURE 7.20 The trajectories of reference $r(t)$, state $x(t)$, and estimated state $\hat{x}(t)$ of the robot $\alpha_{1,5}$ by the traditional PID computed torque controller without FTC [30].

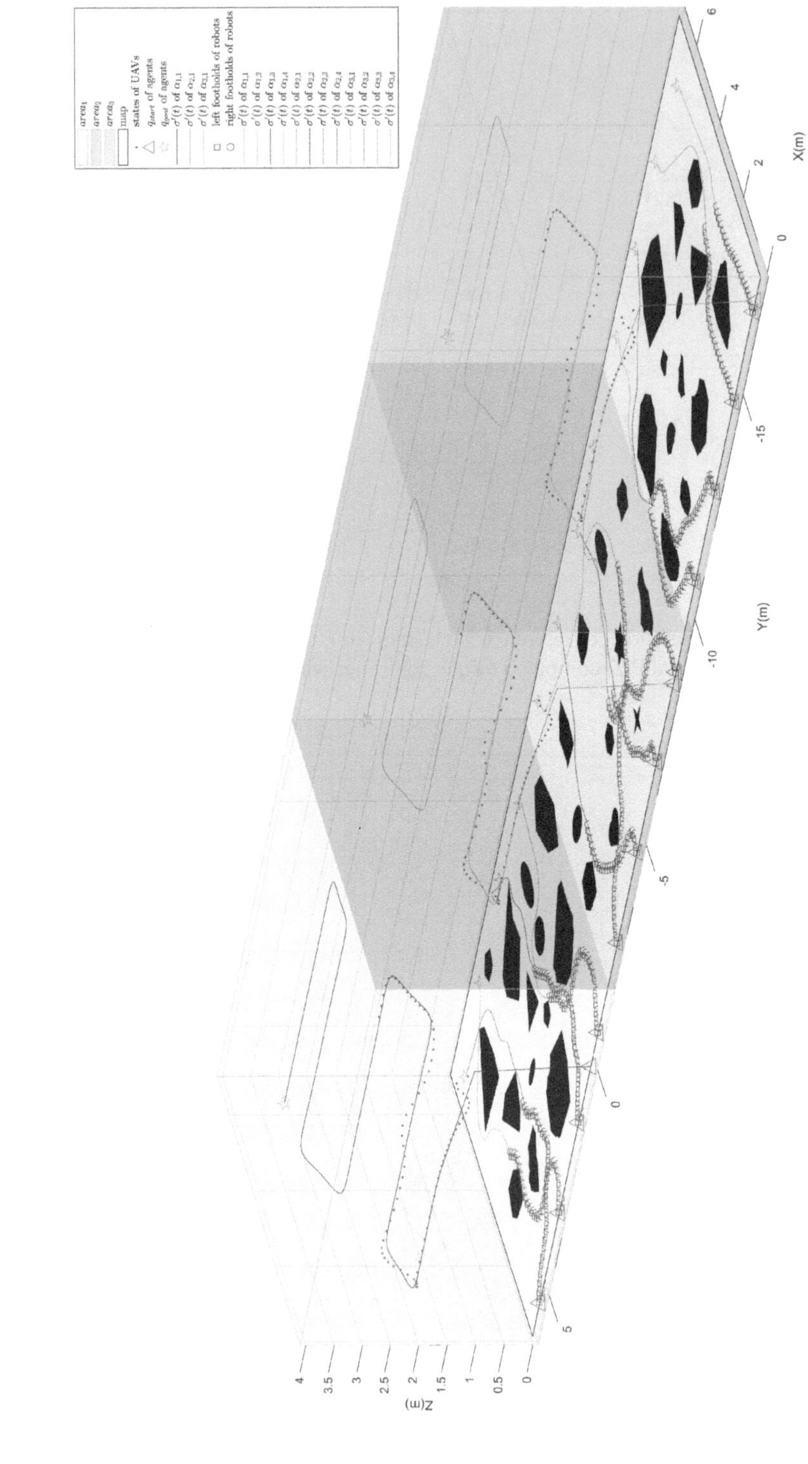

FIGURE 7.21 The simulation of team tracking of URTS for $team_i, i = 1,2,3$ when they perform search tasks.

Although it is still stable, the tracking performance has deteriorated significantly compared to Figure 7.11 and Figure 7.15, respectively.

We use three teams $team_i, i = 1,2,3$ to show the team tracking results of hybrid URTS, as shown in Figure 7.21.

As shown in Figure 7.6, the UAV and robot in URTS will be released along the direction of the road (*Y*-axis) and each team is responsible for an area. First, the respective tasks of each agent are determined by the task allocation algorithm. According to the task content, each agent will be determined a goal q_{goal}. As shown in Figure 7.7, this simulation assumes that $team_i, i = 1,2,3$ are assigned with search tasks. Subsequently, the path (σ_n) for each UAV is assigned and the collision-free path (σ_n) for each robot is planned by the path planning algorithm to reach the goal configuration q_{goal}. Depending on the terrain environment, an appropriate behavior (β_n) is decided to enable the agent to follow the path (σ_n). A reference path $r[k]$ corresponding to a behavior is planned via local motion planning to enable an actual mechanical body to perform the behavior. Since this chapter only gives the motion planning method of flying (for UAV) and walking (for biped robot), it is assumed that the appropriate behaviors from q_{start} to q_{goal} are all flying or walking behaviors in this simulation. In this figure, we draw the smoothed path (σ_n') of each agent, which is the intermediate result of local motion planning as shown in Figure 7.4. It represents the path that the agent is going to reach in the task space. According to $r[k]$ and the dynamic model of UAV and robot, the reference trajectory $r(t)$ of each agent is planned. Finally, through the decentralized H_∞ observer-based feedforward reference tracking FTC scheme proposed in this chapter, the team tracking results of the agents in $team_i$, where $i = 1,2,3$ in URTS are shown. Note that the above process is dynamic. If a higher block in Figure 7.1 makes a new decision that produces a new reference planning, the lower blocks must recalculate based on it. Relatively, this simulation is the static result, that is, there is no re-decision from q_{start} to q_{goal}. The real-time simulation of reference planning and team formation tracking control of URTS is given in Reference [31].

For further verification, we visualized the real-time simulation of reference planning and team formation tracking control in hybrid URTS and the configuration trajectory of biped robot on the online resource [31]. The results again demonstrate the effectiveness of the proposed H_∞ decentralized observer-based feedforward reference tracking FTC method for agents in URTS.

7.6 CONCLUSION

In this chapter, a system architecture of performing S&R tasks for each agent in hybrid URTS is given. The task allocation problem and path planning problem are investigated and unified to the integration of local motion planning and robust H_∞ decentralized feedforward reference tracking FTC for hybrid URTS. By decomposing the path planning process into three subprocess, that is, global path planning, behavior decision and local motion planning, some common roadmap-based path planning algorithm can be applied in the global path planning of URTS. Through the behavior decision, the agent can decide the appropriate behavior according to the terrain environment. Next, we focus on the local motion planning of UAV flying and robot walking behavior. Besides, the bridging method between the reference path planned by the local motion planning block and the reference trajectory to be followed in the reference tracking control block is also given. By a general nonlinear agent dynamic model, the tracking control problem of UAV and robot can be analyzed together. Through a novel feedforward linearization control strategy, the nonlinear tracking control problem with external disturbances is transformed to a regulation problem with fault signals. Then, a smoothing signal model is introduced to embed the fault signals into the state vector to avoid the corruption on the agent dynamic model. After that, a robust H_∞ decentralized observer-based feedforward reference tracking FTC strategy is proposed for each agent in the hybrid URTS. To solve the robust H_∞ decentralized observer-based feedforward reference tracking FTC problem, we transform it into an LMI-constrained optimization problem by a two-step design procedure, which can be effectively solved by MATLAB LMI toolbox. A simulation example is given to illustrate more concretely how the proposed hybrid URTS architecture actually works. Finally, the effectiveness of the proposed robust H_∞ decentralized observer-based feedforward reference tracking FTC method is also verified by the simulation results.

REFERENCES

[1] L. Kloeker, T. Moers, L. Vater, A. Zlocki, and L. Eckstein, "Utilization and potentials of unmanned aerial vehicles (UAVs) in the field of automated driving: A survey," in *2021 5th International Conference on Vision, Image and Signal Processing (iCVISP)*, Kuala Lumpur, Malaysia 2021, pp. 9–17.

[2] A. Dorri, S. S. Kanhere, and R. Jurdak, "Multi-agent systems: A survey," *IEEE Access*, vol. 6, pp. 28 573–28 593, 2018.

[3] F. Chen, W. Ren, et al., "On the control of multi-agent systems: A survey," *Foundations and Trends® in Systems and Control*, vol. 6, no. 4, pp. 339–499, 2019.

[4] W. Cheng, K. Zhang, B. Jiang, and S. X. Ding, "Fixed-time fault-tolerant formation control for heterogeneous multi-agent systems with parameter uncertainties and disturbances," *IEEE Transactions on Circuits and Systems I: Regular Papers*, vol. 68, no. 5, pp. 2121–2133, 2021.

[5] G. M. Skaltsis, H.-S. Shin, and A. Tsourdos, "A survey of task allocation techniques in mas," in *2021 International Conference on Unmanned Aircraft Systems (iCUAS)*, Athens, Greece 2021, pp. 488–497.

[6] H. Y. Zhang, W. M. Lin, and A. X. Chen, "Path planning for the mobile robot: A review," *Symmetry*, vol. 10, no. 10, p. 450, 2018.

[7] J. N. Yasin, S. A. S. Mohamed, M.-H. Haghbayan, J. Heikkonen, H. Tenhunen, and J. Plosila, "Unmanned aerial vehicles (UAVs): Collision avoidance systems and approaches," *IEEE Access*, vol. 8, pp. 105 139–105 155, 2020.

[8] B. Arbanas, A. Ivanovic, M. Car, M. Orsag, T. Petrovic, and S. Bogdan, "Decentralized planning and control for UAV–UGV cooperative teams," *Autonomous Robots*, vol. 42, no. 8, pp. 1601–1618, 2018.

[9] L. Li, Y. Miao, A. H. Qureshi, and M. C. Yip, "MPC-MPNet: Model-predictive motion planning networks for fast, near-optimal planning under kinodynamic constraints," *IEEE Robotics and Automation Letters*, vol. 6, no. 3, pp. 4496–4503, 2021.

[10] T. Olcay and A. Özkurt, "Design and walking pattern generation of a biped robot," *Turkish Journal of Electrical Engineering and Computer Sciences*, vol. 25, no. 2, pp. 761–769, 2017.

[11] Y. Sabri, N. El Kamoun, and F. Lakrami, "A survey: Centralized, decentralized, and distributed control scheme in smart grid systems," in *2019 7th Mediterranean Congress of Telecommunications (CMT)*, Fez, Morocco, 2019, pp. 1–11.
[12] C. Wang, J. Wang, P. Wu, and J. Gao, "Consensus problem and formation control for heterogeneous multi-agent systems with switching topologies," *Electronics*, vol. 11, no. 16, p. 2598, 2022.
[13] D. Rotondo, F. Nejjari, and V. Puig, "Passive and active FTC comparison for polytopic LPV systems," in *2013 European Control Conference (ECC)*, Zurich, Switzerland, 2013, pp. 2951–2956.
[14] B. Paden, M. Čáp, S. Z. Yong, D. Yershov, and E. Frazzoli, "A survey of motion planning and control techniques for self-driving urban vehicles," *IEEE Transactions on Intelligent Vehicles*, vol. 1, no. 1, pp. 33–55, 2016.
[15] A. Khamis, A. Hussein, and A. Elmogy, "Multi-robot task allocation: A review of the state-of-the-art," *Cooperative Robots and Sensor Networks 2015*, pp. 31–51, 2015.
[16] M. Elbanhawi and M. Simic, "Sampling-based robot motion planning: A review," *IEEE Access*, vol. 2, pp. 56–77, 2014.
[17] Y. Liu and R. Bucknall, "A survey of formation control and motion planning of multiple unmanned vehicles," *Robotica*, vol. 36, no. 7, pp. 1019–1047, 2018.
[18] J. Yu and S. M. LaValle, "Multi-agent path planning and network flow," in *Algorithmic Foundations of Robotics X*, Springer, Berlin, Heidelberg, 2013, pp. 157–173.
[19] F. Sabatino, "Quadrotor control: Modeling, nonlinearcontrol design, and simulation," KTH, School of Electrical Engineering (EES), Automatic Control, Stockholm, Sweden, 2015.
[20] B. S. Chen, Y. Y. Tsai, and M. Y. Lee. *Supplementary File*. Hsinchu, Taiwan [Online]. Available: https://www.dropbox.com/ s/qiv63y5hw7vrubp/SupplementFileRobot.pdf
[21] S. Kajita, O. Matsumoto, and M. Saigo, "Real-time 3d walking pattern generation for a biped robot with telescopic legs," in *Proceedings 2001 ICRA. IEEE International Conference on Robotics and Automation (Cat. no. 01ch37164)*, Seoul, Korea (South), vol. 3, 2001, pp. 2299–2306.
[22] Q. Huang, K. Yokoi, S. Kajita, K. Kaneko, H. Arai, N. Koyachi, and K. Tanie, "Planning walking patterns for a biped robot," *IEEE Transactions on Robotics and Automation*, vol. 17, no. 3, pp. 280–289, 2001.
[23] S. Kajita, F. Kanehiro, K. Kaneko, K. Fujiwara, K. Harada, K. Yokoi, and H. Hirukawa, "Biped walking pattern generation by using preview control of zero-moment point," in *2003 IEEE International Conference on Robotics and Automation (Cat. No.03CH37422)*, Taipei, Taiwan, vol. 2, 2003, pp. 1620–1626.
[24] F. L. Lewis, D. Vrabie, and V. L. Syrmos, *Optimal Control*, John Wiley & Sons, New York, 2012.
[25] B. S. Chen and J. H. Lin, "Robust collaborative team formation control of hybrid teams of biped robots and wheeled vehicles under external disturbance and communication interactions," *IEEE Access*, vol. 10, pp. 77 633–77 657, 2022.
[26] J. Yang, C. Liu, M. Coombes, Y. Yan, and W.-H. Chen, "Optimal path following for small fixed-wing UAVs under wind disturbances," *IEEE Transactions on Control Systems Technology*, vol. 29, no. 3, pp. 996–1008, 2021.
[27] B. S. Chen, C. C. Wu, and Y. W. Chen, "Human walking gait with 11-dof humanoid robot through robust neural fuzzy networks tracking control," *International Journal of Fuzzy Systems*, vol. 15, no. 1, 2013.
[28] B. S. Chen, M. Y. Lee, W. Y. Chen, and W. Zhang, "Reverse-order multiobjective evolution algorithm for multi-objective observer-based fault-tolerant control of t-s fuzzy systems," *IEEE Access*, vol. 9, pp. 1556–1574, 2021.

[29] S. Boyd, L. El Ghaoui, E. Feron, and V. Balakrishnan, *Linear Matrix Inequalities in System and Control Theory*. SIAM, Philadelphia, PA, 1994.

[30] E. Rastogi and L. B. Prasad, "Comparative performance analysis of PD/PID computed torque control, filtered error approximation based control and NN control for a robot manipulator," in *2015 IEEE UP Section Conference on Electrical Computer and Electronics (uPCON)*, Allahabad, India, 2015, pp. 1–6.

[31] B. S. Chen and T. W. Hung. Hsinchu, Taiwan [Online]. Available: https://www.dropbox.com/scl/fo/5n6y25k3i8kjxsx9efxa3/h?dl=0&rlkey=x5p4stwtuf2gmch6l07etvz4b

Index

For Product Safety Concerns and Information please contact our
EU representative GPSR@taylorandfrancis.com Taylor & Francis
Verlag GmbH, Kaufingerstraße 24, 80331 München, Germany